Application and Analysis in Fluid Power Systems

Application and Analysis in Fluid Power Systems

Editors

Paolo Casoli
Massimo Rundo

MDPI • Basel • Beijing • Wuhan • Barcelona • Belgrade • Manchester • Tokyo • Cluj • Tianjin

Editors
Paolo Casoli
University of Parma
Italy

Massimo Rundo
Politecnico di Torino
Italy

Editorial Office
MDPI
St. Alban-Anlage 66
4052 Basel, Switzerland

This is a reprint of articles from the Special Issue published online in the open access journal *Energies* (ISSN 1996-1073) (available at: https://www.mdpi.com/journal/energies/special_issues/aa_fps).

For citation purposes, cite each article independently as indicated on the article page online and as indicated below:

LastName, A.A.; LastName, B.B.; LastName, C.C. Article Title. *Journal Name* **Year**, *Volume Number*, Page Range.

ISBN 978-3-0365-7866-8 (Hbk)
ISBN 978-3-0365-7867-5 (PDF)

Contents

About the Editors

Paolo Casoli

Paolo Casoli is a Professor and lecturer of Fluid power system at the University of Parma. Relevant topics in his research curriculum are: simulations and experiments on positive displacement pumps and hydraulic components; condition monitoring applied to fluid power pumps, solutions for reducing fuel consumption in mobile machinery.

Massimo Rundo

Massimo Rundo is associate professor and lecturer of Fluid Power courses at the Politecnico di Torino. His research activity is mainly focused on modelling, simulation and testing of fluid power components and systems, with specific focus on positive displacement machines.

energies

MDPI

Article

Energy Savings in Hydraulic Hybrid Transmissions through Digital Hydraulics Technology

Israa Azzam, Keith Pate, Jose Garcia-Bravo and Farid Breidi *

School of Engineering Technology, Purdue University, West Lafayette, IN 47906, USA; iazzam@purdue.edu (I.A.); pate5@purdue.edu (K.P.); jmgarcia@purdue.edu (J.G.-B.)
* Correspondence: breidi@purdue.edu; Tel.: +1-(765)-496-5140

Abstract: Hydraulic hybrid drivetrains, which are fluid power technologies implemented in automobiles, present a popular alternative to conventional drivetrain architectures due to their high energy savings, flexibility in power transmission, and ease of operation. Hydraulic hybrid drivetrains offer multiple environmental benefits compared to other power transmission technologies. They provide heavy-duty vehicles, e.g., commercial transportation, construction equipment, wagon handling, drilling machines, and military trucks, with the potential to achieve better fuel economy and lower carbon emissions. Despite the preponderance of hydraulic hybrid transmissions, state-of-the-art hydraulic hybrid drivetrains have relatively low efficiencies, around 64% to 81%. This low efficiency is due to the utilization of conventional variable displacement pumps and motors that experience high power losses throughout the drive cycle and thus fail to maintain high operating efficiency at lower volumetric displacements. This work proposes and validates a new methodology to improve the overall efficiency of hydraulic hybrid drivetrains by replacing conventional pump/motor units with their digital counterparts. Compared to conventional pump/motors, the digital pump/motor can achieve higher overall efficiencies at a wide range of operating conditions. A proof-of-concept digital pump/motor prototype was built and tested. The experimental data were integrated into a multi-domain physics-based simulation model of a series hydraulic hybrid transmission. The proposed methodology permits enhancing the overall efficiency of a series hydraulic hybrid transmission and thus allows for energy savings. Simulating the system at moderate load-speed conditions allowed achieving a total efficiency of around 89%. Compared to the average efficiency of the series hydraulic hybrid drivetrains, our simulation results reveal that the utilization of the state-of-the-art digital pump enables improving the total efficiency of the series hydraulic hybrid drivetrain by up to 25%.

Keywords: fluid power; hydraulic hybrid drivetrains; digital hydraulics; energy savings; efficiency

Citation: Azzam, I.; Pate, K.; Garcia-Bravo, J.; Breidi, F. Energy Savings in Hydraulic Hybrid Transmissions through Digital Hydraulics Technology. *Energies* **2022**, *15*, 1348. https://doi.org/10.3390/en15041348

Academic Editors: Paolo Casoli, Massimo Rundo and Helena M. Ramos

Received: 6 January 2022
Accepted: 10 February 2022
Published: 13 February 2022

1. Introduction

Fluid power has a vast area of engineering technology, where it uses pressurized fluids to generate, monitor and control, and transmit power [1]. Hydraulic oil is typically used due to its low freezing point, high boiling point, relatively large bulk modulus, and self-lubrication characteristics [2]. Fluid power has a well-defined research area and scholarly activities, which serves widespread industries like agriculture, construction, transportation, aerospace, marine, manufacturing, and many other industries that require high power to weight ratios [3]. In addition to its capability of achieving high power densities, fluid power is characterized by its flexibility in transmission, ease of operation, relatively low cost, controllability, and manageability [4]. These advantages make fluid power technology convenient for performing crucial operations that require high power density and reliability, like heavy-duty applications, e.g., construction vehicles, material handling equipment, and military operations [5].

The fluid power sector is a Workhorse of the United States economy [6]. According to the National Fluid Power Association (NFPA), thousands of companies in the United

States utilize hydraulic power systems. Such companies are renowned as the top industrial U.S. firms, involving more than 845,000 employees for an annual payroll of more than $60 billion [7]. The fluid power sector has multiple application areas. Four main application areas comprise mobile hydraulics, industrial hydraulics, pneumatics, and aerospace [8].

Hydraulic hybrid technology, a subset of mobile hydraulics, is considered one of the pioneer frameworks in automobiles. It is a desirable research topic widely examined by scholars to improve the overall efficiency of power transmission between a prime mover and the wheels. This technology allows for high fuel efficiency and offers energy savings and multiple environmental benefits, becoming a popular alternative to conventional drivetrain configurations [9]. Hydraulic hybrid drivetrains offer multiple benefits over the conventional manual/automatic transmissions and the continuously variable transmissions (CVTs). Unlike conventional drivetrains and CVTs, hydraulic hybrid drivetrains comprise two power sources: the prime mover (an internal combustion engine or electric motor) and a positive variable displacement pump/motor installed with a high-pressure storage device (accumulator) [10]. The variable displacement pump allows achieving variable speeds by regulating the flow rate, which prevents the gearshift discontinuities in conventional drivetrains (manual/automatic) during speed regulation. CVTs do not experience excessive gearshift discontinuities during speed regulation. However, CVTs have limited belts, also known as chain strengths, which cause torsional moments, generating excessive frictional forces and torsional moments [11]. Besides this advantage, utilizing a hydraulic accumulator in hydraulic hybrid drivetrains allows for better power management [12]. The accumulator enables storing and discharging energy and thus achieving higher overall efficiencies and energy savings while attaining lower carbon emissions and better fuel economy [13].

In addition to its preponderance over manual/automatic drivetrains and CVTs, the hydraulic hybrid drivetrains have many advantages over electric hybrid transmission [14,15]. Although electric hybrid systems also utilize a transmission source and an energy storage source (batteries), the batteries in electric systems can only charge and discharge around 20% of energy available during braking [16]. However, the accumulators employed in the hydraulic hybrid transmissions can store and deliver up to 80% of the braking energy [17], depending on the duty cycle. This operation is desirable during vehicle acceleration/deceleration to improve fuel economy and reduce emissions [18,19]. It permits transmitting a large amount of energy rapidly, thus attaining higher power densities and power to weight ratios [20].

These advantages make the hydraulic hybrid technology more convenient for heavy-duty operations [21], like heavy equipment manufacturing, road appliances, heavy load material handling equipment, and many other types of equipment that require high power densities and energy capture. For instance, commercial vehicles, such as heavy load vehicles undergoing high dynamic-drive-cycle operations, have started to rely on this transmission technology. They carry bulky loads while experiencing continual stops, requiring more braking and high energy capture [15]. Besides heavy-duty vehicles, some utility vehicles designed to carry light weights also rely on hydraulic hybrid technology. For example, a sport utility vehicle was developed in a research laboratory to demonstrate the concept of a hydraulic hybrid transmission in a passenger vehicle. The vehicle was designed to operate the system in four different modes: hydrostatic driving mode, hybrid driving mode, a combination of hydrostatic and hybrid driving modes, and braking mode [22].

This paper is organized as follows: Section 2 exhibits a review of the currently existing hydraulic hybrid transmissions. It introduces the different available hydraulic hybrid drivetrain configurations: series, parallel, and series-parallel, and it discusses the overall operation of each. Section 3 presents a series of notable applications on hydraulic hybrid drivetrains. Section 4 formulates the problem statement of this work, elucidating the inadequacies of the currently existing hydraulic hybrid drivetrains. Moreover, it highlights and introduces the adopted methodology to overcome such inadequacies. Section 5 gives an overview of digital hydraulics technology in fluid power and presents the design of the state-of-the-art digital pump, its basic operation, digital motoring/pumping data, and the capability to achieve high efficiencies at different pump displacements. Section 6 demonstrates the hydraulic hybrid transmission efficiency improvement after integrating digital hydraulics by simulating a physical series hydraulic hybrid architecture using the intended digital motoring/pumping data. It presents the adopted circuit schematic, the comprehensive simulation model, and the resulting outcomes. Section 7 concludes with a brief discussion of the main ideas learned from the work.

2. Existing Hydraulic Hybrid Transmission Configurations

Three main architectures of hydraulic hybrid transmission drivetrains are currently available: the parallel configuration, series configuration, and the series-parallel configuration. These hydraulic hybrid transmission configurations employ a variable displacement pump/motor, a high-pressure reservoir for energy storage (typically a nitrogen accumulator), and a low-pressure reservoir (fluid tank) [23]. The variable displacement pump/motor acts as a motor when converting the hydraulic energy into mechanical energy to drive the vehicle. It acts as a pump when converting the mechanical energy into hydraulic energy stored in the accumulator to be delivered when needed.

The following sections exhibit a brief review of existing relevant hydraulic hybrid drivetrains (parallel, series, and series-parallel) and discuss their overall operation.

2.1. Parallel Hydraulic Hybrid Configuration

A parallel hydraulic hybrid has a simple architecture, design, and implementation compared to other configurations [24]. It involves two sub-systems: a conventional mechanical transmission directly connected to the output shaft via a differential gearbox and a hydraulic transmission [25]. The sub-systems are coupled in parallel using a mechanical power transmitter like a gearbox, a chain, or a belt. Figure 1 shows a parallel hydraulic hybrid configuration utilizing a belt mechanical power transmitter. As shown in Figure 1, the drivetrain consists of mechanical transmission (7), a clutch (2), and the prime mover (1), typically an internal combustion engine. The mechanical transmission (7) is directly connected to the engine (1) via the clutch component (2) and to the differential gearbox (8) on the drive shaft (9) to provide the needed power to each wheel. The hydraulic drivetrain comprises a variable displacement pump/motor (4), an accumulator (5), a fluid tank (6), and other hydraulic components, e.g., valves and fittings. The hydraulic pump/motor (4) is connected in parallel to the mechanical transmission (7) via the belt linkage (3). The clutch (2) allows engaging and disengaging the engine shaft from the rest of the system based on the accumulator's stored energy, thus controlling the engine on/off signal [26]. For instance, decoupling the engine from the system occurs when enough energy is in the accumulator to boost the wheels during acceleration. However, engaging the engine shaft to the transmission shaft is required to drive the wheels at low accumulator storage levels.

After introducing the basic concept beyond the parallel hybrid configuration, now, we briefly discuss its overall operation. The transmission duty cycle constitutes three main stages: acceleration, cruising, and braking. During acceleration, the accumulator acts as a hydraulic launch assist. It discharges its pressurized fluid to drive the hydraulic pump/motor (4) (operating as an energy source) and thus transforms the high-pressure energy into mechanical energy at the motor's shaft. The mechanical belt (3) transmits the generated mechanical power to the drive shaft (9), thus boosting the motion at the wheels. At the start of the cruising phase, the accumulator's pressure decreases below its pre-charged pressure, forcing the clutch to engage the engine shaft with the rest of the system to power the drive shaft (9) and drive the wheels. During the last stage, i.e., braking, the clutch (2) disengages the engine shaft from the drive shaft (9), the vehicle slows down. The rotary motion at the wheels is transmitted via the belt (3) to drive the hydraulic pump/motor (4) that will start acting as a pump to charge the accumulator. This cycle repeats every braking operation.

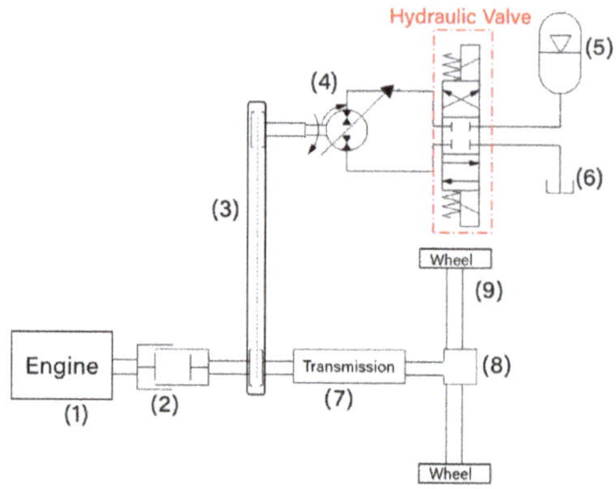

Item	Description
1	Engine
2	Mechanical Clutch
3	Mechanical Transmission Belt
4	Variable Displacement Motor/Pump
5	Gas Accumulator
6	Reservoir
7	Mechanical Transmission
8	Differential Gear Box
9	Output Shaft

Figure 1. Schematic for the parallel hydraulic hybrid configuration.

2.2. Series Hydraulic Hybrid Configuration

Unlike the parallel hydraulic hybrid configuration, the series hydraulic hybrid configuration does not utilize a mechanical transmission. Instead, it employs a variable displacement pump directly connected to the prime mover via a clutch [27]. As shown in Figure 2, in the series configuration, the mechanical transmission is replaced with a hydraulic variable displacement pump (3) connected in series to the hydraulic drivetrain consisting of a variable displacement pump/motor (6), accumulator (5), fluid tank (4), and other hydraulic components.

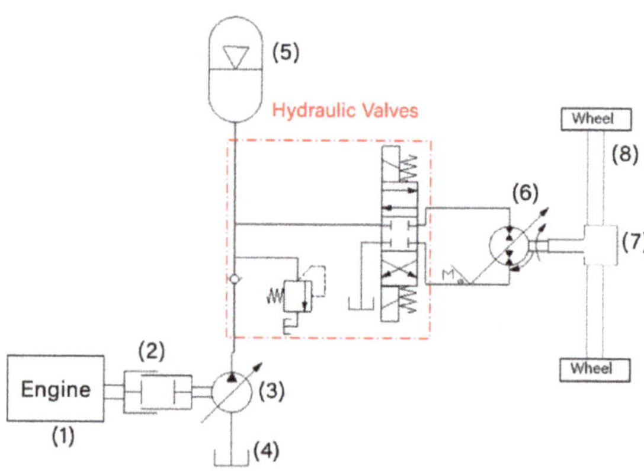

Item	Description
1	Engine
2	Mechanical Clutch
3	Variable Displacement Pump
4	Reservoir
5	Gas Accumulator
6	Variable Displacement Motor/Pump
7	Differential Gear Box
8	Output Shaft

Figure 2. Schematic for the series hydraulic hybrid configuration.

Similar to the parallel architecture, the operation comprises three main phases: acceleration, cruising, and braking. During the acceleration phase, the clutch (2) disengages the engine shaft from the pump shaft, and the pressurized hydraulic fluid stored in the accumulator (5) is delivered into the hydraulic pump/motor (6), allowing it to act as a motor. This operation converts the hydraulic energy in the accumulator into mechanical energy at the output shaft (8). After acceleration, cruising starts, and the fluid inside the accumulator drops to its minimum level. Its pressure decreases below the nitrogen's pre-charge pressure, forcing the engine (1) to operate and the clutch (2) to engage. Accordingly, the engine's energy is transmitted to the hydraulic pump (3), allowing the pump to deliver the fluid from the low-pressure reservoir (4) to the hydraulic pump/motor (6) to drive the wheels. The excess high-pressure fluid produced by the hydraulic pump is stored in the accumulator to be delivered when needed. During braking, the engine shuts down, the clutch disengages, and the hydraulic pump/motor (6) starts acting as a pump to push the fluid from the low-pressure reservoir into the accumulator. Thus, the kinetic energy from the wheels is converted into hydraulic energy stored in the accumulator. This stored energy will be used to accelerate the vehicle when needed.

2.3. Series-Parallel Hydraulic Hybrid Configuration

The series-parallel hydraulic hybrid system, also known as power-split, comprises the series and the parallel hydraulic hybrid layouts [28]. It integrates the two hydraulic hybrid configurations in one design [29]. Like the parallel configuration, the power-split architecture, presented in Figure 3, utilizes two sub-systems coupled via a mechanical transmitter, for example, a belt (3): a conventional mechanical drivetrain and a hydraulic drivetrain. The mechanical drivetrain employs a mechanical transmission (8) directly connected to the output shaft (10) via a differential gearbox (9) and the engine (1) via a mechanical clutch (2). The hydraulic drivetrain comprises two variable displacement pump/motors (4 & 5), accumulator (6), low-pressure reservoir (7), and other hydraulic components. Similar to the series architecture, the hydraulic transmission is directly connected to the output shaft (10).

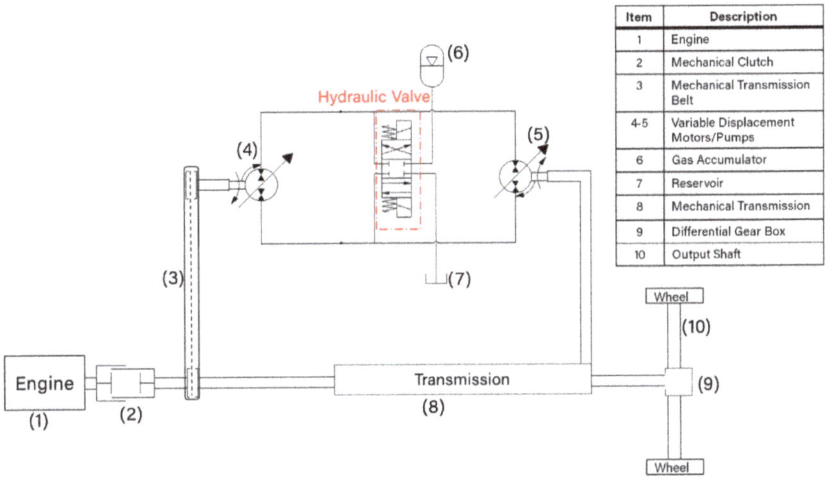

Figure 3. Schematic for the series-parallel hydraulic hybrid configuration.

This configuration enables operating the system in two different modes: pure hydraulic mode and hybrid mode. The pure hydraulic mode offers a simultaneous operation of the variable displacement pump/motors (4 & 5). However, the hybrid mode allows a seamless operation of the mechanical transmission (8) and the hydraulic drivetrain [30]. The seamless integration between the mechanical drivetrain and the hydraulic system allows transmitting the power from the engine to the wheels while maintaining optimal engine operation. This mechanism permits achieving high mechanical transmission performance in addition to the high-efficiency energy storage and regeneration, and thus operating the system at higher overall efficiency and lower fuel consumption. However, the design comes at the expense of increased system complexity, weight, and cost.

3. Summary of Notable Mobile Applications Using Hydraulic Hybrid Drivetrains

This section briefly reviews notable applications and proposed models for hydraulic hybrid drivetrains.

A comprehensive work presented by Stecki et al. [31] presented a review of the most notable computer simulation models of hydraulic hybrids available in the literature before 2005 and summarized the resulting simulation outcomes. The comprehensive report addresses the design, implementation, and simulation of more than eight prototype vehicles using CVT transmissions combined with hydrostatic regeneration. Most of the vehicles studied were on-road vehicles. The authors mentioned ADVISOR (National Renewable Energy Laboratory, Golden, CO, USA), the Advanced Vehicle Simulator developed by the National Renewable Energy Laboratory based in the US Department of Energy's (DOE's), as a simulation tool for developing hybrid drivetrain models. One of the vehicles discussed focused on a tactical vehicle, developed in MATLAB/Simulink which is created by MathWorks (an American privately held corporation based in California) and implemented in ADVISOR [32]. In this work, a Permo-Drive regenerative energy management system for an army tactical vehicle was modeled and simulated on MATLAB/Simulink. The model was then incorporated into the vehicle simulator (ADVISOR), a simulation tool for developing hybrid drivetrain models. The authors concluded their review by stating that extra fuel trails are required for examining the accuracy of the adopted fuel map and thus providing further validation of the model. Thy concluded that the accumulators might present a safety issue and recommended that their implementation require further accumulator technologies.

Besides the work of Stecki et al., the work of Stelson et al. [10] presents a simulation study of various hydraulic hybrid drivetrains comparing the three hydraulic hybrid configurations (series, parallel, and power-split). The prime mover is emulated as a combustion engine throughout their simulation studies. A dynamic programming approach is adopted, in which the fuel consumption was used as the optimization variable. The developed models reveal that the three configurations of the presented on-road vehicles (series, parallel, and power-split) improve fuel consumption for the two cycles: urban drive cycle and highway drive cycle. For the urban drive cycle, the three hydraulic hybrid configurations (series, parallel, and power-split) show an improvement of 14.7%, 53.2%, and 10.0%, respectively.

Regarding the highway drive cycle, the series, parallel, and power-split architectures were enhanced by 8.9%, 11.9%, and 8.5%, respectively. Furthermore, throughout the simulation, the employed pump/motors were oversized. This forced the units to operate at partial displacements and thus led to a low overall hydraulic efficiency.

In 2004, Wu et al. [33] utilized the dynamic programming optimization methodology to optimize the accumulator's charging/discharging process for a Class VI parallel hydraulic hybrid delivery truck. The researchers used a 99.9 L accumulator operating at a maximum pressure of 36 MPa throughout this work. The proposed control algorithm involves two optimization parameters: the fuel consumption variable as the primary control parameter and the state of charge of the accumulator (SOC) as the secondary optimization parameter. The researchers estimated the fuel economy between 28% and 47%, depending on the efficiency of the hydraulic components. In 2009, Van Batavia [34] proposed a new control approach to monitor and regulate the hydraulic launch assist operation. The adopted strategy relies on operating the engine at its peak Torque-RPM performance. An approximate range of operation comprised between 1000 RPM to 5500 RPM and 125 N.m to 200 N.m for a Diesel engine rated for 110 kW capacity. The work claims that the proposed control strategy improved the fuel economy by 50% and reduced emissions by 40%. The proposed configuration used a variable displacement bent axis pump and a separate variable displacement bent axis motor. The bent axis units are known to have high efficiencies at 100% displacements of up to 98% [35], yet this efficiency is only achievable at high displacements and is substantially lower at low displacements.

Buchwald et al. [36] introduced a Hydraulic—Internal combustion hybrid concept and four topologies for a passenger bus in Denmark. Besides the latter work, in 2008, Eaton Hydraulics [37] presented its hydraulic launch assist system for parallel drivetrain configurations. The objective was to improve power savings by supplementing engine power while launching the vehicle from a stopped position. The system was conceived for vehicles operating with heavy stop-and-go cycles, like refuse trucks, delivery vehicles, and buses. This system, developed in collaboration with the U.S. Environmental Protection

Agency, depended on using a state-of-the-art pump/motor unit specifically designed for a parallel drivetrain configuration using an axial piston design.

Tollefson et al. [38] examined a new configuration of internal combustion engine/hydro-pneumatic accumulator energy storage for automobiles. The proposed architecture requires one single hydrostatic pump-motor (P/M) unit. Computer-based simulations were conducted over a wide range of parameter variations using a suitable control strategy. The simulation results revealed that the novel concept allowed significant fuel savings. The proposed hydraulic accumulator passenger car could achieve up to 65 mpg during city driving conditions, unlike the present automobiles that can attain only 25–30 mpg under similar conditions.

Similarly, Kepner [39] presents a Hydraulic Power Assist (HPA) demonstration on a full-size sport utility vehicle. It exhibits a description of the vehicle, the resulting fuel economy data, and performance and noise outcomes. The vehicle is constructed as a collaboration between Ford Motor Company, United States Environmental Protection Agency (EPA), and Advanced Technology Division, in the context of a Cooperative Research and Development Agreement (CRAD). It utilizes a valve block provided by Ifield Technologies, a pump/motor connected in parallel to the driveshaft with the conventional powertrain, and carbon fiber accumulators acting as HPA. Results show that the utilization of the HPA improved the Environmental Protection Agency (EPA) and the Heavier Acceleration city cycles by 23.6% and 35.5%, respectively. Bozic [40] presented a hybrid configuration comprising a combustion engine as the prime mover, hydrostatic transmission, and a coexisting vehicle electric generation. According to the literature consulted, it is evident that the implementation of hybrid powertrains significantly contributes to energy savings, in some cases, as high as 50% [33,34]. Yet, the system's efficiency may be compromised by the efficiency of the heart of the hybrid transmission; that is, the pump, the motor, or the pump/motor units. These units can operate at an efficiency as high as 98%, provided the units are operating at 100% displacement, but this may be the case of only a fixed segment of a duty cycle and may not be true at all operating conditions.

4. Inadequacies of the Existing Hydraulic Hybrid Drivetrains

Despite the prominence of fluid power transmission, the existing state-of-the-art fluid power drivetrains are ineffective [3]. A study published by the United States Department of Energy shows that, among all industries, fluid power drives have low efficiencies between 9% and 60%, depending on the application, with an average efficiency not exceeding 22% [8]. The insufficiency in the fluid power transmissions' efficiencies is notably associated with the performance of the employed hydraulic units. The variable displacement pump/motor is an essential hydraulic unit employed in all hydraulic applications; it is the heart of fluid power systems [41]. Its efficiency significantly affects the performance and overall efficiency of fluid power transmission [42,43].

The conventional variable displacement pump/motors, i.e., existing state-of-the-art hydraulic pumps, are efficient only when operating at full displacements, whereas at partial displacements, their performance decreases until operating at low efficiencies down to 30% [42,44,45]. The insufficiency of the conventional pumps at partial or low displacements affects the system's overall efficiency due to the high pressure that may remain in the displacement chamber, in addition to leakage and other mechanical losses [46]. Given the limited efficiency of the conventional variable displacement pump/motors, the overall efficiency of the hydraulic hybrid drivetrains is relatively low, around 64–81% [47]. This critical problem has motivated many researchers in multiple fluid power applications to propose, design, and implement new techniques and methodologies to improve the performance of hydraulic hybrid transmissions by enhancing the maximum efficiency of variable displacement vane and piston pumps [48–55]. For instance, the work in [49] addresses the pre-defined problem by proposing a dynamic seven-stage model that predicts the thickness of the fluid film between the valve plate and cylinder block in axial piston pumps to get accurate information on lubrication. In [51], relying on a coating methodology, the authors improved the overall efficiency map of axial piston pumps by minimizing the friction losses by applying nanocoatings to the pump's slippers. Besides these works, the work in [54] provides research results for a previously executed research study. The resulting outcomes can be used to enhance the efficiency of many technological systems in power engineering and transport involving hydraulic pumps and motors. Similarly, the work in [55] focuses on improving the numerical algorithm used for solving the optimal control problem for the three-dimensional heat equation, which can improve many technological systems like hydraulic units.

All the proposed techniques in the pre-listed works still result in low overall efficiencies when operating at partial volumetric displacements. There is no noticeable improvement in the pumps' lowest efficiency at low volumetric displacements [56]. Accordingly, the work presented in [42] proposed integrating digital technologies into fluid power systems through designing digital variable displacement pump/motors that allow maintaining high efficiencies over a wide range of displacements.

Building on the results from [41,42,57,58], this work aims at enhancing the performance and efficiency of a hydraulic hybrid transmission by simulating the implementation of an experimentally-validated digital pump/motor into a hydraulic hybrid transmission. As will be proved through a physical-based simulation, unlike the ordinary variable displacement pump/motors, the adopted digital pump/motors pursue to overcome the efficiency limitation in hydraulic hybrid drivetrains by increasing the maximum overall efficiency of the transmission.

5. Digital Hydraulics

Digital Hydraulics is a pioneering sub-field of fluid power, which relies on advanced on/off control technologies to enhance hydraulic systems' performance, reinforce advancements in fluid power applications, and minimize energy consumption [59]. This technology can replace conventional fluid power components, such as directional control valves, with multiple switching on/off valves in a parallel configuration to improve efficiency [41]. Such valves' configurations promote a superior design in this application because they use a poppet-style actuation that does not allow leakage across the valve, commonly seen in spool valves. In turn, this configuration reduces the system's losses, enables more control, increases the performance and efficiency, and reduces the cost of maintaining components [60–62]. This technology has also been utilized in various other configurations to improve the efficiency of conventional systems, including replacing traditional check valves on fixed displacement pumps. It also enables the opportunity to limit and divert flow in conventional fixed displacement pumps allowing variable displacements [44].

Some conventional piston pump/motors, e.g., reciprocating piston pumps, utilize mechanical check valves on the inlet and outlet ports of the pump to separate the system's high- and low-pressure sides. During pumping, this configuration allows fluid to enter through the low-pressure check valve when the pump is on its intake stroke. When the pump exhausts its cylinder, pressure closes the intake check valve and opens the exhaust check valve, displacing fluid to the high-pressure side of the system. This configuration limits the pump to fixed displacements since the pressure differential in the chamber is the only method of controlling the actuation of the intake and exhaust valves. Thus, the work in [42] proposes an electrically actuated digital inline 3-piston pump that replaces the conventional valves with digital switching on/off valves. The proposed digital pump/motor is a 28cc/rev unit comprising three displacement chambers, each with two high-speed switching on/off valves. Figure 4 shows the pump/motor's configuration for a single displacement chamber. The figure shows that Valve 1 controls the flow at the low-pressure side, and Valve 2 controls the flow at the high-pressure side. The proposed configuration allows opening or closing the valves at any time throughout the pumping/motoring cycle. This configuration adds a degree of controllability and flexibility to the system, enabling various digital pumping/motoring techniques, referred to as flow diverting and limiting operating strategies [44,63].

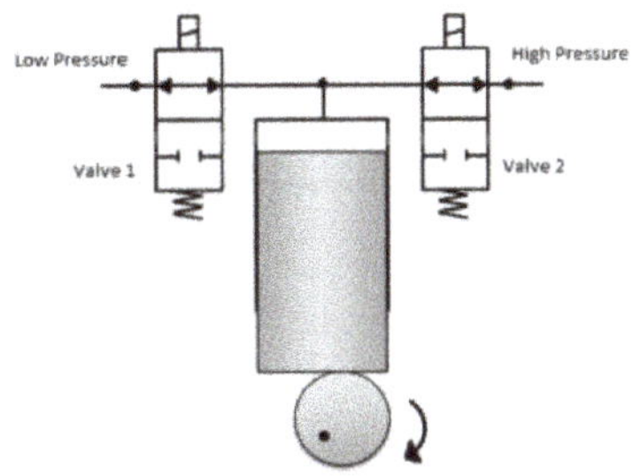

Figure 4. Single-piston digital pump/motor configuration [64].

Besides the electrically actuated digital pump, our team designed and imple-mented a mechanically actuated digital pump, which can also achieve high efficiencies over a wide range of displacements. The following sub-sections exhibit an overview of the internal mechanical structure of the mechanically actuated digital pump, intro-duce the electrically actuated pump's most efficient operating strategy, and present the experimental testing and validation on the electrically actuated pump.

5.1. *Internal Mechanical Structure of the Mechanically Actuated Digital Pump*

This subsection gives a brief introduction to the mechanical structure of the mechanically actuated digital pump, which can achieve efficiencies higher than the ones reported in this paper. The internal mechanical structure of the intended digital pump, shown in Figure 5, comprises three main sub-systems: (1) gear-cam unit, (2) cam-follower unit, and the (3) valve manifold. The gear-cam unit utilizes three main half-masking cams, where each cam consists of two sub-cams that are phased relative to one another to allow achieving different cam profiles. This unit is responsible for the phasing mechanism of the camshafts, where it allows varying the cams' orientation relative to the shaft during rotation. The cam-follower unit consists of the three followers, their mount bracket, and the three valves to be actuated via the cam-follower mechanism. This unit allows converting the rotational motion at the cam to a linear motion at the follower and thus controlling the valves' opening and closing. Therefore, varying the cam profile using the phasing mechanism permits varying the opening and closing of the valves and thus enables achieving variable pump displacement. A detailed description of the mechanical structure and the pumping mechanisms is presented in [65].

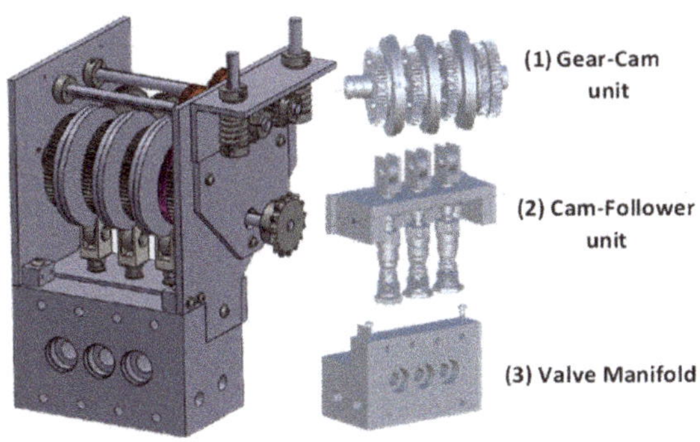

Figure 5. CAD drawing showing the basic units of the mechanically actuated digital pump.

5.2. *Operating Strategies of the State-of-the-Art Electrically Actuated Digital Pump*

The electrically actuated digital variable displacement pump/motor [42] relies on the control flow entering and leaving the displacement chambers. The pump/motor displacement variation can be achieved by controlling the actuation of two on/off valves on the intake and exhaust ports of the pump. The control and timing of the valves' actuation enable various pumping/motoring operating strategies. Four different operating strategies have been explored and tested [42] utilizing the state-of-the-art digital pump. The four different operating modes are Partial Flow-Diverting (PFD), Partial Flow-Limiting, Sequential Flow-Diverting (SFD), and Sequential Flow-Limiting (SFL). The most efficient operating strategy discovered is the SFL [63]. SFL is a technique that permits the states of each cylinder to be enabled or disabled, sending either 100% or 0% displacement to the output valve, respectively. This method can be used for both pumping and motoring cycles.

An example of a 100% displacement pumping cycle is shown in Figure 6. This cycle achieves similar functionality as the mechanical check valve. During pumping, the intake valve is opened for the entire stroke (A) and closed when the exhaust stroke begins (B). The exhaust valve is then opened and held until the exhaust stroke is completed (C). This transfers 100% of the fluid entering the piston chamber on the intake stroke, disregarding the dead volume in the cylinder.

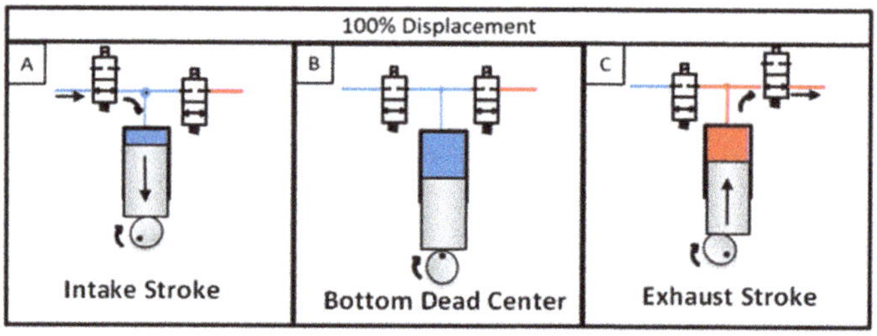

Figure 6. Enabled chamber operating at 100% displacement during SFL mode.

An example of 0% displacement for pumping is shown in Figure 7. During this cycle, all intake and exhaust valves remain closed for the entire process, allowing the fluid to be compressed and decompressed. As the cylinder begins its intake stroke, the piston starts to decompress the fluid in the dead volume of the cylinder (A) until the cylinder reaches the Bottom Dead Center (B), which allows depressurizing the fluid and creating a partial vacuum in the chamber. Once the piston begins the exhaust stroke (C), the fluid is compressed back to its original state. Since the intake and exhaust valves remained closed during this cycle, 0% displacement is delivered by the cylinder. Thus, this cycle allows the chamber's dead volume fluid to be depressurized and pressurized, and no fluid is delivered to the system.

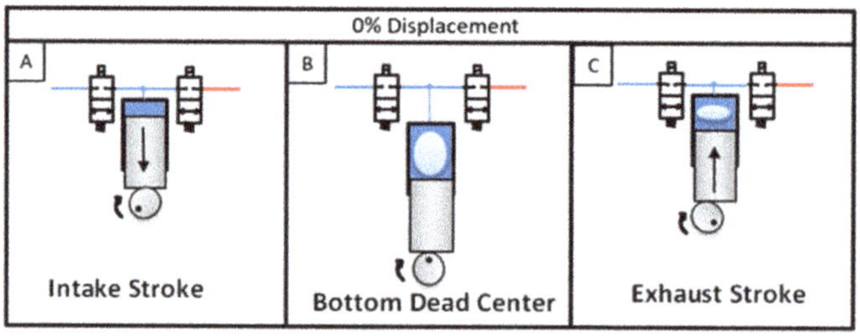

Figure 7. Disabled chamber operating at 0% displacement during SFL mode.

The partial vacuum created during the intake stroke is called chamber voiding. The chamber voiding phenomenon was experimentally inspected by [66]. It was experimentally proved that the gas created during the chamber voiding phase, i.e., (B), would dissolve entirely as the fluid is re-compressed.

The central concept of this strategy relies on varying the state of the chambers sequentially to change the 'ump's displacement. To illustrate, a three-piston pump configuration is shown below in Figure 8. To achieve 66% pumping displacement, two chambers (1 & 2) would operate at 100% displacement, while the third chamber (3) would run at 0% displacement.

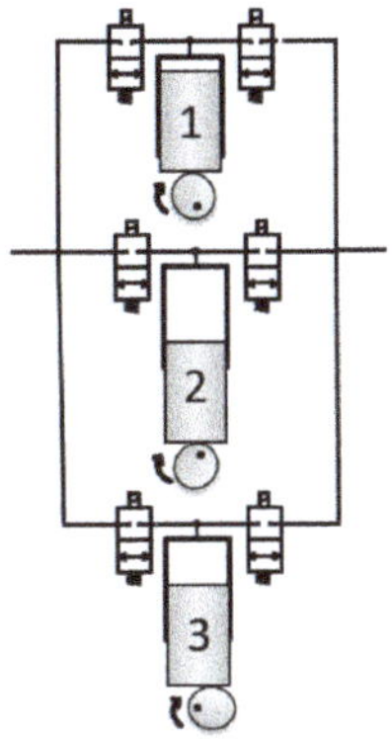

Figure 8. Three-piston digital inline piston pump configuration.

Sequencing is averaged over various pump revolutions to achieve a wide range of displacements between 0, 33%, 66%, and 100% [42]. As a case in point, to achieve 11.11% displacement, the first cycle of the pump would turn on one cylinder (1) and turn off two cylinders (2 & 3). The following two pump revolutions would sequence all cylinders (1, 2 & 3) off. This operation achieves 1/9th, or 11%, displacement over the three cycles. A ~1% displacement resolution can be achieved by averaging the sequences over 32 shaft revolutions.

5.3. Experimental Validation of the Electrically Actuated Digital Pump/Motor

The experimental test stand used for testing the SFL operating strategy is shown in Figure 9. The test stand involves the intended 28 cc/rev digital pump, two accumulators (one at the suction side and one at the discharge side), manifold housing with six on/off valves (two for each piston chamber), and other hardware used to complete the hydraulic circuit, data acquisition, and control. The two accumulators were added to the system to reduce pressure and flow ripples and avoid cavitation problems during the suction phase. For instance, if the hydraulic system experienced sudden stops during suction, rapid changes in the system pressure will generate pressure ripples causing shock waves. In this case, the system's pressure will go above the accumulator's pre-charged pressure. The accumulator at the discharge port will act as a shock cushioning. It will allow the excessive flow to enter, compressing the gas cushion (nitrogen bladder) and thus cushion and minimize the shock. The second accumulator (one located at the suction port) is required because the pump is bi-directional, i.e., the pump can discharge fluid from either port. A detailed description of the hardware used is presented in [66].

The experimental testing of the SFL pumping/motoring strategy was conducted under steady-state conditions at a temperature T = 50 °C and different pump displacements, 25%, 50%, 75%, and 100% displacement. Moreover, the SFL pumping and motoring cycles were examined at different system pressures ranging from 4 MPa to 18 MPa and shaft speeds up to 700 RPM. Operating the digital pump at shaft speeds beyond 700 RPM requires high-speed switching valves, affecting the valves' variability and causing non-linearity and errors in the valves' timing (opening and closing time). A timing valve correction algorithm was developed in [67] and utilized throughout the experimental testing to overcome this problem and account for the valve timing error. The developed algorithm obtains the delay in the valve timing using the pressure readings at the low- and high-pressure ports, thus regulating the valves' opening and closing time. Although the adopted algorithm enabled high-speed switching, the valves' opening and closing speeds are not high enough. Thus, the maximum operating shaft speed was limited to 700 RPM throughout the research study.

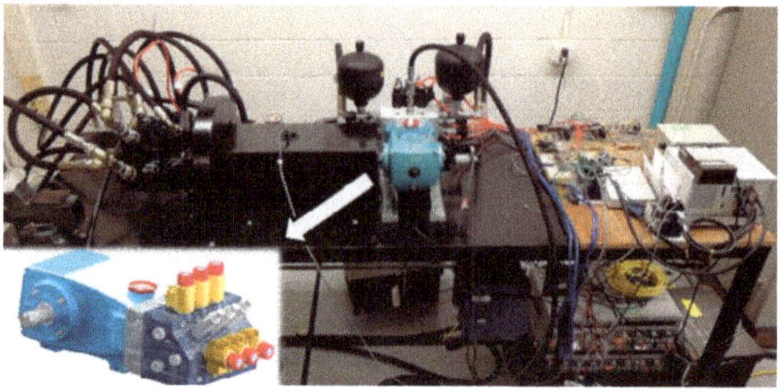

Figure 9. The experimental test stand for testing different operating strategies of the electrically actuated digital 3-piston pump [42].

After executing the experimental testing, the required data (torque, pressure, flow, etc.) were acquired at various pump displacements (25%, 50%, 75%, and 100% displacement), system pressures (4 MPa, 10 MPa, and 18 MPa), and shaft speeds (300 RPM, 500 RPM, and 700 RPM). Then, the efficiencies were calculated according to the obtained data.

The experimental outcomes resulting from operating the SFL strategy at 100% displacement during the pumping and motoring cycles are shown in Figures 10 and 11. Figure 10 illustrates the calculated efficiency curves across various system pressures and shaft speeds at 100% displacement. As shown, at high shaft speeds and low pressures, the pump can achieve efficiencies of at least 78%, while at low shaft speeds and high pressures above 16 MPa, the pump can achieve high efficiencies up to 92%. Furthermore, the system pressure significantly impacts the overall pump's efficiency, where the pump's efficiency significantly increases as pressure increases.

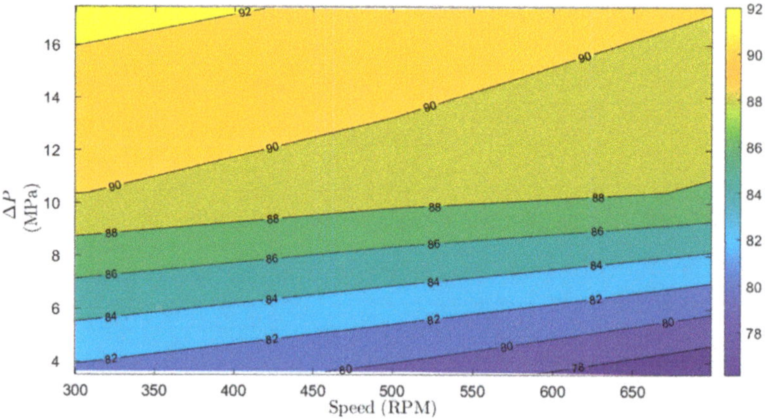

Figure 10. Efficiency map of the electrically actuated digital pump/motor operated in the SFL pumping strategy at 100% displacement.

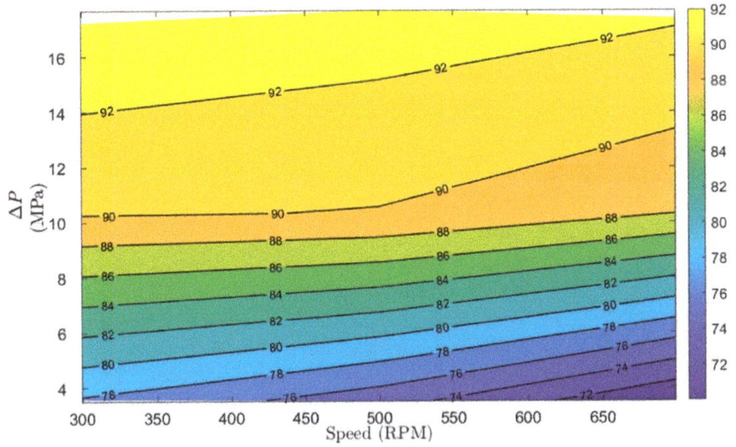

Figure 11. Efficiency map of the electrically actuated digital pump/motor operated in the SFL motoring strategy at 100% displacement.

Similarly, the efficiency map corresponding to operating the pump with the SFL strategy at 100% displacement during the motoring mode is shown in Figure 11. As shown in the figure, the motoring mode enables efficiencies of at least 72% and up to 92%. Unlike the results of the pumping mode, in motoring mode, the efficiency curves are almost parallel to the x-axis, which means that the shaft's speed has a negligible impact on the pump's efficiency. For instance, we can still achieve efficiencies from 72% to 92% at any RPM, depending on the system's pressure.

6. Simulation Model: Hydraulic Hybrid Drivetrains Utilizing Digital Pump/Motors

This section presents a physics-based series hydraulic hybrid drivetrain simulation conducted on MATLAB Simscape, a graphical programming software created by MathWorks [68]. The simulated MATLAB Simscape model mimics regular terrain hydraulic vehicles' series hydraulic hybrid transmission. The goal of the model is to investigate the improvement of the hydraulic hybrid transmission performance after replacing the conventional pump/motors with the digital pump/motor units. The intended state-of-the-art digital units are emulated by applying the digital motoring/pumping data acquired at the tested displacements (25%, 50%, 75%, and 100% displacement), pressures (4 Mpa, 10 MPa, and 18 MPa), and shaft speeds (300 RPM, 500 RPM, and 700 RPM) to the physical simulation components (pump and motor). The digital data are applied to the physical components of the simulation model by converting it into lookup tables imported to the physical pump and motor, employed throughout the simulation.

The following sub-sections present the designed circuit schematic, display the simulation model, design, and functionality during the three transmission stages (acceleration, cruising, and braking), and discuss the resulting outcomes.

6.1. Circuit Schematics

The adopted series hydraulic transmission, illustrated in Figure 12, comprises a variable displacement pressure compensated pump (5) driven by an electric motor (2), a mechanical clutch (3) to engage and disengage the motor's shaft from the rest of the system, a pressurized gas accumulator (1) for energy storage-boost purposes, a low-pressure reservoir (8), and a variable displacement pump/motor (11) for power transmission. The pump/motor (11) operates as a motor during acceleration and cruising, i.e., motoring mode, and works as a pump during braking, i.e., pumping mode. A check valve (6) is added to block the backflow (flow from and into the accumulator) from entering the pump's discharge port. Two 4/3 way (9 & 10) and one 3/2 way (7) directional flow control valves allow a seamless operation during the motoring and pumping modes. The 4/3 way valves (9 & 10) are closed-center valves with spring return having two directly actuated solenoid coils, while the 3/2 way valve (7) is a spring return valve with one actuated solenoid coil. As demonstrated through the simulation model, valve (9) is employed during the motoring mode, i.e., acceleration and cruising, where the user controls the valve based on the direction of motion, i.e., forward or backward drive. However, valve (10) is used during the pumping mode, i.e., regenerative braking, and it is controlled based on the direction of rotation received by the unit's shaft (11). Valve (7) allows switching between valves (9) and (10) based on the system's operation mode. During the motoring mode, i.e., acceleration and cruising, valve (7) is set at its un-actuated position, allowing pumping the fluid into the motor (11) through valve (9). However, during the pumping mode, i.e., at braking, the mechanical braking signal will be converted into an electrical signal at valve (7), forcing it to switch to the other position and thus allowing the motor (11) to pump the flow from the tank into the accumulator through the valve (10).

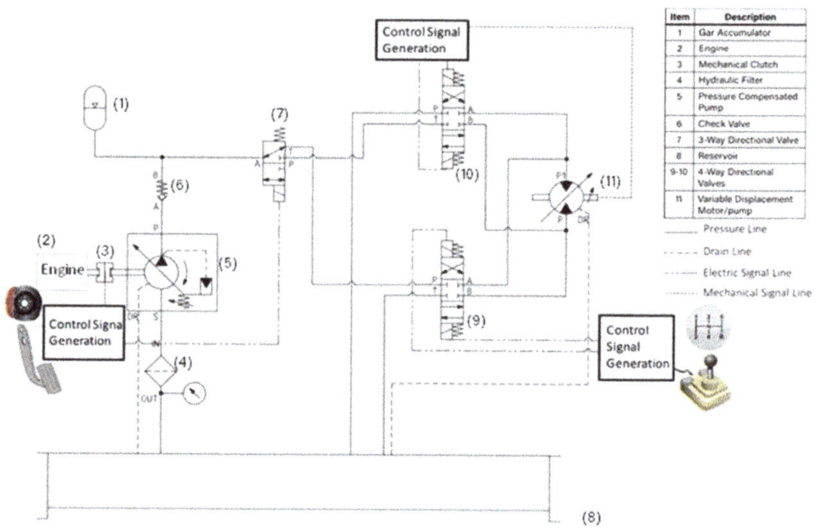

Figure 12. A circuit illustrating the simulated series hydraulic hybrid transmission.

6.2. Simulation Model, Design, and Operation

Figure 13 exhibits the physical model for the series hydraulic hybrid transmission illustrated in Figure 12. The adopted system is modeled and simulated using Simscape Fluids and Drivetrains libraries on MATLAB-Simulink. The simulation model is constructed by integrating the required numerical block diagrams with the physical paradigms. All the hydraulic physical components, i.e., pump, motor, valves, and other hydraulic units, are compiled using interfaceable physical connections. The figure shows that the model involves a pressure compensated pump and a variable displacement motor that operates as a motor when pumping high-pressure fluid into one of its ports and as a pump when receiving a mechanical signal at its output shaft. Furthermore, the model involves physical hydraulic sensors to sense the flow and the pressure at specific points in the system. For instance, pressure-flow sensors are installed at the pump's discharge port, accumulator's port, and motor's inlet and outlet ports. The accumulator's flow sensor is installed into the accumulator. Then, a positive flow rate at the accumulator's port designates flow into the accumulator; during charging. However, a negative flow denotes flow out of the accumulator; during discharging. Besides the pressure-flow sensors, mechanical sensors, e.g., rotational motion and torque sensors, are added at the pump's and motor's shaft to measure the resulting shaft's speed and torque. Acquiring this data (torque and speed) allowed us to compute the input and output mechanical powers. The power data was then exported into the MATLAB workspace for calculating the average input and output powers. The average input and output powers were individually calculated over the entire cycle using the mean function in MATLAB.

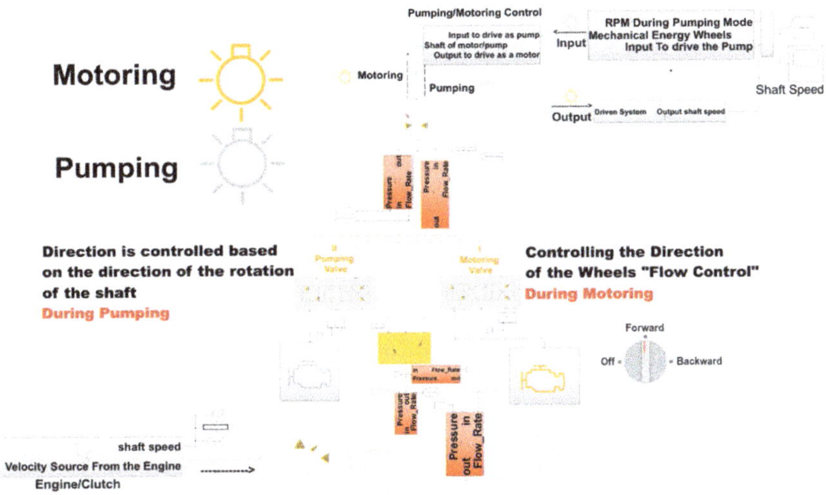

Figure 13. A MATLAB Simscape simulation model for the series hydraulic hybrid transmission.

The digital pumping/motoring data presented in Section 5.2 are converted into lookup tables and then imported to the physical pumping/motoring units (the utilized pressure-compensated pump and the variable displacement motor) to incorporate the pre-introduced digital pump/motor. The pumping data is imported into the physical motor (11) during braking as a lookup table. Also, it is imported into the physical pump (5) during cruising, knowing that the pump will be off during acceleration. Similarly, the motoring data is applied to the physical motor as a lookup table during acceleration and cruising. Thus, the physical pump and motor (pressure-compensated pump and the variable displacement motor) are sized based on the imported lookup tables, which allows setting the maximum and minimum pressures in the system in addition to the required flow rate. Then, the accumulator and other hydraulic components, e.g., valves, fittings, hoses, etc., are sized and selected based on the acquired system's pressure and flow rate. Based on the digital pumping/motoring data, the hydraulic units (5) and (11) will operate between 300 RPM and 700 RPM with a maximum displacement of 28 cc/rev. Furthermore, the hydraulic units will work under a volumetric efficiency up to 94% and overall efficiency of at least 78% and up to 90% during pumping. However, unit (11) will operate during motoring under volumetric and total efficiencies up to 99% and 92%, respectively. Furthermore, the simulation model accounts for the system's internal losses, e.g., losses in hydraulic pipelines, fittings, and valves. These internal losses are simulated by adding hydraulic resistance, capacitance, and inductance to the distributed parameter model using the simscape block models for hydraulic pipelines, fittings, and hoses.

During the series hydraulic hybrid transmission, the gas accumulator serves as a hydraulic lunch assist device to provide auxiliary flow during high demand requirements (within the acceleration stage), forcing the gas inside the accumulator to expand and compress under adiabatic conditions. Based on the intended application, the accumulator is sized using the following equations [69]:

$$P_1 V_1^k = P_2 V_2^k \tag{1}$$

$$V_1 = \frac{V_x(P_3/P_1)^{1/k}}{1 - (P_3/P_2)^{1/k}} \quad (2)$$

where P_1, P_2, and P_3 in Pa denote the accumulator's pre-charged pressure, system maximum operating pressure, and system minimum operating pressure, respectively, set based on the experimentally validated data of the digital pump. V_1 and V_2 in L designate the nitrogen's volume at P_1 and P_2, respectively, V_x in L denotes the required fluid volume, and $k = 1.4$ is the nitrogen constant during adiabatic operation. To prevent the accumulator's poppet (poppet that opens and closes, allowing fluid to enter and leave) from fully closing during the expansion of the nitrogen bladder, the accumulator's pre-charged pressure P_1 has to be set between 60% to 80% of the minimum system pressure [70,71]. Thus, for P_2 = 1.9e + 7 Pa, P_3 = 5.3e + 6 Pa and $P_1 = 0.75P_3$, we get V_1 = 36.3 L and V_x = 18 L.

The adopted physical model comprises two emulation options, options one and two shown in Figure 14, for simulating the intended system. Option one allows for a user interface simulation. The user can interact with the transmission model by switching between the motoring and the pumping modes, thus mimicking the driving and braking situations. Given this option, the user generates the duty cycle by choosing when to drive and brake. On the other hand, option two enables a complete auto-simulation over the three transmission stages (acceleration, cruising, and braking). The simulation will be done automatically based on the provided duty cycle for this option.

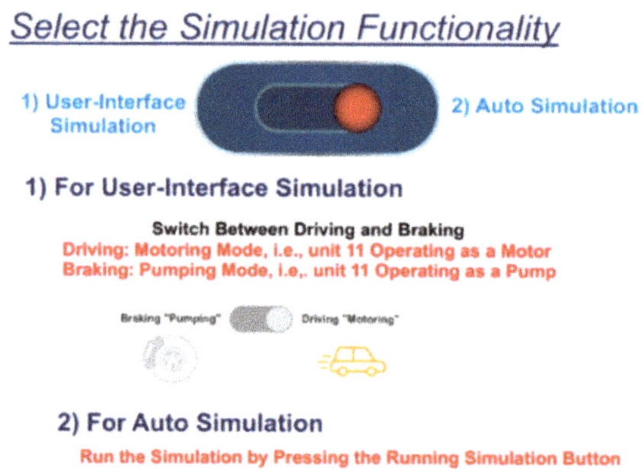

Figure 14. User interface illustrating the functionality of the simulation model.

Relying on option two, several drive cycles were adopted to test the simulation model's capability over the three transmission stages. The total time frame of the tested drive cycles ranged from 200 s/cycle to 300 s/cycle, where the acceleration and braking phases required long times up to 60 s and 40 s, respectively. The acceleration and braking timing is due to the accumulator's charging and discharging time, which was selected based on the experimentally acquired pump data that was limited with a maximum and minimum pressure around 19.1 MPa and 5.3 MPa, respectively.

One of the tested duty cycles is shown in Figure 15. The figure shows that the physical system is simulated for 400 s for two consecutive cycles. Each cycle comprises the three stages of the series hydraulic hybrid transmission (acceleration, cruising, and braking). During the first cycle, the acceleration phase occurs between 0 and 60 s, the cruising starts at time = 60 s and continues until time = 140 s, and the re-regenerative braking takes place within 140 and 180 s. The second cycle starts 20 s after the first cycle, which accounts for the complete stopping of the vehicle. Similarly, the three transmission stages (acceleration, cruising, and braking) occur for 60, 80, and 40 s, respectively, during the second cycle.

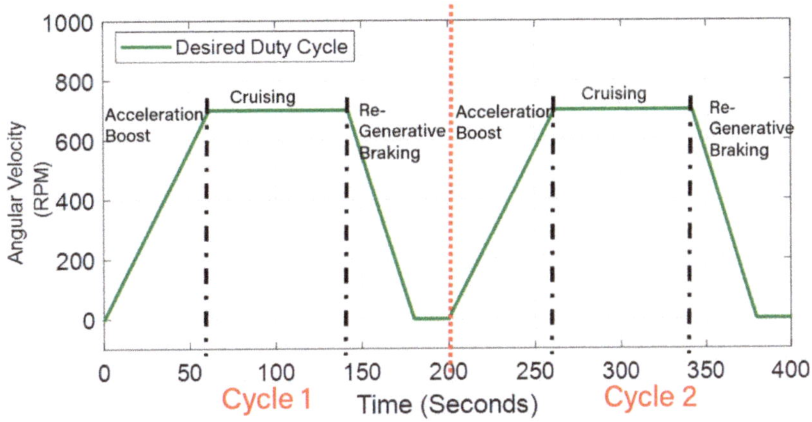

Figure 15. The desired duty cycle.

To fully simulate the hydraulic hybrid series transmission system for achieving the desired duty cycle in Figure 15, it is required to apply the desired operating angular velocity at the system's input port and specify its load. The system's operating RPM is simulated by applying an input signal builder using a physical velocity source. The transmission's total mass was included within the physical paradigms of the utilized hydraulic units and all components in the Simscape environment. The transmission's load is simulated using four Simscape Tire blocks, which simulate the tires by their radius and ideal inertia. In [72], Valeski approximates the average tire's weight and diameter to be around 10 kg and 40 cm, respectively. Thus, the calculated moment of inertia on each tire was around 0.37 kg·m^2. Given the calculated moment of inertia on each tire, a moderate physical torque source with an ideal moment of inertia of 1.5 kg·m^2 was applied at the motor's shaft. Now, we explain the operation of the simulation model during each stage.

6.2.1. Acceleration

During acceleration, the accumulator is fully-charged with pressurized fluid, and its pre-charged pressure is greater than the system's operating pressure. At this phase, the pump is off, and thus the accumulator acts as an auxiliary power supply. The accumulator discharges its stored energy into the motor (11), operating in its motoring mode, to accelerate the vehicle. The valves' opening/closing operation is controlled based on the accumulator's pre-charged pressure. For instance, during this phase, because the pre-charged pressure of the nitrogen inside the accumulator is greater than the system's pressure, no signal is generated at the valve (7). Then, valve (7) is kept at its un-actuated position to pass the flow through the valve (9) and block it at valve (10).

6.2.2. Cruising

At the end of the acceleration phase, the fluid inside the accumulator starts decreasing, and the pre-charged pressure becomes below the system's pressure. Namely, the accumulator discharged almost all its stored energy, and thus the system requires an additional hydraulic power supply to drive the wheels. At this stage (cruising), the drop in the pre-charged pressure triggers an electric signal at the engine and forces the clutch to engage the engine to the rest of the system. The pump starts pumping the fluid from the low-pressure reservoir (hydraulic tank) into the hydraulic motor (11) that is still operating as a motor to execute the motion at the wheels. Like the acceleration phase, the fluid is delivered to the motor through the valve (9), where valve (10) is still closed.

6.2.3. Braking

During braking, the motor (11) switches to the pumping mode, and the clutch decouples the engine from the rest of the system, forcing the pump (5) to shut down. The pre-charged pressure of the nitrogen gas inside the accumulator drops below the system's pressure. At this stage, an electric signal is triggered at valve (7), forcing it to switch to the second position and thus pumping the fluid through valve 10 and blocking it at valve 9. The motor utilizes the braking kinetic energy at its output shaft to pump fluid from the tank into the accumulator converting the mechanical energy into hydraulic energy stored in the accumulator to be delivered during acceleration.

7. Resulting Outcomes

After introducing the simulation model and discussing its fundamental operation, throughout the following sections: Sections 7.1 and 7.2, we present the simulation outcomes resulting from applying the digital motoring/pumping data into units (5) and (11) at moderate load-speed conditions (1.5 kg·m^2 and 700 RPM).

7.1. Pressure-Flow Outcomes

Figures 16 and 17 illustrate the variations in the system's pressure and flow throughout the proposed duty cycle. They show the pressure-flow outcomes at different ports in the system during the three transmission stages (acceleration, cruising, and braking). As shown in the figures, during the acceleration phase (0 to 60 s in cycle one and 200 to 260 s in cycle two), the pump's shaft angular velocity is kept at zero, then triggered after 40 s. The pump started operating after 40 s because the accumulator had enough stored energy at the start of the acceleration phase (from 0 to 40 s). Thus, the clutch disengages the engine's shaft from the pump's shaft, causing a zero RPM at the pump's input shaft. The accumulator acts as a hydraulic launch assist and discharges its stored energy into the motor. Thus, the flow rate at the accumulator's port flows in the negative direction, i.e., from the accumulator into the system, as shown in Figure 16. A pressure drop of about 10 MPa is created between the motor's inlet and outlet ports, as shown in Figure 17, allowing it to convert the hydraulic energy into mechanical energy enabling the wheels to accelerate until the output shaft speed reaches 600 RPM. As the accumulator delivers the auxiliary power, the fluid's pressure at the accumulator's port, shown in Figure 16, decreases because the accumulator's poppet is still open to deliver fluid to the system. After the accumulator discharges about 80% to 85% of its stored energy, i.e., at time = 40 s, the pump starts operating via the clutch-engine mechanism maintaining the system's output angular velocity at about 600 RPM, where its discharge pressure, captured in Figure 16, increases up to 6–7.5 MPa. After 20 s, at time = 60 s, the vehicle undergoes the cruising stage. The pump operates at its maximum speed (700 RPM) with an operating pressure of around 7.5 to 8 MPa. The motor's output shaft operates at almost moderate speed (500 RPM). During this stage, the pressure at the accumulator's port, shown in Figure 16, is the same as the system's pressure because the accumulator's poppet is closed. Thus, the accumulator is neither charging nor discharging. After 80 s, the vehicle undergoes the braking phase, in which the clutch starts disengaging the engine shaft from the pump shaft resulting in a decrease in its RPM. The motor (11) switches to its pumping mode during this phase, pumping the fluid from the tank into the accumulator. The pressure at the motor's outlet port, shown in Figure 17, increases up to 17.5 MPa causing a high pressure at the accumulator's port and the pump's discharge port, shown in Figure 16. The accumulator starts charging, where its flow rate, shown in Figure 16, flows in the positive direction, i.e., from the system into the accumulator.

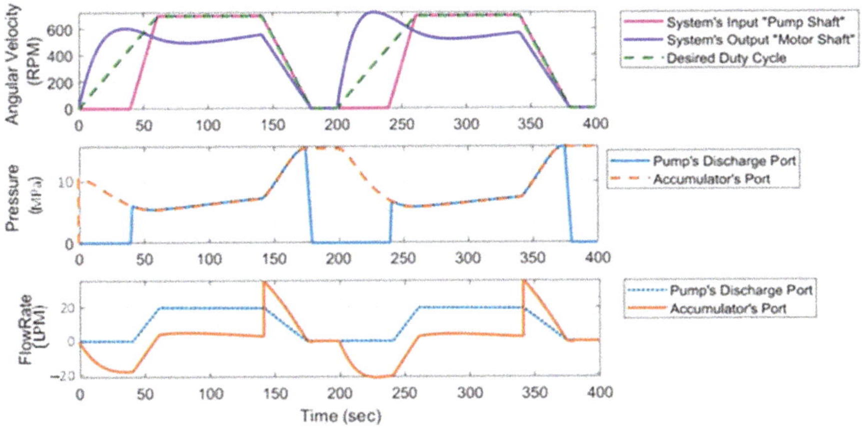

Figure 16. The resulting pressure and flow simulation outcomes at the pump's discharge and accumulator's ports throughout the desired duty cycle.

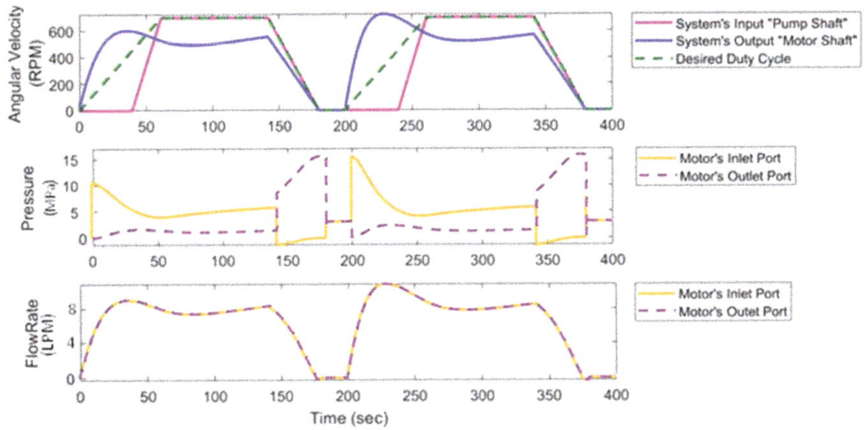

Figure 17. The resulting pressure and flow simulation outcomes at the Motor's inlet and outlet ports throughout the desired duty cycle.

Figures 18 and 19 show the variations in the accumulator's fluid volume and the motor's pressure drop, respectively, throughout the desired duty cycle for the three transmission stages (acceleration, cruising, and braking). As discussed previously, during the acceleration phase, the accumulator discharges its stored energy into the motor's inlet port to accelerate the motion at the wheels. Thus, the motor operates in its motoring mode. The accumulator's discharging process causes a decrease in fluid volume, shown in Figure 18, and a positive pressure drop of about 10 MPa at the motor, shown in Figure 19. As shown in Figure 19, the 10 MPa pressure drop across the motor slightly decreases until 5–6.5 MPa. Then, because the accumulator is neither charging nor discharging during the cruising phase, its fluid volume, shown in Figure 18, is kept almost constant. During this phase, the pressure drop across the motor, shown in Figure 19, is maintained at around 7.5 MPa. During braking, the accumulator's fluid volume, shown in Figure 18, increases because the accumulator is charging. A negative pressure drop is created across the motor, as shown in Figure 19, where the motor is operating in its pumping mode.

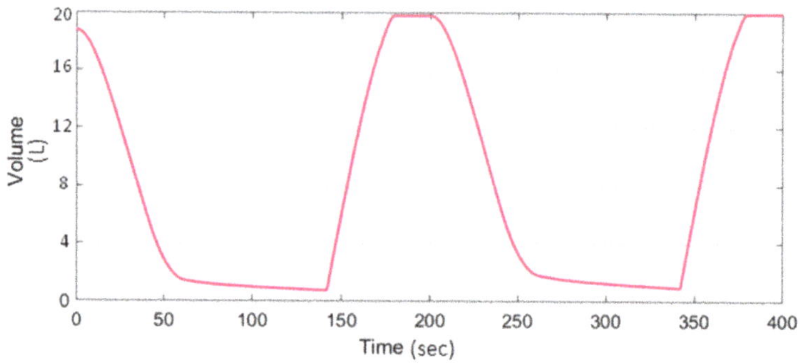

Figure 18. Simulation outcomes for the fluid volume inside the accumulator.

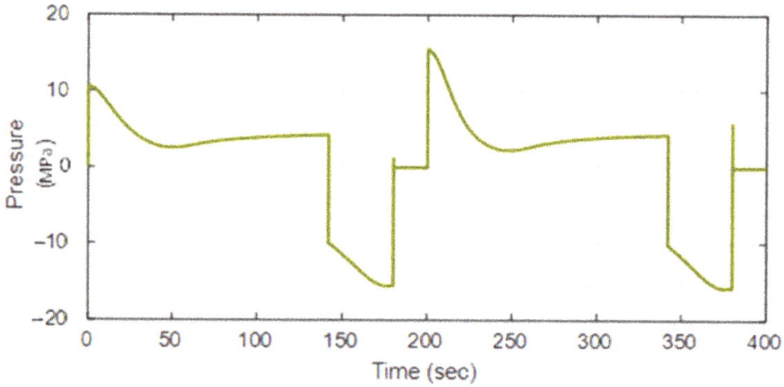

Figure 19. Simulation results for the pressure drop across the motor, i.e., the pressure difference between the inlet and outlet motor's pressure.

7.2. Power-Efficiency Outcomes

Figure 20 shows the load-speed outcomes and the resulting mechanical power at the pump and motor shafts, i.e., the system's input and output shafts, respectively, throughout the three transmission stages (acceleration, cruising, and braking). During acceleration, the pump is not operating, and thus the accumulator is the only power supply in the system. The accumulator discharges its stored energy into the motor. The motor converts the hydraulic energy into mechanical energy resulting in an output mechanical power of around 1300 W in cycle one and about 2000 W in cycle two. The difference in the mechanical powers in cycles one and two is due to the accumulator's initial stored energy. During cruising, the pump starts operating at around 700 RPM, producing an input mechanical power of about 2100 W. This results in an output mechanical power of around 1000 W at the motor's shaft due to the 1.5 kg·m^2 inertia added at the wheels. During braking, the motor switches to its pumping mode. It utilizes the braking energy and converts it into hydraulic energy stored in the accumulator.

Simulating the system to achieve the desired duty cycle at moderate load-speed conditions (1.5 kg·m^2 and 700 RPM) results in average input power of 876 W at the system's input shaft. This results in an average output power of 780 W at the system's output shaft, which outputs an overall efficiency of around 89%. This result reveals a significant improvement in the efficiency of the series hydraulic hybrid drivetrains. Due to the rated efficiency of the conventional hydraulic pump/motors, the total average efficiency of the series hydraulic hybrid transmission is around 64% to 81%, which is relatively low [47]. Therefore, the utilization of the state-of-the-art multi-piston digital pump reinforced the system's performance by increasing the overall efficiency at least 8% and up to 25%.

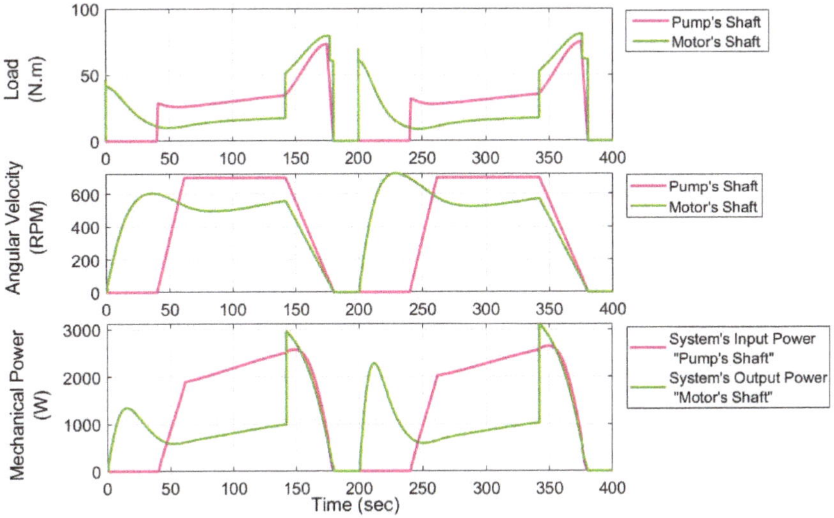

Figure 20. Load-Speed and mechanical power simulation results.

8. Conclusions

The overall average efficiency of series hydraulic hybrid drivetrains is comparatively low, between 64% and 81%, due to the utilization of conventional variable displacement pumps that become inefficient when operating at partial displacements. This work estimates and addresses the efficiency improvement of a series hydraulic hybrid transmission by integrating hydraulic hybrid technology with digital hydraulics. It proposes replacing the conventional piston pump/motor employed in series hydraulic hybrid drivetrains with the multi-piston digital pump designed and implemented in preceding works. The proposed methodology is tested by conducting a physical-based simulation of a series hydraulic hybrid drivetrain on MATLAB Simscape. The state-of-the-art multi-piston digital pump is simulated by applying its digital motoring/pumping data into the power supply units, i.e., pump and motor utilized throughout the simulation model. The adopted simulation model permitted evaluating the performance of the series hydraulic hybrid drivetrain, and it allowed examining the improvement of the system's overall efficiency. The final simulation outcomes resulted in 89% overall efficiency and thus exceeded the conventional series hydraulic hybrid drivetrain (64% to 81%). This resulting efficiency reveals that the utilization of the state-of-the-art digital pump enables enhancing the total efficiency of the hydraulic hybrid drivetrains by at least 8% and up to 25%. The attained 89% efficiency can also be better enhanced using a more refined digital unit. The digital pump utilized throughout the simulation model was a proof of concept prototype limited to 700 RPM shaft speed. Thus, a refined one would run at a broader range of inputs and achieve even better results. As future work, we aim to transform the proof-of-concept digital pump into a higher-speed refined unit to test the proposed transmission experimentally.

Author Contributions: Conceptualization, F.B. and I.A.; Investigation, J.G.-B. and F.B.; Methodology, I.A. and F.B.; Supervision, J.G.-B. and F.B.; Validation, J.G.-B.; Visualization, I.A. and F.B.; Writing—original draft, I.A., K.P. and J.G.-B.; Writing—review & editing, I.A. and F.B. All authors have read and agreed to the published version of the manuscript.

Funding: This research received no external funding.

Data Availability Statement: The data that support the findings of this study are available on request from the corresponding author, F.B.

Conflicts of Interest: The authors declare no conflict of interest.

References

1. McCloy, D.; Martin, H.R. *Control of Fluid Power: Analysis and Design*, 2nd ed.; Ehlh: Chichester, UK, 1980. Available online: https://ui.adsabs.harvard.edu/abs/1980ehlh.book.....M/abstract (accessed on 10 October 2021).
2. Durfee, W.; Sun, Z. *Fluid Power System*; A National Science Foundation Engineering Research Center: Minneapolis, MN, USA, 2009; pp. 1–48.
3. Vacca, A. Energy Efficiency and Controllability of Fluid Power Systems. *Energies* **2018**, *11*, 1169. [CrossRef]
4. Konami, S.; Nishiumi, T. *Hydraulic Control Systems*; John Wiley & Sons: Singapore, 2016; pp. 187–299. [CrossRef]
5. Fabis-Domagala, J.; Domagala, M.; Momeni, H. A Concept of Risk Prioritization in FMEA Analysis for Fluid Power Systems. *Energies* **2021**, *14*, 6482. [CrossRef]
6. What is Fluid Power? National Fluid Power Association. 2021. Available online: https://www.nfpa.com/home/AboutNFPA/What-is-Fluid-Power.htm (accessed on 1 December 2021).
7. U.S. Fluid Power Industry Brief. 2021. Available online: https://www.nfpa.com/home/industry-stats/Industry-Brief.htm (accessed on 1 December 2021).
8. Love, L.; Lanke, E.; Alles, P. Estimating the Impact (Energy, Emissions and Economics) of the US Fluid Power Industry. In Oak Ridge National Laboratory (Issue December). 2012. Available online: http://www.osti.gov/servlets/purl/1061537/ (accessed on 1 December 2021).
9. Kwon, H.; Sprengel, M.; Ivantysynova, M. Thermal modeling of a hydraulic hybrid vehicle transmission based on thermodynamic analysis. *Energy* **2016**, *116*, 650–660. [CrossRef]
10. Stelson, K.A.; Meyer, J.J.; Alleyne, A.G.; Hencey, B. Optimization of a passenger hydraulic hybrid vehicle to improve fuel economy. *Proc. JFPS Int. Symp. Fluid Power* **2008**, *2008*, 143–148. [CrossRef]
11. Chen, J.-S. Energy Efficiency Comparison between Hydraulic Hybrid and Hybrid Electric Vehicles. *Energies* **2015**, *8*, 4697–4723. [CrossRef]
12. Gary, J. Hydrid Transmission. *Acad. Manag. Rev.* **2006**, *31*, 386–408.
13. Zhang, Z.; Chen, J.; Wu, B. The control strategy of optimal brake energy recovery for a parallel hydraulic hybrid vehicle. *Proc. Inst. Mech. Eng. Part D J. Automob. Eng.* **2012**, *226*, 1445–1453. [CrossRef]
14. Rydberg, K.-E. Energy Efficient Hydraulic Hybrid Drives. In Proceedings of the 11th Scandinavian International Conference on Fluid Power, SICFP'09, Linköping, Sweden, 2–4 June 2009.
15. Tvrdić, V.; Podrug, S.; Šuljić, I.; Matić, B. Hydraulic hybrid vehicle configurations and comparison with hybrid electric vehicle. In Proceedings of the Comtemporary Issues in Economy & Technology—CIET 2018, Split, Croatia, 1–2 June 2018; pp. 548–556.
16. Florida, N.; López, C.; Pocomucha, V. CORE View metadata, citation and similar papers at core. ac. uk. *J. Dedik. Pendidik.* **2012**, *2*, 35–43.
17. Shan, M. Modeling and Control Strategy for Series Hydraulic Hybrid Vehicles. Ph.D. Thesis, The University of Toledo, Toledo, OH, USA, December 2009.
18. Tavares, F.; Johri, R.; Salvi, A.; Baseley, S.; Filipi, Z. *Hydraulic Hybrid Powertrain-In-the-Loop Integration for Analyzing Real-World Fuel Economy and Emissions Improvements*. SAE Technical Paper. 2011. Available online: https://doi.org/10.4271/2011-01-2275 (accessed on 20 October 2021).
19. Midgley, W.; Cebon, D. Comparison of regenerative braking technologies for heavy goods vehicles in urban environments. *Proc. Inst. Mech. Eng. Part D J. Automob. Eng.* **2012**, *226*, 957–970. [CrossRef]
20. Li, C.-T.; Peng, H. Optimal configuration design for hydraulic split hybrid vehicles. In Proceedings of the 2010 American Control Conference, Baltimore, MD, USA, 30 June–2 July 2010; pp. 5812–5817. [CrossRef]
21. Kim, N.; Rousseau, A. A Comparative Study of Hydraulic Hybrid Systems for Class 6 Trucks. *SAE Tech. Pap.* **2013**, 2. [CrossRef]
22. Sprengel, M.; Bleazard, T.; Haria, H.; Ivantysynova, M. Implementation of a Novel Hydraulic Hybrid Powertrain in a Sports Utility Vehicle. *IFAC-PapersOnLine* **2015**, *48*, 187–194. [CrossRef]
23. Bottiglione, F.; Mantriota, G.; Valle, M. Power-Split Hydrostatic Transmissions for Wind Energy Systems. *Energies* **2018**, *11*, 3369. [CrossRef]
24. Jae, Y.; Filipi, Z.; Kim, Y.J.; Filipi, Z. Simulation Study of a Series Hydraulic Hybrid Prop System for a Light Truck. *SAE Trans.* **2021**, *116*, 147–161.
25. Hui, S.; Junqing, J. Research on the system configuration and energy control strategy for parallel hydraulic hybrid loader. *Autom. Constr.* **2010**, *19*, 213–220. [CrossRef]
26. Zhou, S.; Walker, P.; Tian, Y.; Zhang, N. Mode switching analysis and control for a parallel hydraulic hybrid vehicle. *Veh. Syst. Dyn.* **2020**, *59*, 928–948. [CrossRef]
27. Sim, T.P.; Li, P.Y. Analysis and Control Design of a Hydro-Mechanical Hydraulic Hybrid Passenger Vehicle. In Proceedings of the Dynamic Systems and Control Conference, Hollywood, CA, USA, 12–14 October 2009; pp. 667–674. [CrossRef]
28. Cheong, K.L. Design and Analysis of Hydraulic Hybrid Passenger Vehicles. Ph.D. Thesis, University of Minnesota, Minneapolis, MN, USA, 2015.
29. Ramdan, M.I.; A Stelson, K. Optimal design of a power-split hybrid hydraulic bus. *Proc. Inst. Mech. Eng. Part D J. Automob. Eng.* **2016**, *230*, 1699–1718. [CrossRef]
30. Sprengel, M.; Ivantysynova, M. Recent Developments in a Novel Blended Hydraulic Hybrid Transmission. *SAE Tech. Pap.* **2014**, *2014*. [CrossRef]

31. Stecki, J.; Matheson, P. Advances In Automotive Hydraulic Hybrid Drives. *Proc. JFPS Int. Symp. Fluid Power* **2005**, *2005*, 664–669. [CrossRef]
32. Matheson, P.; Stecki, J. Development and simulation of a hydraulic-hybrid powertrain for use in commercial heavy vehicles. *SAE Tech. Pap.* **2003**, *112*, 114–123. [CrossRef]
33. Wu, B.; Lin, C.-C.; Filipi, Z.; Peng, H.; Assanis, D. Optimal Power Management for a Hydraulic Hybrid Delivery Truck. *Veh. Syst. Dyn.* **2004**, *42*, 23–40. [CrossRef]
34. Van Batavia, B.L. *Hydraulic Hybrid Vehicle Energy Management System*; SAE Technical Paper; SAE International: Warrendale, PA, USA, 2009. [CrossRef]
35. Achten, P.A.J.; Schellekens, M.P.A.; Murrenhoff, H.; Deeken, M. Efficiency and Low Speed Behavior of the Floating Cup Pump. *J. Commer. Veh.* **2004**, *113*, 366–376. [CrossRef]
36. Buchwald, P.; Christensen, G.; Larsen, H.; Sunn Pedersen, P. Improvement of citybus fuel economy using a hydraulic hybrid propulsion system—A theoretical and experimental study. *SAE Tech. Pap.* **1979**, *88*, 1042–1056. [CrossRef]
37. Hydraulic Launch Assist. 2013. Available online: https://www.eaton.com/SEAsia/ProductsSolutions/Hydraulics/ProductsServices/HydraulicLaunchAssist/index.htm (accessed on 1 October 2021).
38. Tollefson, S.; Beachley, N.H.; Fronczak, F.J. Studies of an Accumulator Energy-Storage Automobile Design with a Single Pump/Motor Unit. *SAE Trans.* **1985**. [CrossRef]
39. Kepner, R.P. Hydraulic power assist—A demonstration of Hydraulic Hybrid Vehicle regenerative braking in a road vehicle application. *SAE Tech. Pap.* **2002**, *111*, 826–833. [CrossRef]
40. Bozic, A. Introducing Hydraulic-Electric Synergy into Hybrid Transmission Using the Free-piston Engine Technology. In Proceedings of the Commercial Vehicle Engineering Congress and Exhibition and Powertrain & Fluid Systems Conference, Rosemont, IL, USA, 29 October–1 November 2007. [CrossRef]
41. Breidi, F.; Garrity, J.; Lumkes, J. Design and Testing of Novel Hydraulic Pump/Motors to Improve the Efficiency of Agricultural Equipment. *Trans. ASABE* **2017**, *60*, 1809–1817. [CrossRef]
42. El-Breidi, F. Investigation of digital pump/motor control strategies. Ph.D. Thesis, Purdue University, West Lafayette, IN, USA, 2016.
43. Yang, H.-Y.; Pan, M. Engineering research in fluid power: A review. *J. Zhejiang Univ. A* **2015**, *16*, 427–442. [CrossRef]
44. Merrill, K.J.; Breidi, F.Y.; Lumkes, J. Simulation based design and optimization of digital pump/motors. In *Fluid Power Systems Technology*; American Society of Mechanical Engineers: New York, NY, USA, 2013.
45. Li, M.; Foss, R.; Stelson, K.A.; Van De Ven, J.D.; Barth, E.J. Design, Dynamic Modeling, and Experimental Validation of A Novel Alternating Flow Variable Displacement Hydraulic Pump. *IEEE/ASME Trans. Mechatron.* **2019**, *24*, 1294–1305. [CrossRef]
46. Wilhelm, S.R.; Van De Ven, J.D. Synthesis of a Variable Displacement Linkage for a Hydraulic Transformer. In Proceedings of the International Design Engineering Technical Conferences and Computers and Information in Engineering Conference, Washington, DC, USA, 28–31 August 2011. [CrossRef]
47. Chen, Y.L.; Liu, S.A.; Jiang, J.H.; Shang, T.; Zhang, Y.K.; Wei, W. Dynamic analysis of energy storage unit of the hydraulic hybrid vehicle. *Int. J. Automot. Technol.* **2013**, *14*, 101–112. [CrossRef]
48. Ivantysynova, M. Innovations in pump design-what are future directions? *Proc. JFPS Int. Symp. Fluid Power* **2008**, *2008*, 59–64. [CrossRef]
49. Zhang, C.; Huang, S.; Du, J.; Wang, X.; Wang, S.; Zhang, H. A new dynamic seven-stage model for thickness prediction of the film between valve plate and cylinder block in axial piston pumps. *Adv. Mech. Eng.* **2016**, *8*. [CrossRef]
50. Shang, L.; Ivantysynova, M. A temperature adaptive piston design for swash plate type axial piston machines. *Int. J. Fluid Power* **2016**, *18*, 38–48. [CrossRef]
51. Rizzo, G.; Massarotti, G.; Bonanno, A.; Paoluzzi, R.; Raimondo, M.; Blosi, M.; Veronesi, F.; Caldarelli, A.; Guarini, G. Axial piston pumps slippers with nanocoated surfaces to reduce friction. *Int. J. Fluid Power* **2015**, *16*, 1–10. [CrossRef]
52. Inaguma, Y.; Hibi, A. Reduction of friction torque in vane pump by smoothing cam ring surface. *Proc. Inst. Mech. Eng. Part C J. Mech. Eng. Sci.* **2007**, *221*, 527–534. [CrossRef]
53. Wieczorek, U.; Ivantysynova, M. Computer Aided Optimization of Bearing and Sealing Gaps in Hydrostatic Machines—The Simulation Tool Caspar. *Int. J. Fluid Power* **2002**, *3*, 7–20. [CrossRef]
54. Sazonov, Y.A.; Mokhov, M.A.; Gryaznova, I.V.; Voronova, V.V.; Tumanyan, K.A.; Frankov, M.A.; Balaka, N.N. Development and Prototyping of Jet Systems for Advanced Turbomachinery with Mesh Rotor. *Emerg. Sci. J.* **2021**, *5*, 775–801. [CrossRef]
55. Kostikov, Y.A.; Romanenkov, A.M. Approximation of the Multidimensional Optimal Control Problem for the Heat Equation (Applicable to Computational Fluid Dynamics (CFD)). *Civ. Eng. J.* **2020**, *6*, 743–768. [CrossRef]
56. Manring, N.D. Valve-Plate Design for an Axial Piston Pump Operating at Low Displacements. *J. Mech. Des.* **2003**, *125*, 200–205. [CrossRef]
57. Helmus, T.; Breidi, F.; Lumkes, J. Simulation of a variable displacement mechanically actuated digital pump unit. In Proceedings of the Eight Workshop on Digital Fluid Power, Tampere, Finland, 24–25 May 2016; pp. 95–106.
58. Chehade, A.; Breidi, F.; Pate, K.S.; Lumkes, J. Data-driven Adaptive Thresholding Model for Real-time Valve Delay Estimation in Digital Pump/Motors. *Int. J. Fluid Power* **2019**, *20*, 271–294. [CrossRef]

59. Winkler, B. Recent Advances in Digital Hydraulic Components and Applications. In Proceedings of the The Ninth Workshop on Digital Fluid Power, Aalborg, Denmark, 7–8 September 2017. Available online: https://www.et.aau.dk/digitalAssets/377/377525_id117-recent-advances-in-digital-hydraulic-components-and-applications.pdf (accessed on 1 December 2021).
60. Brandstetter, R.; Deubel, T.; Scheidl, R.; Winkler, B.; Zeman, K. Digital hydraulics and "Industrie 4.0". *Proc. Inst. Mech. Eng. Part I J. Syst. Control. Eng.* **2016**, *231*, 82–93. [CrossRef]
61. Peng, S.; Branson Iii, D.T.; Guglielmino, E.; Boaventura, T.; Caldwell, D.G. Performance Assessment of Digital Hydraulics in a Quadruped Robot Leg. 2012. Available online: http://asmedigitalcollection.asme.org/ESDA/proceedings-pdf/ESDA2012/44861/227/4451304/227_1.pdf?casa_token=fWco0NtKGfUAAAAA:9c3YrOxP0rRp-PmWLcv8DUFtas4Qe6HnyCq02vGig-lSARq94FAcQnGOpL9W-XpcM-sdwGk (accessed on 5 November 2021).
62. Pinto, L.P.G.; Belan, H.C.; Locateli, C.C.; Krus, P.; De Negri, V.J.; Lantto, B. New perspectives on digital hydraulics for aerospace applications. In Proceedings of the Aerospace Technology Congress, Solna, Stockholm, 11–12 October 2016.
63. Breidi, F.; Helmus, T.; Lumkes, J. The impact of peak-and-hold and reverse current solenoid driving strategies on the dynamic performance of commercial cartridge valves in a digital pump/motor. *Int. J. Fluid Power* **2015**, *17*, 37–47. [CrossRef]
64. Breidi, F.; Chehade, A.; Lumkes, J. Monitoring Digital Technologies in Hydraulic Systems Using CUSUM Control Charts. In *Fluid Power Systems Technology*; American Society of Mechanical Engineers: New York, NY, USA, 2019.
65. Helmus, T. Investigation and Implementation of Mechanically Actuated Valves for Digital Hydraulic Units. Ph.D. Thesis, Purdue University, West Lafayette, IN, USA, 2017.
66. Holland, M.A. Design of Digital Pump/Motors and Experimental Validation of Operating Strategies Doctor of Philosophy. 2012. Available online: http://www.purdue.edu/policies/pages/teach_res_outreach/c_22.html (accessed on 1 December 2021).
67. Breidi, F.; Garrity, J.; Lumkes, J., Jr. Investigation of a real-time pressure based valve timing correction algorithm. In *Fluid Power Systems Technology*; American Society of Mechanical Engineers: New York, NY, USA, 2017.
68. Simscape—MATLAB & Simulink. 2021. Available online: https://www.mathworks.com/products/simscape.html (accessed on 6 October 2021).
69. Sizing Tool—Reasontek Corp. 2021. Available online: https://www.reasontek.com/sizing-tool/ (accessed on 23 November 2021).
70. Cheema, J. Air and Oil Do Mix: Selecting the Right Accumulator Is Critical for Wind Farm Applications. Hydraulics and Pneumatic. 2013. Available online: http://ezproxy.derby.ac.uk/login?url=http://search.ebscohost.com/login.aspx?direct=true&db=edsbl&AN=RN332050940&site=eds-live (accessed on 28 November 2021).
71. Xu, Z.; Li, W.; Liu, X.; Chen, Z. Dynamic characteristics of coupling model of valve-controlled cylinder parallel accumulator. *Mech. Ind.* **2019**, *20*, 306. [CrossRef]
72. Valeski, B. Average Tire Weight (With 10 Examples)—Survival Tech Shop. 2021. Available online: https://www.survivaltechshop.com/tire-weight/ (accessed on 1 December 2021).

energies

MDPI

Article

Energy Comparison between a Load Sensing System and Electro-Hydraulic Solutions Applied to a 9-Ton Excavator

Paolo Casoli [1,*], Fabio Scolari [1], Carlo Maria Vescovini [1] and Massimo Rundo [2]

[1] Department of Engineering and Architecture, University of Parma, 43124 Parma, Italy; fabio.scolari@unipr.it (F.S.); carlomaria.vescovini@unipr.it (C.M.V.)
[2] Department of Energy, Politecnico di Torino, Corso Duca degli Abruzzi 24, 10129 Turin, Italy; massimo.rundo@polito.it
* Correspondence: paolo.casoli@unipr.it

Abstract: With the increasingly stringent regulations on air quality and the consequent emission limits for internal combustion engines, researchers are concentrating on studying new solutions for improving efficiency and energy saving even in off-road mobile machines. To achieve this task, pump-controlled or displacement-controlled systems have inspired interest for applications in off-road working machines. Generally, these systems are derived from the union of a hydraulic machine coupled to an electric one to create compact components that could be installed near the actuator. The object of study of this work is a 9-ton excavator, whose hydraulic circuit is grounded on load sensing logic. The validated mathematical model, created previously in the Simcenter Amesim© environment, represents the starting point for developing electro-hydraulic solutions. Electric components have been inserted to create different architectures, both with open- and closed-circuit layouts, in order to compare the energy efficiency of the different configurations with respect to the traditional load sensing system. The simulations of a typical working cycle show the energy benefits of electro-hydraulic solutions that allow for drastically reducing the mechanical energy required by the diesel engine and, consequently, the fuel consumption. This is mainly possible because of the elimination of directional valves and pressure compensators, which are necessary in a load sensing circuit, but are also a source of great energy dissipations. The results show that closed-circuit solutions produce the greatest benefits, with higher energy efficiencies than the open-circuit solution. Furthermore, closed-circuit configurations require fewer components, allowing for more compact and lighter solutions, as well as being cheaper.

Keywords: excavator; electro-hydrostatic actuator; load sensing; energy saving; pump-controlled systems

Citation: Casoli, P.; Scolari, F.; Vescovini, C.M.; Rundo, M. Energy Comparison between a Load Sensing System and Electro-Hydraulic Solutions Applied to a 9-Ton Excavator. *Energies* **2022**, *15*, 2583. https://doi.org/10.3390/en15072583

Academic Editors: Rafael J. Bergillos, Helena M. Ramos and Abdessattar Abdelkefi

Received: 10 February 2022
Accepted: 31 March 2022
Published: 1 April 2022

1. Introduction

The study of new solutions for improving energy efficiency and fuel saving applied to mobile off-road work machines is also due to the increasingly stringent regulations on emissions from internal combustion engines. Among the different proposals, some researchers have studied hybrid architectures to improve the energy efficiency of the transmission system. In the literature [1,2], solutions are presented to increase the energy efficiency of hydraulic systems, introducing architectures for energy recovery. In the case of an excavator, the most commonly used strategy is generally the recovery of kinetic energy given by the rotation of the swing and the potential energy of the arms; a complete discussion of these solutions can be found in [3–5].

Nowadays, in many engineering fields, the interest of researchers is aimed at the introduction of electro-mechanical actuation systems, replacing the traditional hydraulic architectures controlled by valves [6]. The distributed hydraulic solutions, replacing the centralized one, are derive from the aeronautical sector, where the thought of "more electric aircraft (MEA)" [7] is predominant, and is also spreading in various fields of application such as aeronautics [7], submarine [8], and land vehicles [9].

Electro-hydraulic systems could be achieved using electro-hydrostatic actuators (EHA) that combine compact dimensions and high energy efficiencies, while also offering the advantage of allowing plug-and-play installations. Many scientific papers compare EHA systems to traditional valve-controlled hydraulic architectures. Schmidt et al. [10] propose various solutions that provide pump control also with load holding characteristics. The distribution of energy and the possibility of energy recovery are described for all of the studied solutions, comparing the results with those of a traditional hydraulic system controlled by valves. Padovani et al. [11] suggest another electro-hydraulic drive system applied to single-rod cylinders with a passive load holding capability. Applying this architecture to a single-arm crane, the authors highlight how the error on the final position of the arm remains within ± 2 mm, while the overall energy efficiency reaches about 60% during handling. Among the various applications in the aeronautical field, Takahashi et al. [12] have proposed an electro-hydraulic system applied to a single-rod double-acting actuator for the landing gear. Ketelsen et al. [13] carry out a detailed review of the scientific literature, including electro-actuated systems with variable displacement hydraulic pumps and solutions with variable speed electric motors, comparing the advantages and disadvantages of each architecture. The authors report the progressive trend towards solutions based on electric motors, moving further and further away from diesel engines. The EHA architecture also offers the possibility of recovering energy, with the further advantage of sharing it subsequently on several actuators.

Abekawa et al. [14] state that excavators are responsible for about 60% of the CO_2 emissions produced by construction machinery. It follows that the implementation of new solutions in order to reduce the emission of greenhouse gases by excavators makes it possible to significantly decrease the amount of global CO_2 emissions. Niraula et al. and Zhang et al. [15,16], focusing on 1-ton mini excavators, showed that an electro-hydraulic actuation system could lead to a reduction in energy consumption of up to 50% compared to the traditional system based on load sensing circuit architecture. Budden et al. [17] modified a 20-ton excavator by equipping it with a digital displacement pump instead of the swashplate machine. The results show a significant increase in productivity, associated with fuel savings.

This article presents further results of a research activity applied to a 9-ton excavator, on which various electro-hydraulic solutions have been analyzed. The results, in terms of mechanical energy and fuel consumption, have been compared with those obtained by the traditional load sensing architecture. The results have been obtained through simulations carried out using a mathematical model of the excavator created in the Simcenter Amesim© environment. The standard load sensing model of the machine, created with a lumped parameter approach, consists of modeling the diesel engine, the variable displacement main pump, the post compensated flow sharing distributors, and the kinematics of the front equipment. The model has been validated through extensive experimental activity, which made it possible to quantify fuel consumption during the work cycles [18–25].

The mathematical model of the excavator has been modified to study new EHA circuit architectures. In particular, an open-circuit and two closed-circuit solutions have been studied. The energy performance of all configurations has been compared with that of the load sensing system: the results show how the electro-hydraulic circuit solutions offer a significantly higher energy performance than a traditional valve-controlled system, allowing for savings on fuel consumption.

In the previous paper [26], the authors have already investigated an EHA architecture based only on an open-circuit solution; in that solution, the accumulator was connected to one side of the actuators where the lower average pressure occurs, obviously that solution has a drawback because the side at lower pressure can change with the duty cycle. In this paper, to overcome this limit, a shuttle valve has been added to connect the accumulator to the lower pressure side of the actuator [13,27–29]. Furthermore, in this paper, two closed-circuit solutions have been investigated, not considered in [26]. The first closed-circuit presented is similar to the open-circuit in order to make a reliable comparison. The second closed-circuit solution [13] presents a different architecture where a further control of a solenoid valve is requested. The methodology followed is that all the solutions investigated are always compared with the traditional load sensing circuit in terms of both mechanical energy and fuel consumption saving. The contribute of this paper is not in creating new circuit architectures, which are already known, as reported in the cited papers, but in the potential of the mathematical modeling that permits giving reliable information about the application of different solutions in a mobile machine like the excavator considered. This document is structured as follows. Section 2 describes the mathematical model of the standard load sensing hydraulic excavator. Section 3 describes the different circuit solutions based on EHA. The results of the simulations of all architectures are reported in Section 4. Section 5 presents the comparison between the standard model LS and EHA solutions. Finally, Section 6 reports the conclusions and future developments.

2. Mathematical Model of the Standard Hydraulic Excavator

The machine under consideration is a 9-ton hydraulic excavator. The standard version features a 46-kW diesel (ICE) engine that drives an axial piston pump to feed a hydraulic circuit based on a load sensing logic (LS) with flow sharing distributors. The ISO diagram of the circuit layout of the machine is shown in Figure 1, where the several sensors used for the testing activity are also indicated. The mathematical model, based on a lumped parameter approach, has been developed in the Simcenter Amesim® environment. The complete and detailed description of the model can be found in previous publications [18,20,22,23].

As shown in the ISO diagram, the standard hydraulic circuit consists of two pumps. One is a fixed displacement machine with external gears, which feeds the circuit dedicated to pilots. The main pump that supplies the users is a variable displacement axial piston pump, equipped with a pressure compensator, flow compensator, and torque limiter. The pump has been modeled with a gray box approach, where the models of the regulators are white box, while a black box solution has been adopted for the pump flow characteristic, thanks to the use of hydromechanical and volumetric efficiency maps obtained from the experimental activity [25].

The directional valves are post compensated LS flow sharing valves, which allow for extracting the load sensing pressure to be sent to the pump regulators. The main feature of this type of valve is to maintain the same pressure drop in each section, even if the main pump reaches saturation conditions. The distributor model has been validated through the experimental results provided by the tests carried out in the laboratory of the Department of Engineering and Architecture of the University of Parma, Italy, as described in detail in [19,24].

The hydraulic cylinders are single rod double acting actuators. The model of every cylinder includes the effects of linear friction and leakage between the two chambers. The Coulomb and the viscous friction coefficients have been characterized by means of the experimental activity. The hydraulic cylinders have been simulated by applying the continuity equation and the fluid state equation for two control volumes, on the piston side and rod side. The model of the actuators has also remained unchanged in the further circuit solutions proposed in this article.

The kinematic model of the front attachment and of the swing has been developed in order to correctly calculate the forces and torques acting on the hydraulic actuators [21,24]. The front attachment consists of boom, arm, and bucket, which have been modeled as rigid bodies linked together by rotary joints and linear actuators. Characteristic properties such as mass, moment of inertia, and center of gravity have been defined using the CAD geometry of the components. For the swing model, assumed at constant inertia, both the Coulomb friction and viscous friction terms have been considered, defined through experimental tests carried out with fast and slow rotation cycles. In the different circuit solutions presented hereinafter in this article, the kinematic model has remained unchanged.

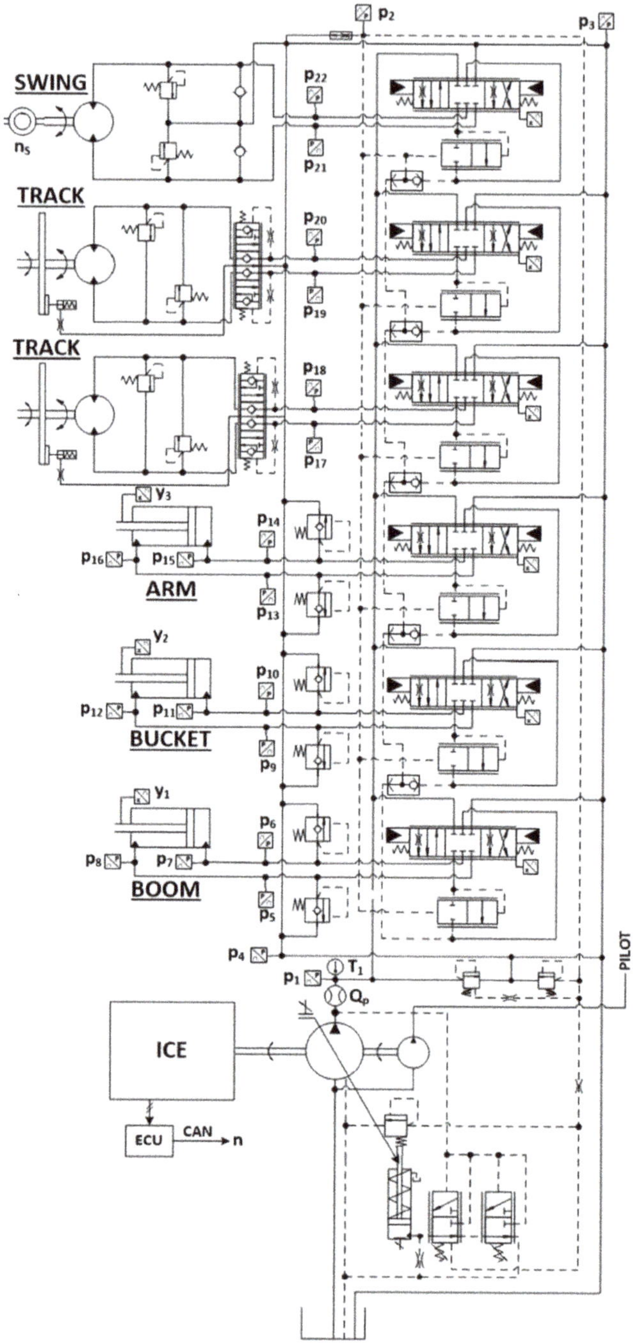

Figure 1. ISO scheme of the standard excavator hydraulic circuit. The sensors used in the test phase are also indicated [23].

The model of the internal combustion engine, developed including some data provided by the manufacturer and other obtained experimentally, allows for calculating the instantaneous fuel consumption, as a function of the torque and of the engine speed, as in Figure 2 [20,22].

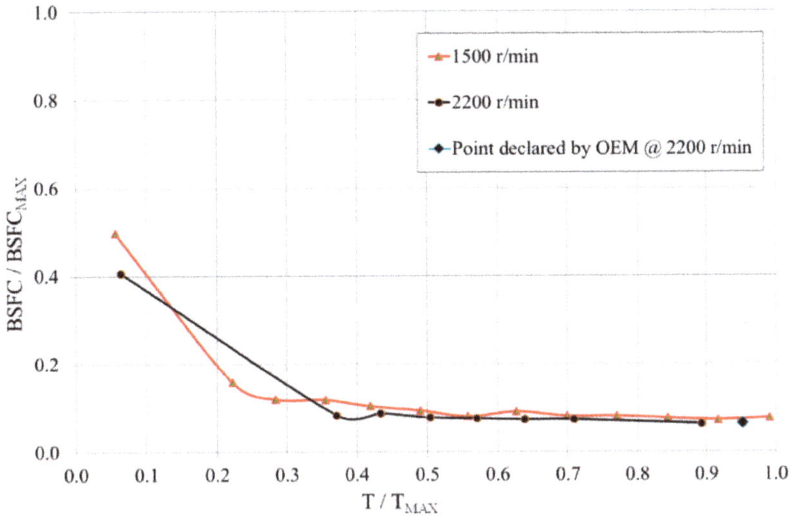

Figure 2. Diesel engine—Brake specific fuel consumption map.

3. Electro-Hydraulics Solutions

The validated model of the excavator represents the starting point from which to develop EHA solutions for carrying out energy comparison. An open-circuit configuration introduced in a previous article [26] is re-proposed in this paper with the name "Open Circuit—Layout 1", with a few changes regarding the position of the accumulator; moreover, two new closed-circuit configurations, called "Layout 2" and "Layout 3", have been developed.

3.1. Open Circuit—Layout 1

The circuit layout is shown in the ISO diagram in Figure 3: for each user, there is a pair of hydraulic machines with fixed displacement, connected to an electric unit, that can work as a pump or as a motor depending on the applied pressure difference. It has been supposed to move the swing by means of only an electric motor. The internal combustion engine, which is the same as in the standard excavator, drives an electric generator that supplies the energy necessary to power the electric motor of each user. The hydraulic machines have been supposed to be external gear units, and the models include maps of volumetric and hydromechanical efficiency as functions of speed and delivery pressure.

The sizing of each hydraulic machine has been set starting from the displacement of the actuator: the ratio between the displacement of the hydraulic machine on the piston side and on the rod side must be similar to the ratio between the active surfaces of the actuator; in general, it is advisable to keep the difference between these ratios below 2% [30,31]. As the displacements of the hydraulic machines are not exactly the same as those of the cylinder and that the internal losses of each component are also considered, a hydraulic accumulator has been introduced to compensate for the differences in flow rate. The accumulator is connected to the side with the lowest pressure by means of a shuttle valve. In case of overrunning load, the hydraulic motor allows for recovering energy, as the electric unit can work as a generator. For this reason, it is necessary to introduce a battery to store the energy produced.

For each side of the actuator, there are electrically controlled on/off valves to block the load, with the pressure drop included. They are powered only when it is necessary to move the cylinder. For safety reasons, the valves are provided with a spring that blocks the load, even in the case of a fault in the electrical power supply system.

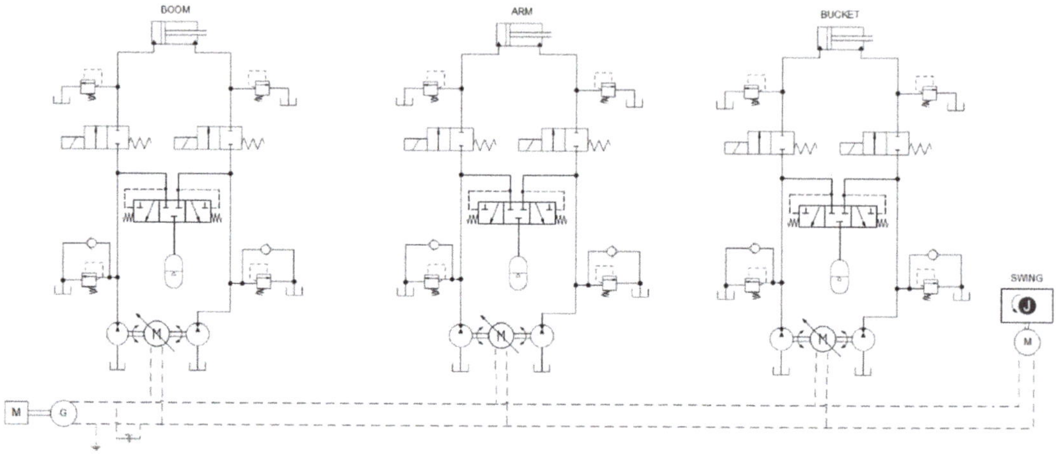

Figure 3. ISO scheme of Open Circuit—Layout 1.

3.2. Closed Circuits

Two different circuit solutions based on EHA architecture are proposed, and the results have been compared with those of the standard LS circuit and of the open-circuit. The proposed configurations are closed-circuit layouts. The feature common to all layouts is the presence of a bidirectional fixed displacement hydraulic pump, driven by an electric motor, and for every cylinder, the connection of the pump delivery port with the rod side or the piston side of the actuator is managed by the direction of rotation of the electric motor. Electric motors are powered by a generator driven by the internal combustion engine, the model of which is the same as in the standard excavator. An electric motor has been inserted in place of the hydraulic motor to move the swing and the tracks have not been considered. All of the proposed solutions permit removing the variable displacement main pump and the directional valves, typical of the standard LS system.

The hydraulic machines are bi-directional external gear pumps and have been modeled considering volumetric and hydromechanical efficiency maps as a function of rotation speed and delivery pressure. The pumps and their electric motors are the same for all proposed configurations.

The presence of different areas between the rod side and the piston side in the actuators makes it essential to balance the differential flow. The two solutions differ in the strategy in order to obtain the flow compensation.

3.3. Closed Circuit—Layout 2

Figure 4 shows the ISO scheme of Layout 1. The solution uses a flushing valve that connects the low-pressure chamber of the cylinder with the accumulator to compensate for the uneven flow rate given by the differential actuator [13]. When the piston side is fed, the higher flow rate required by the pump depressurizes the rod side chamber; the valve switches in order to connect the accumulator with the rod side, which provides the missing flow rate to the pump. Conversely, when the rod side is powered, the excess flow rate from the piston side is introduced into the accumulator.

In the case of an overrunning load, the hydraulic machine works as a motor and drives the electric machine, which, acting as a generator, allows energy to be recovered, storing it in the battery.

Two block valves have been inserted, similarly to the open-circuit layout, to ensure the position of the actuator even in the event of power failure. These on/off valves are opened only when it is necessary to move the cylinder.

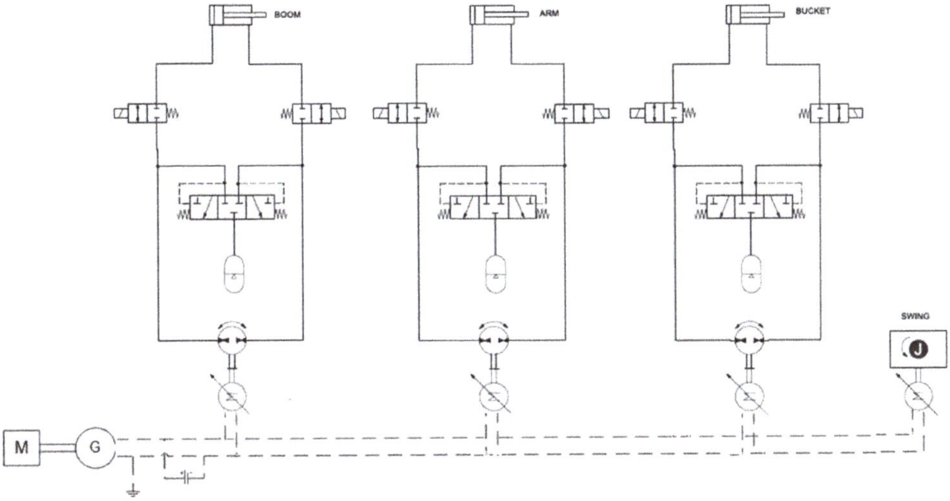

Figure 4. ISO scheme of Closed Circuit—Layout 2.

3.4. Closed Circuit—Layout 3

Figure 5 shows the circuit diagram of Layout 3. This solution has been reported by Padovani et al. [13]. Each user consists of an electric motor and a fixed displacement hydraulic machine driving the double-acting single-rod cylinder, in a closed-circuit layout.

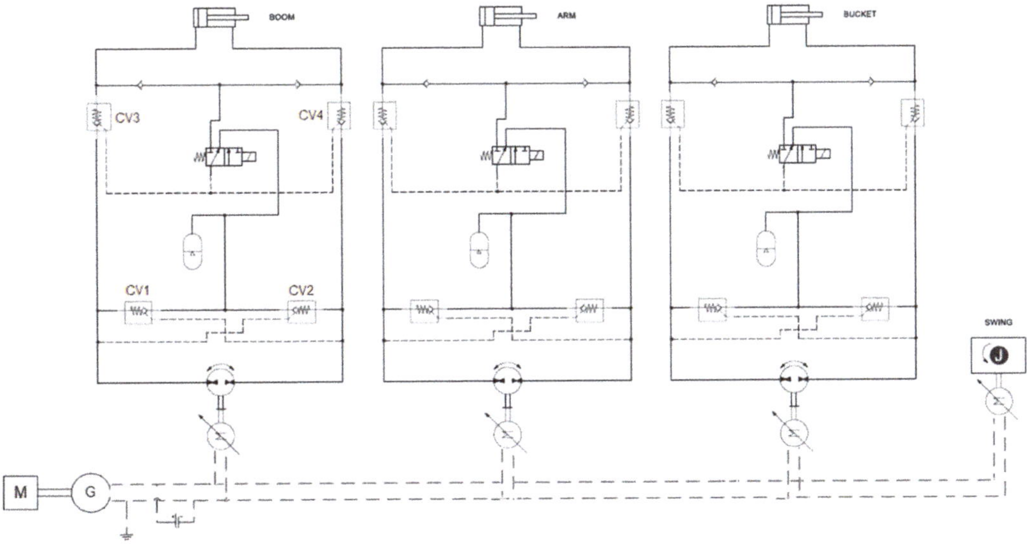

Figure 5. ISO scheme of Layout 3.

The differential flow managed by the low-pressure accumulator is controlled by two piloted check valves (CV1 and CV2). There are also two pilot operated check valves (CV3 and CV4) for load holding, which, unlike Layout 1, are fully hydraulic valves. The actuation of the solenoid valve transmits the signal of the maximum pressure of the actuator to the pilot line to allow for the opening of the two load holding valves to permit the movement of the actuator. When the solenoid valve is not powered, the piloting line of the load holding valves is connected to the low-pressure line and does not allow for their opening. In the event of a power failure, the solenoid valve assumes this latter position by means of a spring, preventing the movement of the load.

The next sections describe the features common to all of the proposed EHA configurations.

3.5. Model of the Operator

The digging duty cycle defined by the JCMAS standard is replicated by the operator model. In the standard model of the excavator, the position of the actuator is compared with the desired position and, consequently, the opening of the directional valve is defined.

In all of the proposed EHA solutions, the position of the actuator depends on the flow rate delivered, given by the rotation speed of the hydraulic machine: the error between the desired position and the actual one controls the speed of the electric motor by means of a PI type controller, as indicated in Figure 6.

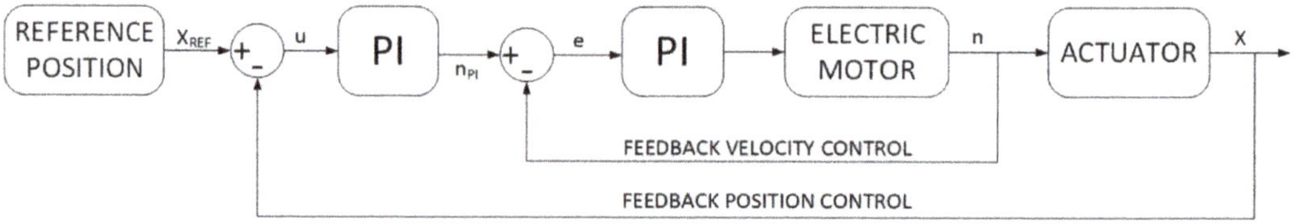

Figure 6. Actuator control scheme.

3.6. Electric Components

The electrical circuit is a feature common to all studied EHA configurations. It consists of three electric machines, a battery, and a generator connected to the diesel engine.

To allow for energy recovery during the overrunning load phases, it is essential to introduce a battery, which has been simulated as an ideal storage system without considering its dynamic behavior or thermal effects. The data shown in the Table 1 refer to a li-ion battery with characteristics that can be found in the scientific literature [32].

The electric machine model present in Amesim© allows for operating in all four quadrants of the torque-speed map: the electric machine can act as a generator, producing electricity that is then stored by the battery. In the operating range of the electric machines, a constant average efficiency has been assumed.

The main characteristics of the electrical components, which are the same for all of the analyzed configurations, are shown in Table 1.

Table 1. Main features of the electric components.

Component	Parameter	Value
Electric motor	Voltage (V)	400
	Max speed (r/min]	4000
	Moment of inertia (kg/m^2]	0.049
Generator	Min speed (r/min]	1000
	Rated speed (r/min]	2200
	Voltage (V]	400
Battery	Specific Energy (Wh/kg]	130
	Nominal voltage (V]	400
	Specific power (W/kg]	2000
	Power (kW]	50

4. Results

This section presents the results obtained with the standard LS circuit and EHA models to compare their performance. Figure 7 shows the position of the actuators of the front attachment during the imposed digging cycle. As can be appreciated from the graphs, all of the analyzed configurations are able to follow the duty cycle, without significant differences compared to the reference one. This result will allow for a reliable comparison of the fuel consumption between the various architectures. Figure 8, obtained from the Amesim© environment, shows the sequence of the positions of the front implement cylinders during the digging cycle, which consists of lowering the arms to dig an earth bucket and, in the subsequent extension, to unload the bucket and return it to the initial position.

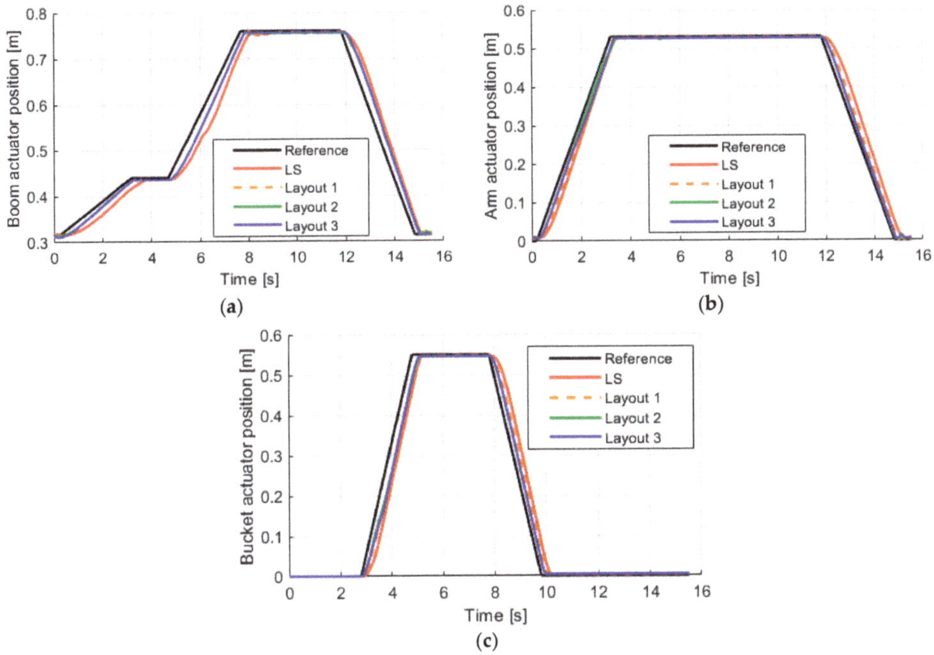

Figure 7. Actuator response to reference position of the digging cycle: simulations results: (**a**) boom, (**b**) arm, and (**c**) bucket.

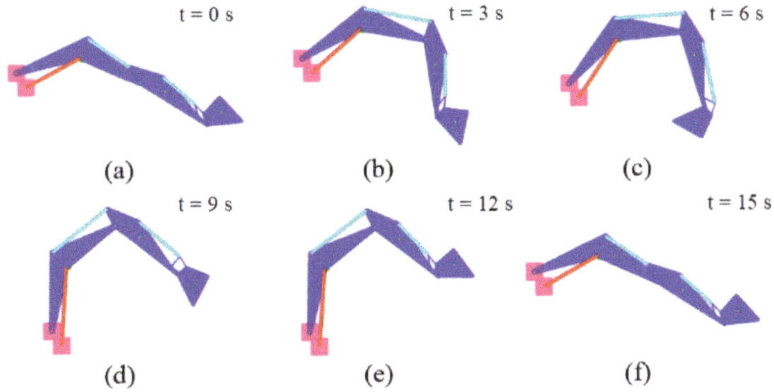

Figure 8. Sequence of front implement movements during the digging cycle.

The simulations have been used to compare the energy demands of the different studied solutions. Table 2 shows the mechanical energy supplied by the diesel engine and required by the pumps for each configuration: for the LS solution, the value represents the energy required by the main pump, while for the EHA solutions, the values are the ones needed by the alternator. The results obtained show that EHA solutions allow for greatly reducing the mechanical energy required, with savings of over 60% for all of the configurations, mainly due to the absence of pressure compensators and directional valves, typical of the load sensing systems. In particular, the closed-circuit solutions allow for obtaining higher improvements than the open-circuit solution. Furthermore, Layout 2 permits a mechanical energy saving of 2% higher than that of Layout 3. These trends are also confirmed by analyzing the fuel savings. Table 3 shows the fuel consumption of the diesel engine for each layout, and this involves considering the engine efficiency map. The savings are also significant in terms of fuel, but with lower percentage values than for the mechanical energy saving: this is because the efficiency of the thermal engine strongly depends on the absorbed torque value. During the cycle, the power required by the diesel engine for EHA solutions is lower than the traditional LS circuit, as shown in Figure 9: in general, the internal combustion engine has lower efficiency if subjected to lower loads. The lower power required by the solutions studied opens up the possibility of downsizing the diesel engine: this solution would allow for even greater fuel saving to be obtained, because it would permit the engine to work at average loads closer to the maximum, where the efficiency is higher. The possibility of downsizing the diesel engine has not been investigated in this paper, but interesting results are reported in [26], applied to an open-circuit EHA solution.

Table 2. Mechanical energy.

Solutions	Mechanical Energy Supplied by Diesel Engine (kJ/Cycle)	Mechanical Energy Saving (%)	Mechanical Energy Supplied to the Pumps (kJ/cycle)	Mechanical Energy Saving (%)
LS	393.6	/	393.6	/
Open Circuit—Layout 1	143.6	63.5%	91.6	76.7%
Closed Circuit—Layout 2	112.2	71.5%	84.7	78.5%
Closed Circuit—Layout 3	136.2	65.4%	109.2	72.3%

Table 3. Diesel engine fuel consumption.

Solutions	Fuel Consumption (g/cycle)	Fuel Saving (%)
LS	34.49	/
Open Circuit—Layout 1	23.92	30.6%
Closed Circuit—Layout 2	22.50	33.1%
Closed Circuit—Layout 3	23.60	31.6%

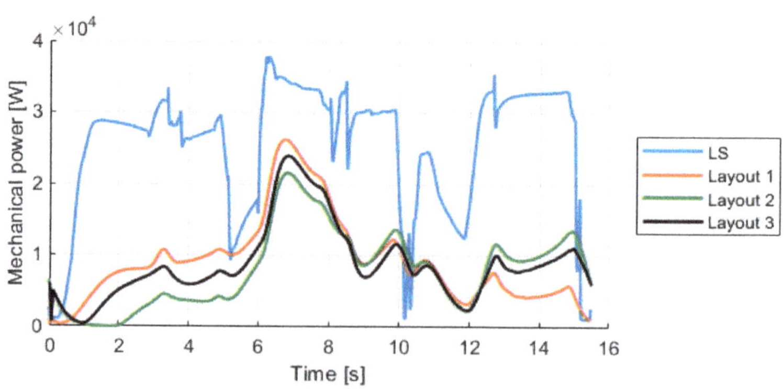

Figure 9. Profile of mechanical power during the digging cycle.

The Figures 10–13 show the Sankey diagrams for all of the studied configurations, which show the distribution of the mechanical input energy on the different components that make up the circuit, allowing for quantifying the energy losses. In the LS standard configuration, in Figure 10, the largest amount of energy loss is attributable to the valve block, which dissipates more than 57% of the energy. The open-circuit EHA solution, in Figure 11, has most of the energy losses concentrated in the group consisting of electric and hydraulic machines. The closed-circuit variants, on the other hand, have similar energy distributions, as seen in Figures 12 and 13, with lower losses than the open-circuit solution. In particular, Layout 3 has less energy losses on hydraulic machines than Layout 2, but at the same time, has much greater losses on the load holding valves, as the piloted check valves generate higher pressure drops than the on/off valves used in Layout 2.

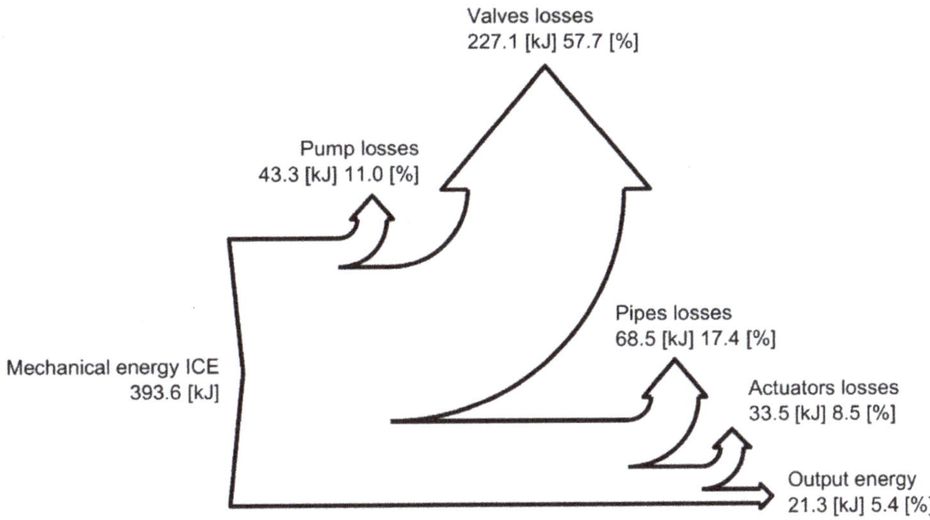

Figure 10. Energy distribution for the LS standard configuration during the digging cycle.

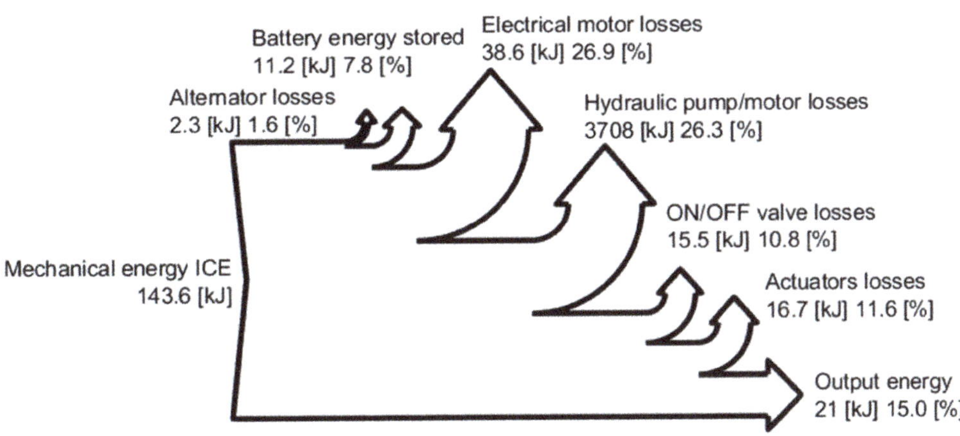

Figure 11. Energy distribution for configuration during the digging cycle for Open Circuit—Layout 2.

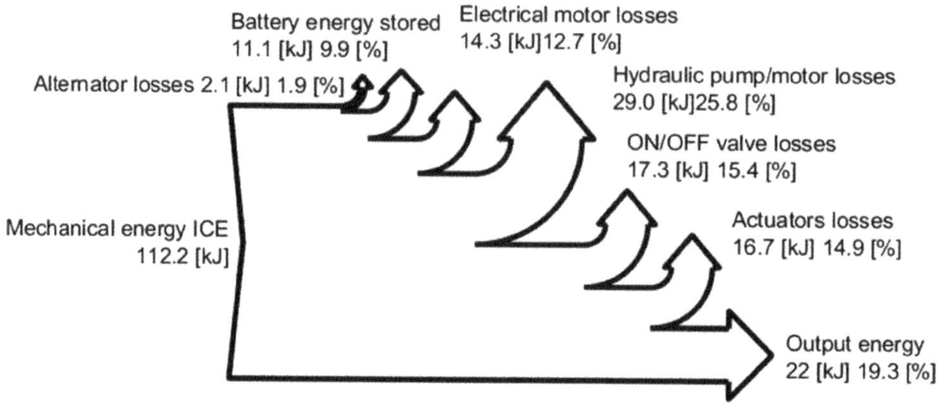

Figure 12. Energy distribution configuration during the digging cycle for Closed Circuit—Layout 2.

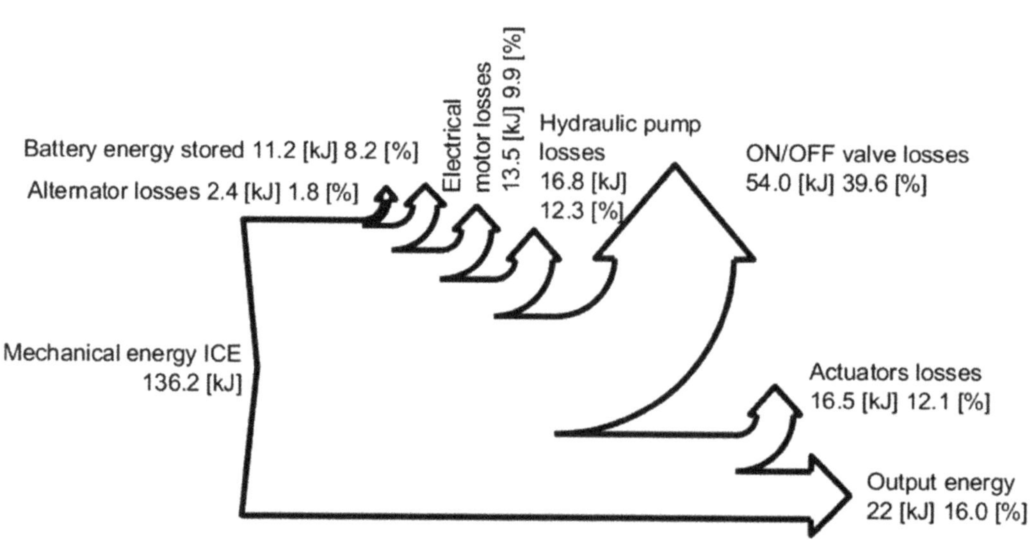

Figure 13. Energy distribution configuration during the digging cycle for Closed Circuit—Layout 3.

5. Discussion

This research activity focused on the application of electro-hydraulic solutions to a 9-ton excavator to achieve energy improvements.

The results are obtained through simulations carried out in the Amesim© environment, starting from a validated mathematical model of an excavator in LS logic, used as a reference. Three EHA configurations have been considered, one in open-circuit configuration and two in closed-circuit configuration.

The energy distributions obtained through the Sankey diagrams for the traditional LS system and for the EHA configurations (open- and closed-circuit) have been determined for a 9-ton excavator. It has been shown that all EHA solutions allow for reducing the mechanical energy requirement by over 60%, mainly due to the elimination of the direction valves required by the traditional load sensing circuit. Furthermore, closed-circuit solutions show greater benefits than the open-circuit solution. The circuit layout therefore strongly influences the efficiency and overall consumption of the system and must be carefully designed. The closed-circuit solutions do not only bring advantages in energy terms. In fact, the open-circuit solution involves the use of two hydraulic machines for each actuator, which involves a much greater cost, size, and weight than the closed-circuit solutions. From these points of view, the Layout 2 solution allows for reducing the number of components, also guaranteeing the best energy performance.

However, although closed-circuit solutions have fewer components, they still suffer from the main drawback typical of all EHA solutions, which is the higher cost. In fact, electric motors, alternators, and a battery (or supercapacitors) are required, as well as electro-systems, to manage the electric drives that make this technology less competitive at present compared with fully hydraulic solutions. Despite this, the advantages presented in this article could help overcome these disadvantages.

The presented results evaluate the energy efficiency of all circuit solutions through the application of characteristic work cycles: during the simulations, the digging cycle defined by the JCMAS standard has been used [33], which takes place without a payload.

This simulation work made it possible to demonstrate the potential of electro-hydraulic solutions by applying them to a medium-sized excavator. Future developments will concern the study of other solutions for compensating the different flow rate given using a differential cylinder.

A further important result is that the mechanical power required by the diesel engine always has lower values than the standard LS circuit. A future development is represented by the study of the possibility of reducing the size of the diesel engine: in this way, the engine would always operate in working points with the highest efficiency, with the advantage of further reducing fuel consumption.

6. Conclusions

This article presents the results from the application of electro-hydraulic actuation (EHA) applied to a 9-ton excavator. A validated mathematical model of the traditional excavator with load sensing circuit has been taken as a reference for subsequent energy analyses. Three EHA configurations have been also considered: the first in open-circuit layout and the other two in closed-circuit layouts. The results of the simulations permit to quantify the possible energy saving given by each studied solution and to characterize in detail the sources of energy losses.

The standard load sensing circuit features directional valves with pressure compensators, which are a source of great energy losses. EHA solutions, on the other hand, make it possible to eliminate these components, significantly reducing hydraulic losses and allowing for considerable fuel savings.

In all of the EHA solutions studied, the mechanical power required by the diesel engine has lower values compared to the LS standard circuit: this interesting result provides the starting point for future developments where it is possible to reduce the displacement of the diesel engine, allowing for increasing its efficiency with the advantage of further reducing fuel consumption.

Author Contributions: Conceptualization, P.C.; methodology, P.C. and M.R.; software, F.S.; validation, P.C. and F.S.; writing—original draft preparation, P.C., F.S. and C.M.V.; writing—review and editing, P.C., F.S., C.M.V. and M.R.; supervision, P.C.; project administration, P.C. All authors have read and agreed to the published version of the manuscript.

Funding: This research received no external funding.

Data Availability Statement: Not applicable.

Acknowledgments: The authors would like to acknowledge the active support of this research by Casappa S.p.A., Parma, Italy.

Conflicts of Interest: The authors declare no conflict of interest.

Nomenclature

Abbreviation	Definition
EHA	Electro-hydraulic actuator
CAD	Computer-aided design
ICE	Internal combustion engine
JCMAS	Japan Construction Machinery Association Standard
LS	Load sensing

References

1. Vukovic, M.; Leifeld, R.; Murrenhoff, H. Reducing Fuel Consumption in Hydraulic Excavators—A Comprehensive Analysis. *Energies* **2017**, *10*, 687. [CrossRef]
2. Inderelst, M.; Weidner, F.D.; Stammen, C. Quantification of Energy Saving Influencers. 21t Excavator Hydraulic System—A Holistic Investigation? In Proceedings of the 11th International Fluid Power Conference, Aachen, Germany, 19–21 March 2018.
3. Joo, C.; Stangl, M. Application of Power Regenerative Boom system to excavator. In Proceedings of the 10. IFK: International Fluid Power Conference, Dresden, Germany, 8–10 March 2016; Volume 3, pp. 175–184.
4. Li, J.; Zhao, J.; Zhang, X. A Novel Energy Recovery System Integrating Flywheel and Flow Regeneration for a Hydraulic Excavator Boom System. *Energies* **2020**, *13*, 315. [CrossRef]
5. Guan, C.; Xu, X.; Lin, X.; Wang, S.-H. Recovering system of swing braking energy in hydraulic excavator. *Zhejiang Daxue Xuebao (Gongxue Ban)/J. Zhejiang Univ. (Eng. Sci.)* **2012**, *46*, 142–149. [CrossRef]
6. Padovani, D.; Rundo, M.; Altare, G. The Working Hydraulics of Valve-Controlled Mobile Machines: Classification and Review. *J. Dyn. Syst. Meas. Control* **2020**, *142*, 070801. [CrossRef]
7. Rosero, J.A.; Ortega, J.A.; Aldabas, E.; Romeral, L. Moving towards a more electric aircraft. *IEEE Aerosp. Electron. Syst. Mag.* **2007**, *22*, 3–9. [CrossRef]
8. Smith, S.; Irving, J.; Irving, J. Electro Hydrostatic Actuators for Control of Undersea Vehicles. In Proceedings of the Joint Undersea Warfare Technology Fall Conference, Groton, CT, USA, 11–14 August 2016.
9. Schneider, M.; Koch, O.; Weber, J. Green Wheel Loader—Improving fuel economy through energy efficient drive and control concepts. In Proceedings of the 10th International Fluid Power Conference, Dresden, Germany; 2016.
10. Schmidt, L.; Ketelsen, S.; Brask, M.H.; Mortensen, K.A. A Class of Energy Efficient Self-Contained Electro-Hydraulic Drives with Self-Locking Capability. *Energies* **2019**, *12*, 1866. [CrossRef]
11. Padovani, D.; Ketelsen, S.; Hagen, D.; Schmidt, L. A Self-Contained Electro-Hydraulic Cylinder with Passive Load-Holding Capability. *Energies* **2019**, *12*, 292. [CrossRef]
12. Takahashi, N.; Kondo, T.; Takada, M.; Masutani, K.; Okano, S.; Tsujita, M. Development of Prototype Electro-Hydrostatic Actuator for Landing Gear Extension and Retraction. In Proceedings of the 7th JFPS International Symposium on Fluid Power, Toyama, Japan, 15–18 September 2008.
13. Ketelsen, S.; Padovani, D.; Andersen, T.O.; Ebbesen, M.K.; Schmidt, L. Classification and Review of Pump-Controlled Differential Cylinder Drives. *Energies* **2019**, *12*, 1293. [CrossRef]
14. Abekawa, T.; Tanikawa, Y.; Hirosawa, A. *Introduction of Komatsu Genuine Hydraulic Oil KOMHYDRO HE*; Komatsu Technical Report; Yumpu: Komatsu, Japan, 2010; Volume 56, No. 163.
15. Niraula, A.; Zhang, S.; Minav, T.; Pietola, M. Effect of Zonal Hydraulics on Energy Consumption and Boom Structure of a Micro-Excavator. *Energies* **2018**, *11*, 2088. [CrossRef]
16. Zhang, S.; Minav, T.; Pietola, M. Improving efficiency of micro excavator with decentralized hydraulics. In Proceedings of the 2017 Bath/ASME Symposium on Fluid Power and Motion Control, FPMC, Sarasota, FL, USA, 16–19 October 2017.
17. Budden, J.J.; Williamson, C. Danfoss Digital Displacement® Excavator: Test Results and Analysis. In Proceedings of the ASME/BATH 2019 Symposium on Fluid Power and Motion Control, FPMC2019, Longboat Key, FL, USA, 7–9 October 2019. [CrossRef]
18. Casoli, P.; Gambarotta, A.; Pompini, N.; Riccò, L. Development and Application of Co-simulation and "Control-oriented" Modeling in the Improvement of Performance and Energy Saving of Mobile Machinery. *Energy Procedia* **2014**, *45*, 849–858. [CrossRef]
19. Casoli, P.; Anthony, A.; Riccò, L. Modeling simulation and experimental verification of an Excavator hydraulic System–Load sensing flow sharing valve model. In Proceedings of the SAE 2012 Commercial Vehicle Engineering Congress, Rosemont, IL, USA, 13–14 September 2012. [CrossRef]
20. Casoli, P.; Gambarotta, A.; Pompini, N.; Riccò, L. Coupling excavator hydraulic system and internal combustion engine models for the real-time simulation. *Control Eng. Pract.* **2015**, *41*, 26–37. [CrossRef]
21. Casoli, P.; Pompini, N.; Riccò, L. Simulation of an Excavator Hydraulic System Using Nonlinear Mathematical Models. *Stroj. Vestn.-J. Mech. Eng.* **2015**, *61*, 583–593. [CrossRef]
22. Casoli, P.; Riccò, L.; Campanini, F.; Lettini, A.; Dolcin, C. Mathematical model of a hydraulic excavator for fuel consumption predictions. In Proceedings of the ASME/BATH Symposium on Fluid Power & Motion Control, Chicago, IL, USA, 12–14 October 2015; Paper No. FPMC2015-9566; ISBN 978-0-7918-5723-6. [CrossRef]

23. Bedotti, A.; Pastori, M.; Casoli, P. Modelling and energy comparison of system layouts for a hydraulic excavator. *Energy Procedia* **2018**, *148*, 26–33. [CrossRef]
24. Casoli, P.; Riccò, L.; Campanini, F.; Bedotti, A. Hydraulic Hybrid Excavator—Mathematical Model Validation and Energy Analysis. *Energies* **2016**, *9*, 1002. [CrossRef]
25. Casoli, P.; Anthony, A.; Rigosi, M. Modeling of an Excavator System—Semi Empirical Hydraulic Pump Model. *SAE Int. J. Commer. Veh.* **2011**, *4*, 242–255. [CrossRef]
26. Casoli, P.; Scolari, F.; Minav, T.; Rundo, M. Comparative Energy Analysis of a Load Sensing System and a Zonal Hydraulics for a 9-Tonne Excavator. *Actuators* **2020**, *9*, 39. [CrossRef]
27. Weber, J.; Beck, B.; Fischer, E.; Ivantysyn, R.; Kolks, G.; Kunkis, M.; Lohse, H.; Lübbert, J.; Michel, S.; Schneider, M.; et al. Novel System Architectures by Individual Drives. In Proceedings of the 10th International Fluid Power Conference, Dresden, Germany, 8–10 March 2016.
28. Michel, S.; Weber, J. Energy-efficient electrohydraulic compact drives for low power applications. In Proceedings of the ASME/BATH 2012 Fluid Power and Motion Control, Bath, UK, 12–14 September 2012; pp. 93–107.
29. Costa, G.K.; Sepehri, N. Four-Quadrant Analysis and System Design for Single-Rod Hydrostatic Actuators. *J. Dyn. Syst. Meas. Control* **2018**, *141*, 021011. [CrossRef]
30. Minav, T.; Zhang, S.; Pietola, M. Eliminating sizing error in direct-driven hydraulics. In Proceedings of the 10th JFPS International Symposium on Fluid Power, Fukuoka, Japan, 24–27 October 2017.
31. Järf, A.; Minav, T.; Pietola, M. Nonsymmetrical flow compensation using hydraulic accumulator in direct driven differential cylinder. In Proceedings of the ASME 2016 9th FPNI Ph.D. Symposium on Fluid Power, FPNI2016, Florianópolis, Brazil, 26–28 October 2016.
32. Rahn, C.D.; Wang, C.-Y. *Battery Systems Engineering*; John Wiley& Sons: New York, NY, USA, 2013; ISBN 9781119979500.
33. *Japanese Technical Standard JCMAS H020: 2007*; Earth-Moving Machinery—Fuel Consumption on Hydraulic Excavator—Test Procedure. JCMAS: Tokyo, Japan, 2007.

energies

MDPI

Article

Effect of Operating Parameters on Efficiency of Swash-Plate Type Axial Piston Pump

Heikki Kauranne [1,2]

[1] Department of Mechanical Engineering, Aalto University, 02150 Espoo, Finland; heikki.kauranne@aalto.fi
[2] Fiellberg Oy, 01720 Vantaa, Finland

Abstract: In an effort to improve the energy economics of hydraulic systems, attention should be paid to reducing power losses in two main entities, energy converting components, and energy controlling and conveying components. Achieving the former requires utilizing components' most energy efficient operating range. The energy converting efficiency of a pump, which is the primary energy converter in a hydraulic system, is determined by several operational factors. Of these, only pressure and rotational speed are normally considered, but also the fluid temperature and derived capacity with variable displacement pumps have a major effect on the efficiency. Omitting these factors may lead to running the pump outside its most efficient operation range and cause high energy losses. Operating the pump in its optimal region calls, however, for detailed knowledge of its performance characteristics, which are not generally made public by the pump manufacturers. This study presents the performance measurement results of a variable displacement axial piston pump in the form of efficiencies as a function of pressure, rotational speed, derived capacity and inlet fluid temperature. The results show that all of these factors have a significant impact on pump's energy conversion efficiency and should, therefore, be taken into account when operating a hydraulic pump.

Keywords: axial piston pump; variable displacement; swash-plate; hydromechanical; volumetric; efficiency; operation point; operation range

Citation: Kauranne, H. Effect of Operating Parameters on Efficiency of Swash-Plate Type Axial Piston Pump. *Energies* **2022**, *15*, 4030. https://doi.org/10.3390/en15114030

Academic Editors: Paolo Casoli and Massimo Rundo

Received: 28 April 2022
Accepted: 24 May 2022
Published: 30 May 2022

1. Introduction

Hydraulic systems are widely used both in stationary and mobile machines because of their high power density and suitability to implement versatile functions. These systems, however, typically have mediocre or even low energy efficiency, especially in applications where the operational circumstances of the system change along with the work cycle, and where the system cannot continuously operate at or close to its optimized operation point. Low energy efficiency, in turn, leads to high energy consumption in relation to a system's energy output and/or performance. This then leads to the need for a cooling system that increases the energy consumption of the system even more.

The energy efficiency of a pump is dependent on the operation point of the pump, i.e., the prevailing pressure difference between the pump's inlet and outlet ports, the rotational speed of the pump, the displacement setting value in the case of a variable displacement pump, the temperature of the fluid and the fluid characteristics like viscosity and density. In relation to all of these, pumps have optimal operation points or ranges where the efficiency of the pump is at its highest value. However, operating the pump in these regions requires knowledge that is not commonly and publicly available as pump manufacturers do not tend to publish comprehensive performance data of their products. For some individual pumps, one may find efficiency data in some research papers, e.g., Innas published [1]. If pump manufacturers publish something more than pure theoretical and lossless values, it usually is 2D efficiency curves as a function of pressure and with one or few selected rotational speeds. This kind of data does not fulfil the needs of a system designer that strives to develop a system, e.g., an electrical motor-run system, where the motor and pump combination is to be used to control the system actuators.

The design of an energy efficient hydraulic system requires comprehensive knowledge of the pump's characteristics and performance over the total ranges of the operational parameters, at the lowest pressure, rotational speed and displacement settings in the case of variable displacement pumps. For pump users, the only way to achieve this seems to be measuring the pump characteristics. This, however, requires a significant investment in research apparatus and environment, and is therefore implemented by very few machine manufacturers.

The aim of this study was to determine the dependence of a swash-plate type axial piston pump's performance on the above-mentioned system parameters.

2. Measurement Systems and Determining Pump's Steady-State Performance

The methods for measuring positive displacement pumps' and motors' characteristics and performance are determined in two ISO standards. ISO 8426:2008 [2] describes the methods for determining the derived capacity of these machines, and ISO 4409:2019 [3] describes the methods for determining their steady state performance. These standards determine the basic structure of the measurement systems (e.g., Figure 1 open circuit for unidirectional pumps), measured quantities, positions of the transducers in relation to the measured machine and each other, system fluid with its filtration and temperature, magnitudes of systematic errors permitted in each measurement accuracy class, and the form of measurement reporting.

When conducting pump measurements, two other ISO standards should also be taken into account. ISO 9110-1:2020 [4] and ISO 9110-2:2020 [5] discuss general measurement related issues as measurement uncertainty, factors affecting it, determination of uncertainty, and calibration of measurement equipment.

Figure 1 presents the basic open circuit measurement system for unidirectional pumps as described in [2,3]. Positions of pressure and temperature measuring devices (hereafter referred as transducers) are determined by the measure of inner diameter of the fluid conduits, but the positions of flow transducers are not so rigorously governed. The recommendation is to place it close to the pump outlet port, and when not possible, as shown in the Figure with the caption "alternative position", the inlet port of the flow transducer has to be equipped with pressure and temperature transducers, and the produced actual pump flow rate is to be calculated with the methods announced in the standards. This calculation is required because the fluid temperature and pressure at the far-placed flow transducer may deviate significantly from the values prevailing at the pump's outlet port. Based on pressure and flow rate measurement at the outlet of the pump, the produced hydraulic power, i.e., the produced output power, can be determined. Determination of the pump's power losses, and thus also its efficiencies, also requires the measurement of the pump's input power. This is a product of the driving rotational frequency and torque at the pump's axle.

When a pump is operated in a real application, its characteristics are affected by the pressure difference prevailing between outlet and inlet ports, the rotational speed of the pump, the pump displacement setting in variable displacement pumps, the characteristics of the fluid used, the impurity level, both the free and dissolved air content of the fluid, and the operational temperatures, i.e., fluid temperature and ambient temperature. In pump measurements, the emphasis is in solving the effects of pressure, rotational speed and displacement on pump characteristics, and, therefore, the effects of all other factors should be eliminated. This calls for adequate filtering and temperature management of the fluid, as well as ambient temperature management. In addition, to be able to produce comparable results between various pumps and measurers, ISO 4409 recommends using fluids of certain viscosity grades and indexes, classes, and densities. Naturally, the fluid used in the measurements has to meet the demands of the pump manufacturer.

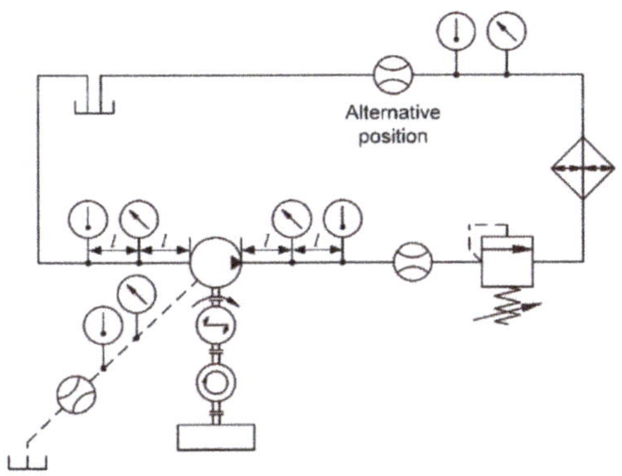

Figure 1. Basic measurement system for determining derived capacity and steady-state performance of a unidirectional positive displacement pump according to standards ISO 8426 and ISO 4409 [2,3].

Control of pressure difference across the pump ports requires some sort of loading device. In the system of Figure 1 this is a simple manually operated pressure relief valve, but when conducting a wide or comprehensive series of measurements with multiple pressure levels, a more user-friendly and accurate device, such as an electrically controlled proportional pressure relief valve or an electrically controlled flow control valve, should be used instead. If the effect of the rotational speed on the pump characteristics lies in the field of interest, it calls for control of the pump's drive motor's rotational speed. This is typically implemented by a frequency converter. In the case of variable displacement pumps, the applied displacement control should be such that it prevents unintentional alterations of setting value during measurements, and should also be equipped with a displacement indicating transducer to verify the value of the displacement setting. The more accurate and easily operated the controls are for all of these available, the shorter time it takes to shift from one measurement point (operating point) to another and conduct a series of measurements.

Determination of the pump's power losses and efficiency by calculation from the measurement data necessitates determination of the pump's derived capacity first. A basic method for this is presented in ISO 8426:2008 [2], but also more advanced methods have been proposed [6]. The ISO 8426 method is based on measurements on one rotational speed and several outlet pressures, whose number is dictated by the measurement class striven for. Measurement results are placed in a q_V/p set of coordinates and the pump output flow at zero pressure difference between the pump's inlet and outlet ports is determined by the least squares method and extrapolation. The value of derived capacity is then determined by dividing this flow rate by the used rotational speed. In more advanced methods, such as those of Toet and Wilson-II [6,7], determining the derived capacity is based on measurements on several pressure and rotational speed levels. In these methods the determination is two-phased, e.g., in the Wilson-II method the first phase is practically similar to ISO 8426, but the q_V/p dependence is determined at several levels of constant rotational speeds, after which the calculated speed-dependent angular coefficients V_i are placed in a V_i/n set of coordinates, and the derived capacity is determined with the least squares method and extrapolated to speed value zero. In the Toet method, the procedure is reversed to that of the Wilson-II method, i.e., first is the determination of the flow rate dependence on rotational speed at several pressures, then placing the solved pressure dependent angular coefficients V_i in a V_i/p set of coordinates, and the derived capacity is determined with the least squares method and extrapolated to pressure value zero.

Figure 2 presents an example of a comparison of derived capacity values calculated with four methods using the same measurement data, and using the value given by the Wilson-II method as a reference. The now replaced ISO 8426 method from 1988 was based on measurements at one low-pressure level and with several rotational speeds, which resulted in a lower than real derived capacity. When used in calculating efficiencies, this would result in higher than real volumetric efficiency values and lower than real hydromechanical efficiency values. The current ISO 8426 method from 2008 results in derived capacity, which is very close to the values given by the Toet and Wilson-II methods even though it is a single-phased determination method. The reason for this is that the volumetric losses of pumps are mainly pressure-related and the effect of rotational speed, which the new ISO 8426 method omits, is minor on the flow losses. Regardless of which one of the three latter methods is used in determining the derived capacity, the data used for it should be selected carefully and any data affected by obvious measurement error should be discarded from calculation, since it might have a strong distorting effect on the resulting value of the derived capacity.

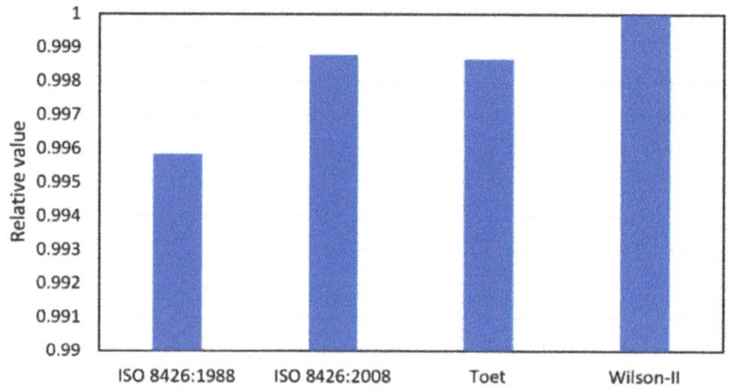

Figure 2. Derived capacities determined with different methods. Values relative to the result given by the Wilson-II method.

The minor deviation between the results of the Toet and Wilson-II methods illustrates the importance of keeping the values of parametrizing quantities accurately at selected levels, since even minor deviations from these will induce errors in calculated angular coefficients in the first calculation phase and therefore also in the second calculation phase. If the deviations in the selected measurement quantity levels were zero, the Toet and Wilson-II methods would give the same result. In the measurements on which the results of Figure 2 are based, the value of pump outlet pressure between separate measurement points has varied relatively more than the rotational speed, thus the minor value of the derived capacity of the Toet method.

Determining the value of derived capacity is of great importance as it has a major impact on the calculated values of power, torque and flow losses, and therefore also on the calculated efficiencies. An incorrect value will lead to either over- or underestimating the losses and efficiencies depending on the direction of the error in derived capacity. Therefore, the measurements should be conducted as accurately as possible.

ISO 4409 classifies the measurements in to three accuracy classes: A, B and C. A certain measurement point in steady-state measurements is acceptable for calculation only when the indicated value of the measurement parameter is within the limits presented in Table 1. In ISO 8426, which determines the conditions for derived capacity measurements, the differences to the values presented in the table are that the pressure boundary value is at 0.15 bar and the allowed temperature variation is half of the value presented in Table 1. In addition, ISO 8426 also determines a permissible variation region for ambient temperature.

Equations for calculating a pump's performance are presented in standard ISO 4391 [8]. These equations are simple to use and require no knowledge of the pump's internal structure. The methods and equations presented in this standard and older version of ISO 4409 are, however, not fully consistent in a methodological sense. Some criticism has been presented towards the equations presented in the standards, e.g., the comparison of pressurized flow rate with zero pressure flow rate, i.e., compressed fluid volume to uncompressed fluid volume, to produce volumetric efficiency.

Table 1. Maximum permissible variations in the values of measurement parameters in each measurement accuracy class when measuring steady state performance of pumps [3].

Measurement Parameter and Unit	Limits of Permissible Variations in Measurement Accuracy Class		
	A	B	C
Rotational frequency, %	±0.5	±1.0	±2.0
Torque, %	±0.5	±1.0	±2.0
Flow rate, %	±0.5	±1.5	±2.5
Pressure ($p < 2$ bar), bar	±0.01	±0.03	±0.05
Pressure ($p > 2$ bar), %	±0.5	±1.5	±2.5
Temperature, °C	±1.0	±2.0	±4.0

Achten et al. [9] pointed out two main deficiencies in the calculation methods: omitting the effect of fluid compressibility and omitting the effect of the pump's dead volume in displacement chambers. Instead of announcing the pump performance in terms of efficiency and dividing the total efficiency into volumetric and hydromechanical efficiency as in ISO standards, Achten et al. proposed presenting the pump performance in terms of power losses and dividing these into overall, volumetric, and hydromechanical power losses. By taking into account the two deficiencies mentioned, they derived new equations for losses using thermodynamical analysis of pump power as a starting point. The heat losses were neglected and the kinetic energies of inlet and outlet flows were assumed to be equal. As a result, Achten et al. proposed two power loss factors, which take into account the fluid's compressibility and the dead volume in the displacement chamber, to be added to the standards' efficiency equations and the loss equations derived from them.

Later, both Li and Barkei [10], and Schänzle and Pelz [11] continued the work on developing more accurate definitions for the efficiencies of hydrostatic machines. Similar to Achten et al., both of them used the thermodynamical analysis of pumping processes as a starting point, but after pointing out some inconsistencies in the work of Achten et al., they resulted in different efficiency definitions, and unlike Achten et al., they were able to present a definition for the volumetric efficiency as well. The efficiency definitions of both Li and Barkei, and Schänzle and Pelz are basically the same, but since the latter researchers chose to make some simplifying assumptions concerning the compressibility of the fluid and the relationship between the density of the fluid and the prevailing pressure, which the former did not make, the work of Schänzle and Pelz resulted in somewhat simpler efficiency definitions compared to Li and Barkei. In this way, Schänzle and Pelz, however, managed to achieve one of their main goals, which was to create practical and easily applicable efficiency definitions. The more complete and accurate Li and Barkei definitions for hydromechanical and total efficiencies include a quantity called volume specific enthalpy, which requires more measurement effort from the user when determining its value for the efficiency equations.

Figure 3 presents factors affecting effective outlet flow in a variable displacement pump. In fixed displacement pump, the control flow can be omitted, and depending on the pump's structure type and implementation, the external leakage may also be omitted.

In Figure 3, the low-pressure flow at the pump inlet port is transformed into high-pressure flow at the outlet port by the displacement elements of the pump. Because of the compressibility of the fluid, the effective outlet flow is decreased in relation to the inlet flow even though the mass flow could be equal if there were no flow losses inside the pump. These losses, however, exist. Part of the fluid volume taken in is used in lubrication of the rotating or reciprocating elements of the pump structure, and this flow of the fluid is pushed back from the high-pressure regions to the low-pressure regions of the structure, or simply put, from the pump outlet either to the pump inlet or/and to the external drain port via the pump casing. Part of the input flow is lost in fluid compression as presented by the above referred researchers, and in variable displacement pumps, a portion of flow is also lost to the displacement control system, if it is hydraulically controlled. The flow rate remaining after these losses is the effective flow rate of the pump.

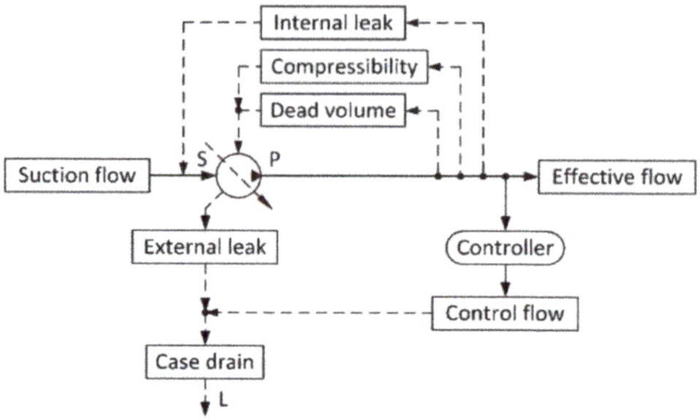

Figure 3. Flow rates of a variable displacement pump.

When determining pump performance, applying the Schänzle and Pelz proposed efficiency determinations, including the averaged isentropic compressibility, is fairly simple, as it only requires knowledge of the bulk modulus of the fluid and its dependency on pressure, temperature, and percentage of dissolved air in the fluid. These determinations, however, also require knowledge of the pump's dead volume, which in turn necessitates knowledge of the pump structure and measures, which are challenging to determine and usually out of reach of the person who does the performance measurements. To be able to apply these efficiency determinations, an experimental method for determining the dead volume should be developed.

The magnitudes and directions of the differences in efficiency values calculated with the ISO standards' determinations and Schänzle and Pelz determinations depend on the magnitude of the pump's dead volume, the value of the averaged isentropic compressibility, and the pressure.

The measurement results, i.e., the calculated efficiencies, shown in this article were determined using the methods presented in the ISO standards. The efficiency determinations presented by Schänzle and Pelz are omitted, as the research plan for measurements and also actual pump measurements were made before the publication of their article [10,11]. The measured pumps and used fluids were no longer available at that point, making, e.g., the determination of bulk moduli of the fluids impossible.

3. Efficiency Characteristics of a Swash-Plate Type Variable Displacement Axial Piston Pump

Measurements in this study were focused on solving the effects of various operational parameters on the performance of the axial piston pump. Figure 4 presents the principle image of the measurement system, where two different sized proportional directional control valves connected in parallel were used as loading devices for accurate control of outlet pressure at low and high flow rates. Individual parameter values and value ranges used in measurements are presented in Table 2. Measurements were carried out using a mineral oil-based fluid of viscosity class 46.

The inlet pressure was maintained stable and at very close to 0 bar (overpressure) by placing the pump below the tank that had an oil column height of one meter, and by using a large diameter inlet tube to achieve very low flow velocity, thus achieving negligible flow friction. The resultant variation in inlet pressure in measurements was of class 0.04 bar. Measurement uncertainties for each measured parameter presented in Table 3 were determined from the data of transducers, amplifiers, and data acquisition units.

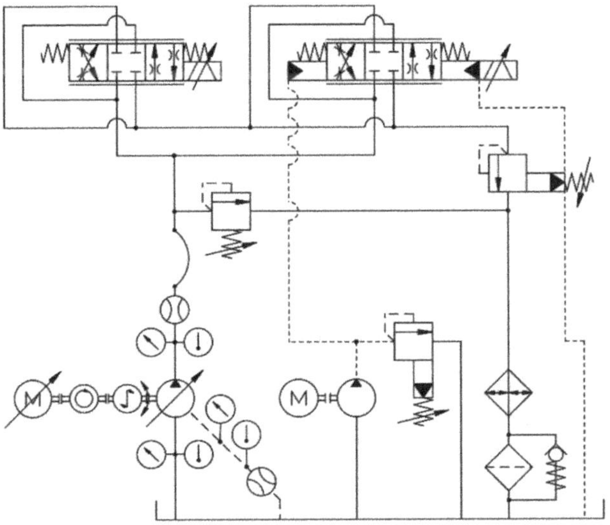

Figure 4. Measurement system.

Table 2. Values of the measurement parameters. Pressure values indicated in relation to atmospheric pressure. Maximum displacement of the measured nine piston pump is 45 cm^3/r, nominal pressure 280 bar, and maximum rotational speed 2600 r/min.

Parameter	Symbol	Value	Step	Unit
Rotational speed	n	500–2000	250	r/min
Inlet pressure	p_1	0	-	bar
Outlet pressure	p_2	25–250	25	bar
Derived capacity	V_i	45, 33, 23, 12	-	cm^3/r
Inlet temperature	θ_i	25, 60	-	°C

Table 3. Uncertainty values of measurement parameters. Coverage factor used in calculating expanded uncertainty $k = 2$.

Parameter	Combined Uncertainty	Expanded Uncertainty	Unit
Rotational speed	±1.70	±3.50	r/min
Torque	±0.41	±0.82	Nm
Flow rate	±0.33	±0.66	L/min
Inlet pressure	$\pm 7.20 \times 10^{-3}$	$\pm 1.40 \times 10^{-2}$	bar
Outlet pressure	±0.60	±1.20	bar
Inlet and outlet temperature	±0.61	±1.22	°C

The main frame of the measurements was to start from the largest value of derived capacity and lowest inlet temperature, followed by unchanged capacity and highest inlet temperature. Then, adjustment of the derived capacity to the next smaller value and repeating the inlet temperature sequence until all the selected derived capacities had been passed. The inner measurement loop at each step of capacity and inlet temperature was at each selected value of rotational speed. The outlet pressure was varied through the selected range at selected steps.

The values of the derived capacity listed in Table 2 are nominal values; the actual values used in calculating pump efficiencies were determined using measurement data and the Wilson-II method. In measurements, the pump was stripped of controllers and the pressure control channel of the pump's displacement control piston was disconnected from the pump's outlet and connected to a zero-pressure tank instead. This, as well as using the two swash-plate angle limiting screws for maximum and minimum displacement to lock the swash-plate in place, inhibited the angle from altering during measurements, thus keeping the value of derived capacity constant at the set value.

In the following, the measurement results are presented from several points of view and commented on briefly. A wider discussion of the results is presented in Section 4.

Figure 5 gives an overall picture of pump efficiencies' behavior and dependence on pressure difference between outlet and inlet ports and rotational speed at two values of derived capacity and inlet fluid temperature. In volumetric efficiency, the high fluid viscosity produces high efficiency values, but in hydromechanical efficiency this phenomenon is achieved only at combinations of high pressures, low rotational speeds, and high values of derived capacity.

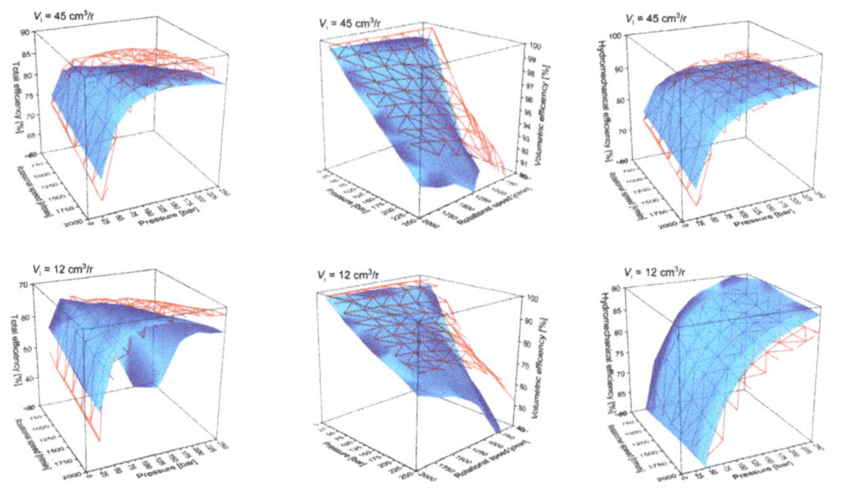

Figure 5. Efficiencies of a variable displacement axial piston pump as a function of pressure and rotational speed with derived capacities of 45 cm^3/r and 12 cm^3/r, and with inlet temperatures of 25 °C [ν = 84 cSt] (red wireframe) and 60 °C [ν = 21 cSt] (blue surface). Note the varying efficiency scales.

Figure 6 presents the dependence of the pump's total efficiency on pressure and rotational speed at four values of derived capacity and two values of fluid temperature at the pump's inlet in the form of contour plots. Reducing the value of derived capacity decreases the efficiency throughout the operational ranges, and rising temperature, i.e., decreasing fluid viscosity, moves the topography of high efficiencies to the direction of lower pressures.

In the following, the general results of Figures 5 and 6, above, are presented in more detail with 2D diagrams. When interpreting the effect of fluid inlet temperature on the pump efficiencies, it must be noted that the fluid viscosities used in the measurements were well within the range recommended by the pump manufacturer. Using very high or very low viscosities will change the behavior of the efficiencies drastically and may lead to damage to the pump. It also must be noted that the presented results are related to one certain variable displacement axial piston pump construction and cannot be generalized to apply to any kind of axial piston pump. The purpose of this presentation is to illustrate the effects of various operational parameters on the performance of a hydraulic pump.

Figures 7–9 present the dependence of the pump's total, volumetric, and hydromechanical efficiency on the pressure difference between outlet and inlet ports. The results are shown for four constant rotational speeds, two derived capacities, and two inlet temperatures (i.e., inlet fluid viscosities).

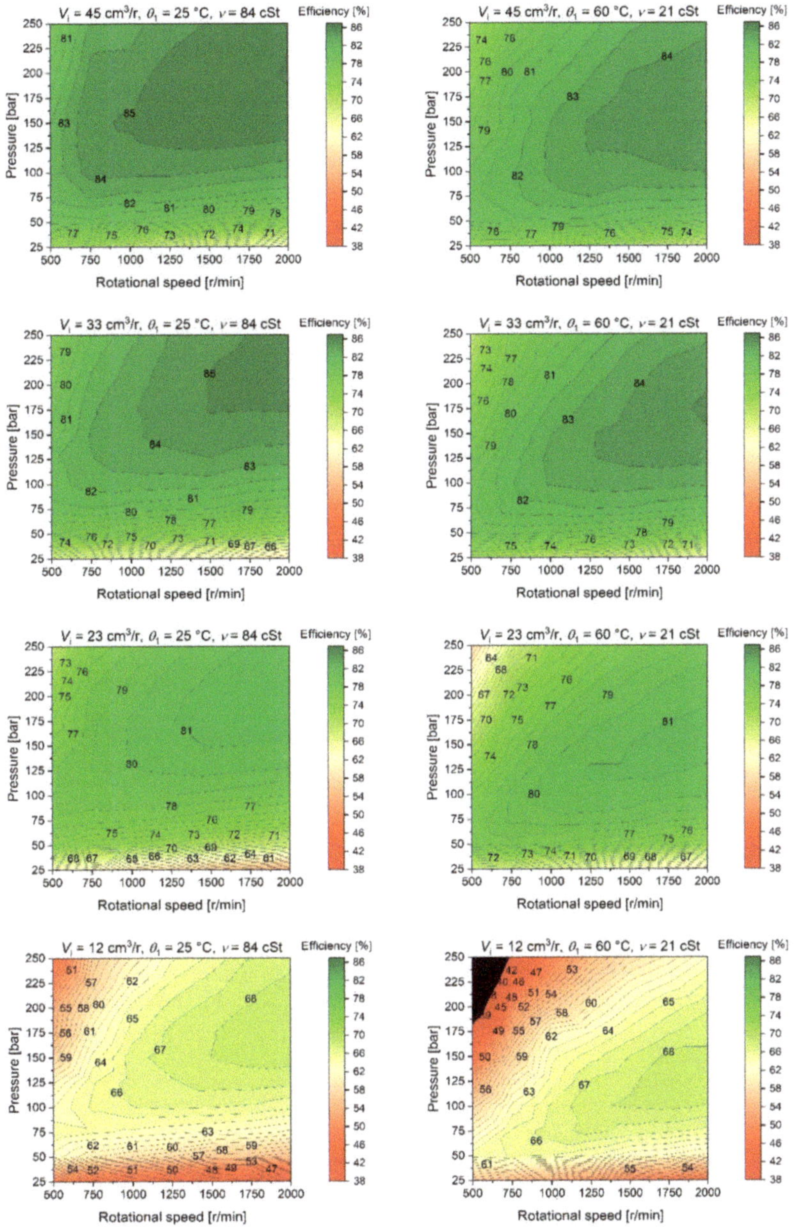

Figure 6. Total efficiency of an axial piston pump as a function of pressure and rotational speed with four values of derived capacities and two inlet temperatures. Contour lines were created from surface plots using the mesh determined by the step values presented in Table 2.

The effect of pressure on total efficiency is twofold. When moving from low pressures to high pressures, the efficiency first increases strongly and then decreases slowly. At low pressures, low rotational speeds result in higher efficiencies than high rotational speeds, but the situation turns the other way at high pressures. Decreasing derived capacity diminishes total efficiency all over, while increasing temperature (i.e., decreasing viscosity) has a milder effect on the total efficiency, increasing the efficiency at lower pressures and decreasing it at higher pressures, the effect being dependent on the rotational speed.

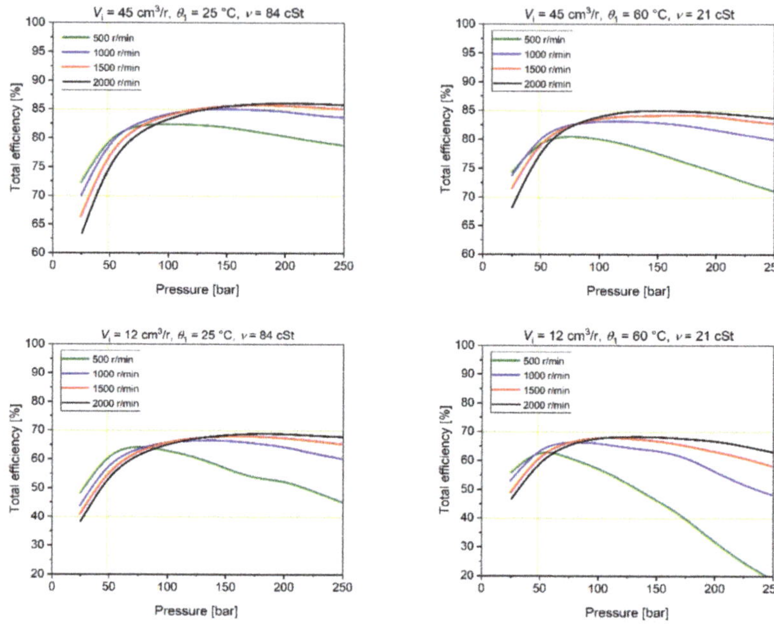

Figure 7. Total efficiency of an axial piston pump as a function of pressure with four constant rotational speeds. The ten-calculation-point efficiency curves have been smoothed with B-splines. Note the varying efficiency scales.

The effect of pressure on volumetric efficiency is highly linear, the value of the efficiency decreases with rising pressure. The effect of the other affecting parameters is conveyed merely in the slope of the efficiency alteration with pressure. Lowering the derived capacity and increasing the temperature (i.e., lowering the viscosity) result in decreased volumetric efficiency.

The effect of pressure on hydromechanical efficiency resembles the behavior of total efficiency, although the decrease in hydromechanical efficiency does not take place at high pressures. Likewise, with volumetric efficiency, the pressure dictates the general behavior of hydromechanical efficiency, while the effects of the other affecting parameters manifest themselves in the vertical positioning of the efficiency curves. Reducing the derived capacity decreases the efficiency, while raising the temperature (i.e., decreasing inlet fluid viscosity) increases the efficiency. Low rotational speeds tend to increase the efficiency, although this effect is lesser in larger derived capacities.

Figures 10–12 present the above shown results from an inverse point of view, showing the dependence of the pump's total, volumetric, and hydromechanical efficiency on rotational speed with selected constant pressures. Results are shown for five pressures, two derived capacities and two inlet temperatures (i.e., inlet fluid viscosities).

The effect of rotational speed on total efficiency depends on the prevailing pressure; at low pressures the efficiency decreases with increasing rotational speed, whilst at high pressures the efficiency rises. As noted above, the decreasing derived capacity results in diminishing efficiency, as does the rise of inlet temperature (i.e., inlet fluid temperature), the effect of the latter being strongest at combinations of high pressures and low rotational speeds.

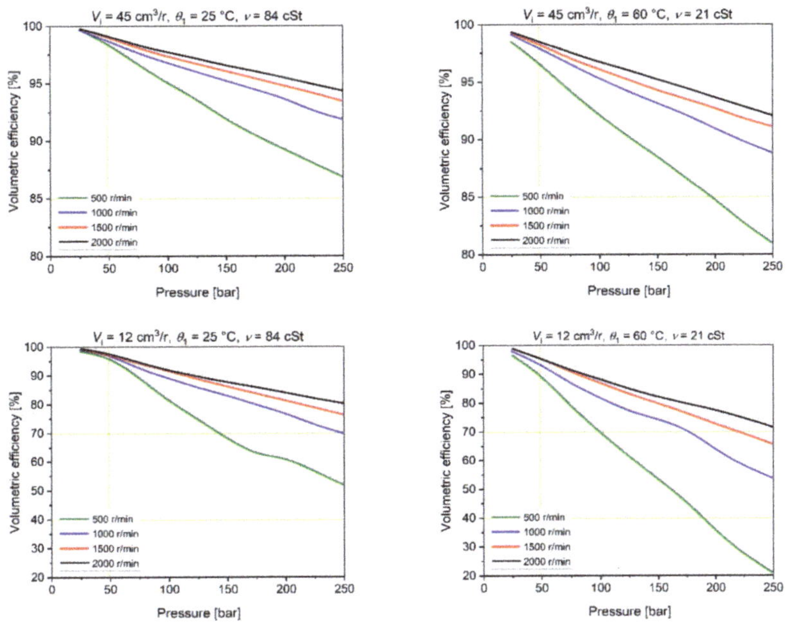

Figure 8. Volumetric efficiency of an axial piston pump as a function of pressure with four constant rotational speeds. The ten-calculation-point efficiency curves have been smoothed with B-splines. Note the varying efficiency scales.

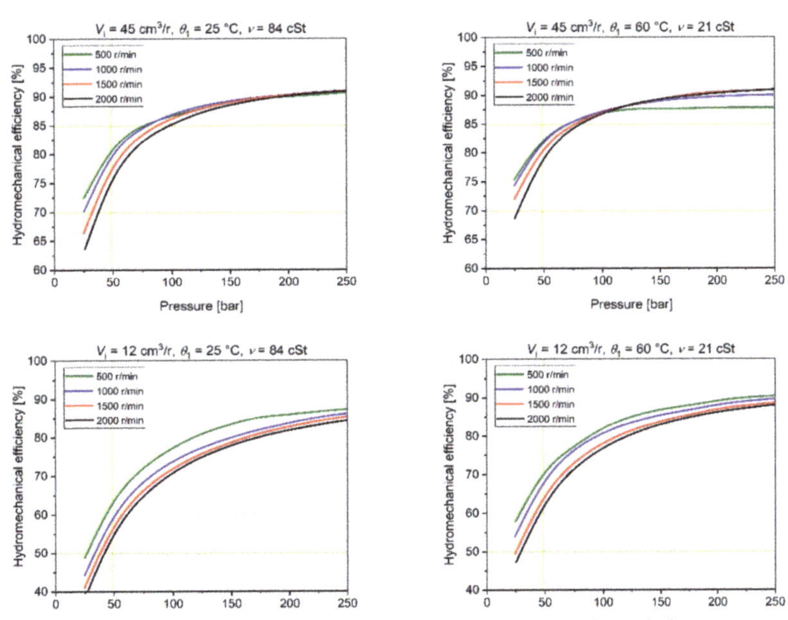

Figure 9. Hydromechanical efficiency of an axial piston pump as a function of pressure with four constant rotational speeds. The ten-calculation-point efficiency curves have been smoothed with B-splines. Note the varying efficiency scales.

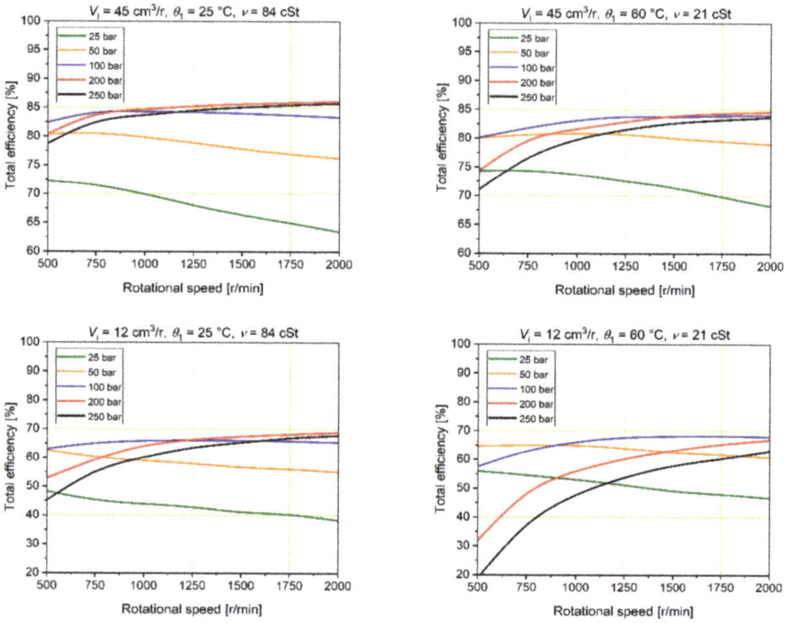

Figure 10. Total efficiency of an axial piston pump as a function of rotational speed with five constant pressures. The seven-calculation-point efficiency curves have been smoothed with B-splines. Note the varying efficiency scales.

The effect of rotational speed on volumetric efficiency is generally elevating. The rate of change in efficiency is the strongest the higher the pressure, although at low pressures the efficiency is already at a high level starting from low rotational speeds. In addition, here, a decrease of the derived capacity leads to diminishing efficiency, and the same effect has also the increasing temperature (i.e., inlet fluid temperature).

The effect of rotational speed on hydromechanical efficiency is generally linear, rising speed results in decreased efficiency, although the effect is strongest at low pressures, whilst at high pressures the effect is minimal or even non-existing. As with the other efficiencies presented above, decreasing the derived capacity leads to diminishing efficiency, and the rate of change is largest at low pressures. The temperature rise (i.e., decreasing inlet fluid viscosity) increases the efficiency in general.

Figures 13–15 present the above shown results from a third angle, showing the dependence of the pump's total, volumetric, and hydromechanical efficiency on derived capacity with selected constant pressures. Results are shown for six pressures, one rotational speed, and two inlet temperatures (i.e., inlet fluid viscosities).

As already demonstrated above, the total efficiency is higher the closer the derived capacity of a variable displacement pump is to its maximum value. Rising pressure basically increases the efficiency, but from the middle pressure region up, the effect is non-existent, and with high temperatures (i.e., inlet fluid temperatures) even the opposite.

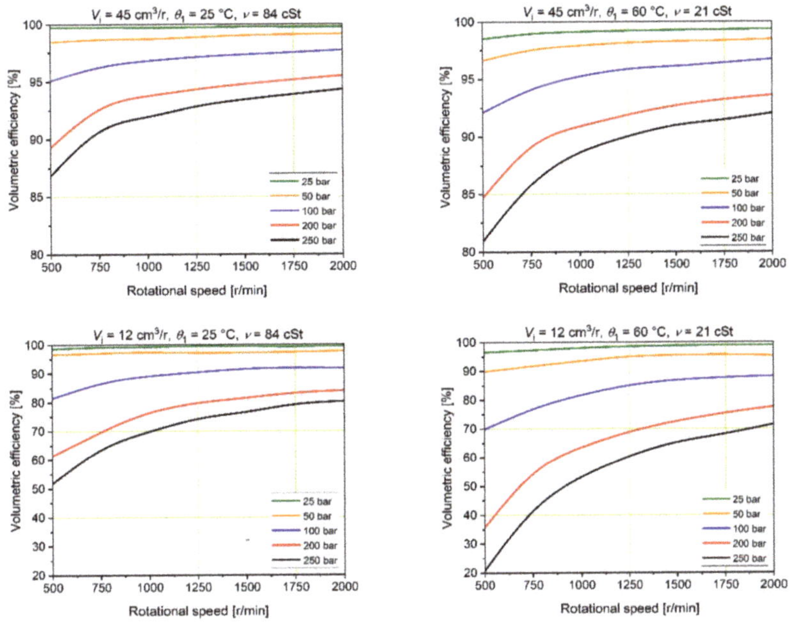

Figure 11. Volumetric efficiency of an axial piston pump as a function of rotational speed with five constant pressures. The seven-calculation-point efficiency curves have been smoothed with B-splines. Note the varying efficiency scales.

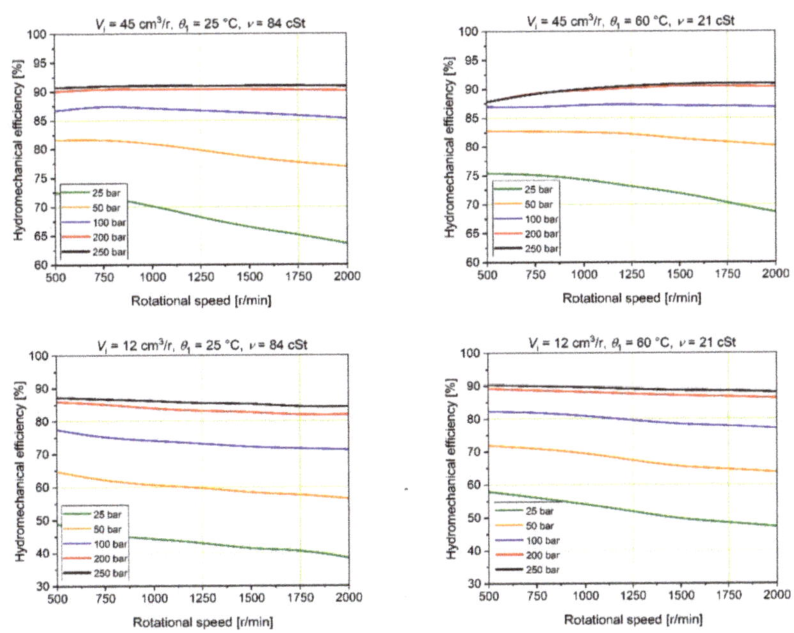

Figure 12. Hydromechanical efficiency of an axial piston pump as a function of rotational speed with five constant pressures. The seven-calculation-point efficiency curves have been smoothed with B-splines. Note the varying efficiency scales.

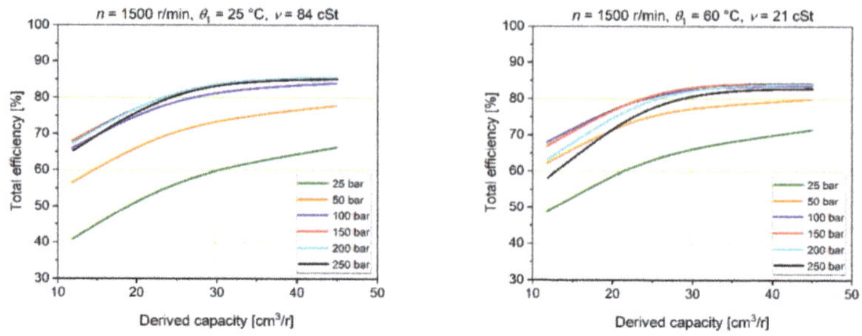

Figure 13. Total efficiency of an axial piston pump as a function of derived capacity with six constant pressures. The four-calculation-point efficiency curves have been smoothed with B-splines.

Increasing derived capacity has a similar rising effect on the volumetric efficiency, but here the efficiency is coherently: the lower the pressure, the higher the efficiency. At low pressures, however, the effect of derived capacity on efficiency is minimal. Rise of temperature (i.e., inlet fluid temperature) decreases the efficiency throughout, but here the effect diminishes with decreasing pressure.

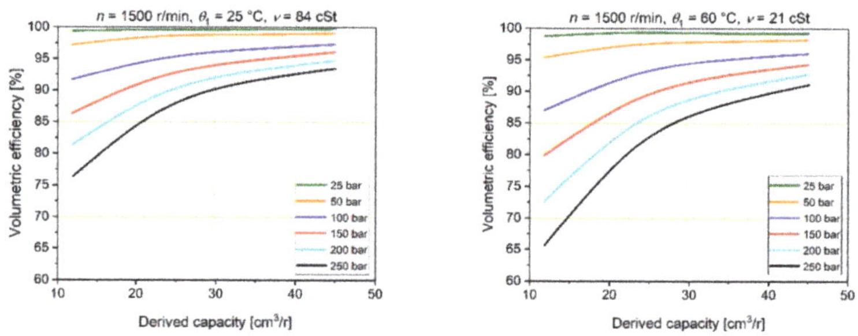

Figure 14. Volumetric efficiency of an axial piston pump as a function of derived capacity with six constant pressures. The four-calculation-point efficiency curves have been smoothed with B-splines.

In the case of hydromechanical efficiency, the effect of increasing derived capacity on the efficiency is not as coherent as with above presented efficiencies. At low pressures the efficiency rises with increasing derived capacity, but at high pressures the efficiency starts to decrease with increasing capacity. Rising temperature (i.e., inlet fluid temperature) increases the efficiency, although this effect diminishes with the combination of rising pressure and increasing derived capacity.

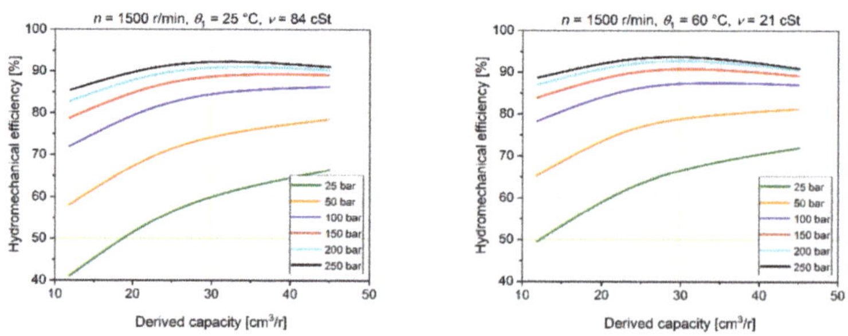

Figure 15. Hydromechanical efficiency of an axial piston pump as a function of derived capacity with six constant pressures. The four-calculation-point efficiency curves have been smoothed with B-splines.

4. Discussion

The pump's performance is dependent on several phenomena, whose combined effect on the pump is typically described in terms of the pump's total efficiency. This, in turn, can be divided into two subefficiencies, namely volumetric efficiency and hydromechanical efficiency. These then are affected by the parameter values prevailing at the operating point of the pump. This point is usually defined by the pressure difference between the pump's inlet and outlet ports, the rotational speed of the pump, and the inlet fluid temperature since it affects the viscosity and density of the fluid. In the case of variable displacement pumps, the derived capacity of the pump becomes one of the factors defining the pump's operating point.

The volumetric efficiency is determined by several types of flow loss (see Figure 3) taking place inside the pump, and hydromechanical efficiency is determined by the minor losses and friction losses taking place inside the pump as well. The minor losses are due to the flow resistance occurring in the pump's complicated internal flow channels, and their magnitude increases with the flow that passes through the pump. Friction losses in turn are due to the parts moving in relation to each other and separated by a lubrication layer, the circumstance of which is defined as viscous friction. Both of these losses are dependent on fluid viscosity and density, i.e., on fluid temperature, besides which the minor losses are also dependent on the flow rate that passes through the pump, while the friction losses are dependent on the pressure that affects the thickness of the lubrication layer, separating the parts that move relative to each other. The higher the pressure, the thicker the lubrication layer, although the thickness of the layer cannot increase unlimitedly due to the physical limitations of the pump structure.

As mentioned, the temperature considered to affect the operating point of a pump is the fluid inlet temperature. This, however, is not the temperature inside the pump, where the minor and friction losses turn into power losses that raise the temperature of both the fluid and pump structure. This in turn leads to alteration of the fluid properties and also of the dimensions of the pump parts, whose effects on the pump performance are virtually impossible to separate from the effects of pressure, rotational speed, and derived capacity. Therefore, these effects induced by the temperature rise in the pump are ignored in this presentation.

Factors that are not commonly, or at all, considered in determining pump performance in an operating point are the properties of fluid other than viscosity and density. For example, the lubrication properties or the pressure durability of the fluid may have a significant impact on the pump performance, either on the overall operating range, or after reaching some limiting operating point in relation to, e.g., pressure or temperature. This should be noted, and the recommendations of the pump manufacturer should be followed considering the properties of the fluid.

In the following, the effects of the main parameters influencing the performance of a variable displacement axial piston pump are discussed in terms of volumetric, hydromechanical, and total efficiency. It must be noted that the conclusions drawn apply only to the measured value ranges of the measurement parameters. In the conducted study, it was not possible to do measurements over the total pressure and rotational speed ranges of the pump because of the underpowered prime mover.

Effects of the studied parameters on volumetric efficiency are coherent, a unidirectional change in a parameter value only either reduces or increases the value of the efficiency. Increasing pressure difference between the inlet and outlet ports of the pump increases flow through various leakage paths in pump and it also increases the compressibility of the fluid thus reducing pump's effective outlet flow rate in relation to pump's theoretical flow delivery resulting in decreasing volumetric efficiency. The change in efficiency with pressure is very linear which implies that the leakage is dominantly laminar in nature. The effect of pressure on the efficiency is dependent on rotational speed in a way that the higher the rotational speed the lesser the relative decrease of efficiency with rising pressure. This implies that the amount of leakage is primarily determined by the pressure, and only remotely affected by the rotational speed, which is confirmed by the behavior of the efficiency as a function of rotational speed. Increasing rotational speed raises the theoretical flow of the pump, and when the rotational speed does not significantly affect the amount of the leakage, but remains close to constant, the raise of rotational speed results into increase of volumetric efficiency. This characteristic is emphasized at high pressures, where the leakages are higher than at low pressures, and where also the changes that take place in the leakage flow/theoretical flow rate proportion are higher when rotational speed is raised. Increase of the derived capacity has similar effect on the efficiency like raising rotational speed, the theoretical flow increases with increasing derived capacity while the pressure-dependent leakage stays close to constant resulting in raising volumetric efficiency. The change in efficiency is the larger the higher is the pressure, since then the portion of the pressure-dependent leakage of the pump's theoretical flow delivery diminishes stronger with the increasing derived capacity than it does at low pressures. Compared to the effects of increasing rotational speed and derived capacity, temperature rise has an opposite effect on the volumetric efficiency, it decreases the viscosity and density of the fluid, which both have increasing effect on the pressure dependent laminar and turbulent leakage thus reducing the volumetric efficiency.

The effects of the studied parameters on hydromechanical efficiency are more complex, since this efficiency is affected by two major phenomenon, the minor losses and the viscous friction losses, whose magnitudes have the above mentioned dependencies on fluid properties, flow rate, and pressure. Raising pressure increases the hydromechanical efficiency, which can be deduced to be due to the reducing viscous friction losses between the moving parts of the pump. The effect of pressure on the minor losses can in turn be assumed to be negligible, as the pressure-related alterations in the dimensions of the pump's flow channels are of minuscule class. In lubrication points, however, the rising pressure increases the height of the lubrication gap and thus also the lubrication layer, resulting in lower viscous friction between the moving parts, which then results in higher hydromechanical efficiency. Increase of the efficiency with the raising of pressure is not linear, which implies that the lubrications gap heights cannot grow limitlessly with rising pressure, but are restricted by the construction and the materials of the pump. Increasing rotational speed is supposed, in general, to have a reducing effect on the hydromechanical efficiency, which is due to the increasing flow through the pump that in turn increases the pump's minor losses. In the studied pump, the efficiency reduction with raising rotational speed is, however, heavily pressure dependent, and is strongest at low pressures, while at high pressures it is very moderate or even non-existent. This implies that in the case of this pump, the rotational speed does not have a major impact on the hydromechanical efficiency, but the deterioration of the efficiency with the increase of speed is primarily due to the viscous friction losses. These are the higher, the lower is the pressure and the higher is the rotational speed, because of the resulting thinner lubrication layer. Increase of the derived capacity increases the hydromechanical efficiency, the phenomenon being strongly pressure dependent. At low pressures the alteration of the efficiency with derived capacity is higher than at high pressures. As with rotational speed, the dependency of the efficiency on derived capacity is mostly related to viscous friction losses while the effect of minor losses is lesser, the latter manifesting themselves only at high flow rates. An example of this is the decrease of the efficiency with increasing derived capacity that takes place at high pressures. A temperature rise that decreases the values of viscosity and density, results in increasing hydromechanical efficiency due to the reducing minor and viscous friction losses. This phenomenon has a limit, especially at some temperatures, e.g., the lubrication capability of the fluid collapses, resulting in high friction losses.

Total efficiency is the product of volumetric and hydromechanics efficiency, and therefore it incorporates the parameter dependencies of both. Which of these has the dominating effect on the total efficiency depends on the operating point. In this context, domination here refers to the effect that significantly decreases the total efficiency.

Regarding the pressure dependency of the total efficiency, the hydromechanical efficiency dominates at low pressures, while the weight of the volumetric efficiency on the total efficiency increases with rising pressure. Rotational speed, derived capacity, or fluid temperature do not have a significant impact on the general dependency of total efficiency on pressure. The highest total efficiency settles somewhere between the lowest and highest value in the pressure range recommended by the pump manufacturer.

The effect of rotational speed on total efficiency is not so clear. Neither of the subefficiencies clearly dominate the total efficiency if the pump is operated at low or medium pressures and at high derived capacities. Significant domination occurs only when the pump is operated at low rotational speeds, in which case the volumetric efficiency has the dominant effect. This is further emphasized at low values of derived capacity. The rise in the temperature decreases the total efficiency consistently throughout the rotational speed range. The highest total efficiency settles at the high end of the rotational speed range recommended by the pump manufacturer.

In addition, the effect of derived capacity on total efficiency is twofold depending on the values of the other affecting parameters. At low pressures, the hydromechanical efficiency strongly dominates the total efficiency, but when pressure is raised, the volumetric efficiency takes over the dominating role. The temperature has similar kind of effect, hydromechanical efficiency dominates at low temperatures, but at higher temperatures the impact difference between the two subefficiencies diminishes. Rotational speed does not affect the relation of the subefficiencies significantly. The highest total efficiency settles in the high end of the derived capacity range.

The efficiency dependencies on various operating point parameters observed in the measurements are collected in the Table 4 in form of a rough generalization.

Table 4. Rough generalization of empirical changes in the efficiencies of the studied swash-plate type axial piston pump in relation to the increase in the operating point parameter values.

Rise of	Volumetric Efficiency	Hydromechanical Efficiency	Total Efficiency
Pressure	Decreases	Increases	Not unambiguous
Rotational speed	Increases	Decreases slightly	Not unambiguous
Derived capacity	Increases	Increases slightly	Increases
Temperature	Decreases	Increases slightly	Decreases

Although this study did not include examination of pump models, the generalized observations presented in Table 4 are quite well aligned with the structures of common and quite simple pump models like the Wilson and Schlösser models and the affecting parameters included in those. Review of these models will be done in following studies.

5. Conclusions

The measurement results presented above manifest the major effect that the operating point has on the performance of a swash-plate type axial piston pump. Pressure, rotational speed, derived capacity and temperature each have their own kind of effect on the volumetric and hydromechanical efficiency, thus also on the total efficiency. In addition, these effects also have an interrelatedness, where the effect of one parameter on the efficiencies is dependent at least on one additional parameter.

Although the dependency of efficiency on operating point parameters is complicated, knowing the pump's efficiency characteristics is essential when striving for higher system energy efficiencies. The pump as an energy exchanger has a major impact on the energy consumption in hydraulic systems, and running it outside its highest efficiency range will lead to low system efficiency even though the system is otherwise optimized for high efficiency.

Results of the study show that the highest values of total efficiencies of the studied pump are found at the high end of the rotational speed range, while their positioning in the pressure range depends on the values of derived capacity and viscosity. Decreasing of both of these shifts the areas of highest efficiencies to the direction of lower pressures. This behavior cannot be generalized to all pump types since it depends heavily on the pump construction type, as shown in the measurement results published by the company Innas.

The results and findings of this study are utilized further in developing methods that enable the definition of the combination of the pump's pressure, rotational speed, and derived capacity that results in the highest possible total efficiency, while meeting the performance requirements of the pump in terms of power and flow rate.

Funding: This research has received funding from Business Finland (Finnish Funding Agency for Innovation), project GOOD [Future electrified mobile machinery for harsh conditions] and School of Engineering, Aalto University, Finland.

Institutional Review Board Statement: Not applicable.

Informed Consent Statement: Not applicable.

Data Availability Statement: Not applicable.

Conflicts of Interest: The author declares no conflict of interest. The funders had no role in the design of the study; in the collection, analyses, or interpretation of data; in the writing of the manuscript; or in the decision to publish the results.

References

1. Performance of Hydrostatic Machines. Extensive Measurement Report. Innas BV. 2020. Available online: https://www.google.com/search?client=firefox-b-d&q=1.+Performance+of+hydrostatic+machines.+Extensive+measurement+report (accessed on 31 January 2022).
2. *ISO 8426:2008*; Hydraulic Fluid Power—Positive-Displacement Pumps and Motors—Determination of Derived Capacity. International Organization for Standardization: Geneva, Switzerland, 2008.
3. *ISO 4409:2019*; Hydraulic Fluid Power—Positive-Displacement Pumps, Motors and Integral Transmissions—Methods of Testing and Presenting Basic Steady-State Performance. International Organization for Standardization: Geneva, Switzerland, 2019.
4. *ISO 9110-1:2020*; Hydraulic Fluid Power—Measurement Techniques—Part 1: General Measurement Principles. International Organization for Standardization: Geneva, Switzerland, 2020.
5. *ISO 9110-2:2020*; Hydraulic Fluid Power—Measurement Techniques—Part 2: Measurement of Average Steady-State Pressure in a Closed Conduit. International Organization for Standardization: Geneva, Switzerland, 2020.
6. Post, W.J.A.E.M. Models for steady-state performance of hydraulic pumps: Determination of displacement. In Proceedings of the Ninth Bath International Fluid Power Workshop, Bath, UK, 9–11 September 1996; pp. 339–352.
7. Toet, G.; Johnson, J.; Montague, J.; Torres, K.; Garcia-Bravo, J. The determination of the theoretical stroke volume of hydrostatic positive displacement pumps and motors from volumetric measurements. *Energies* **2019**, *12*, 415. [CrossRef]
8. *ISO 4391:1983*; Hydraulic Fluid Power—Pumps, Motors and Integral Transmissions—Parameter Definitions and Letter Symbols. International Organization for Standardization: Geneva, Switzerland, 1983.
9. Achten, P.; Mommers, R.; Nishiumi, T.; Murrenhoff, H.; Sepehri, N.; Stelson, K.; Palmberg, J.-O.; Schmitz, K. Measuring the losses of hydrostatic pumps and motors—A critical review of ISO4409:2007. In Proceedings of the ASME/BATH 2019 Symposium on Fluid Power and Motion Control, FPMC2019, Sarasota, FL, USA, 7–9 October 2019.
10. Li, P.Y.; Barkei, J.H. Hydraulic effort and the efficiencies of pump and motors with compressible fluid. In Proceedings of the ASME/BATH 2020 Symposium on Fluid Power and Motion Control, FPMC2020, Virtual, Online, 9–11 September 2020.
11. Schänzle, C.; Pelz, P.F. Meaningful and physically consistent efficiency definition for positive displacement pumps—Continuation of the critical review of ISO 4391 and ISO 4409. In Proceedings of the ASME/BATH 2021 Symposium on Fluid Power and Motion Control, FPMC2021, Virtual, Online, 19–21 October 2021.

energies

MDPI

Article

Graphic Method to Evaluate Power Requirements of a Hydraulic System Using Load-Holding Valves

Luis Javier Berne [1], Gustavo Raush [2,*], Pedro Roquet [3], Pedro-Javier Gamez-Montero [2] and Esteban Codina [2]

1 IHBER, Polígono Malpica, Calle F, Nave 65, 50016 Zaragoza, Spain; ljberne@ihber.com
2 CATMech, Department of Fluid Mechanics, Universitat Politcnica de Catalunya, Colom 7, 08222 Terrassa, Spain; pedro.javier.gamez@upc.edu (P.-J.G.-M.); esteban.codina@upc.edu (E.C.)
3 ROQCAR, Antonio Figueras 68, 08551 Tona, Spain; pereroquet@hotmail.com
* Correspondence: gustavo.raush@upc.edu; Tel.: +34-93-739-82-24

Abstract: It is very well known that the use of a load-holding valve (LHV) in a hydraulic system introduces additional energy consumption. This article presented a simplified graphical method for analyzing the power requirements of hydraulic systems equipped with load-holding valves for overrunning load control. The method helps to understand the performance of load-holding valves during actuator movement. In addition, it allows visualization of the influence on the overall system consumption of the main parameters (pilot ratio, set pressure) and others such as flow rate, back pressure, and load force. The method is attractive because, with only the pressures at the three ports and the valve relief function curve, it is sufficient to evaluate the energy consumption and to define the power ratio as an index indicating the percentage of energy that is to be used to open the LHV valve. The method was applied to real cases, in particular to two types of lifting mobile machines. It was validated following several outdoor tests on two mobile machines where experimental data were obtained. During tests, both machines were equipped with a set of seven different performance LHV valves. The described method could be beneficial for hydraulic machine manufacturers engaged in designing lifting devices when selecting a suitable valve for energy efficiency applications, especially now that the trend towards electrification is a reality.

Keywords: load-holding valve; counterbalance; overcenter; energy balance; hydraulic systems; mobile machinery

Citation: Berne, L.J.; Raush, G.; Roquet, P.; Gamez-Montero, P.-J.; Codina, E. Graphic Method to Evaluate Power Requirements of a Hydraulic System Using Load-Holding Valves. *Energies* **2022**, *15*, 4558. https://doi.org/10.3390/en15134558

Academic Editors: Paolo Casoli and Massimo Rundo

Received: 10 May 2022
Accepted: 16 June 2022
Published: 22 June 2022

1. Introduction

Load Holding Valves (LHV) are simple but critical components of any mobile machine. Basically, they are used in hydraulic circuits to avoid uncontrolled movements of cylinders and motors due to an overrunning or gravity-assisted load, but also are frequently found in positioning circuits and regenerative circuits. The LHVs have had significant technical-commercial impact and are required by law (directives and standards) thanks to their additional safety functionalities. They provide fully secured load holding and no leakage, shock absorption, cavitation protection at load lowering, overload protection, and perform as line rupture safety valves. In addition, they must show some functionalities to guarantee a good dynamic performance of the entire hydraulic system, such as: prevent overflow when applying stepped pilot pressure, quick closing when the pilot pressure disappears, opening action independent of load pressure, and good flow control capabilities. A significant amount of equipment uses LHV valves, such as: excavators, cranes, trucks with one or more arms (aerial platforms, inspection vehicles, concrete pumping equipment), telescopic handlers, forklifts, fruit pickers, recreational facilities, industrial presses, drills, and industrial winches, among many other machines.

LHVs valves are modulating devices that allow free flow from the valve to the cylinder and block reverse flow from the cylinder to the valve until load pressure, pilot pressure, or a combination of both opens them (traditional pressure–spring balance principle, see Figure 1). Though they are simple in design, their application can often frustrate technicians and engineers. The known problems of LHVs are the excessive power consumption associated with over-pressurization of the flow supply and their tendency to introduce instability. The solutions to both issues are opposed to each other since what works well for stability is harmful for energy savings.

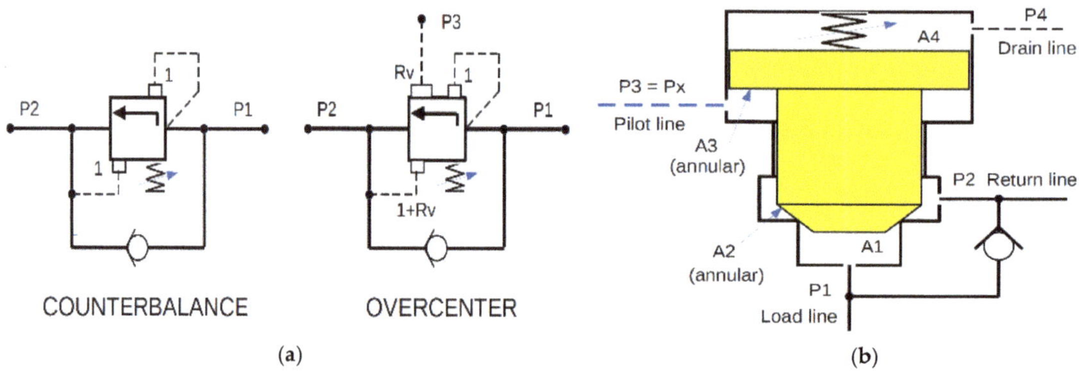

Figure 1. The traditional principle of pressure–spring load-holding valve model. (**a**) ISO 1219 symbols; (**b**) Functional scheme of the load holding valve, LHV.

Any application engineer interested in energy consumption needs to answer the following question: how can the power requirements of a LHV be calculated for a given set of operating conditions? To meet this challenge, the paper attempted to present a simplified graphical method to analyze the power requirements of hydraulic systems equipped with LHVs for overrunning load control. This paper is organized as follows: Section 1 presents the introduction of the work exposed in this paper. Section 2 presents a brief schematic of the load-holding valves and a typical application to control the movement of a linear actuator. Special emphasis is also given to the graphical representation of the characteristic curves of both components. Section 3 is devoted to a short review of state of the art. Section 4 conceptually exposes the graphical method to perform a power balance. Section 5 focuses on describing the experimental tests and shows examples of the use of this analysis methodology. Finally, in Section 6, conclusions on the proposed method are discussed.

2. Review of Load-Holding Valves

It is believed that the first designs were the work of the Vickers engineers staff in the 1930s and the first valves to appear on the market were developed by Racine. Since then, these valves have received different denominations based on their design and functionality: counterbalance valve, overcenter valve, holding valve, load-control valve, pilot-assisted load-control valve, load-holding valve, and motion-control valves, among others.

The main discussion arises in the use of the names counterbalance and overcenter valve. Counterbalance valves are basically pressure relief valves in combination with a check valve to create a unidirectional back pressure in a hydraulic system to prevent the actuator running away with accelerating loads (see Figure 1a left). The valve poppet is balanced by two pressures and a spring. The force generated by the load pressure (P_1) tends to open the valve orifice, while the spring force and the force caused by the back pressure (P_2) tend to close the valve orifice. The overcenter valve, on the other hand, can be considered in a simplified form as a counterbalance valve with an external pilot line (P_3), being more energy efficient in systems with variable loads (see Figure 1a right). Considering the latest innovations in this valve type, the authors are motivated to look for a more generic name, such as load-holding valve, LHV. Figure 1b shows the scheme of a load-holding valve. Usually, the P_4 connects to the P_2, but P_4 is also left directly to atmospheric reference.

According to Figure 1b, where $A_4 = A_1 + A_2 + A_3$, the force balance on the LHV poppet (Equation (1)) allows for calculating the pressure required to open the valve and starts to move the load. In order to hold the load, the valve cracking pressure (or setting pressure) must be set higher than the load pressure. Because of the hysteresis of the moving parts of the valve (mainly due to friction), there is a difference between cracking and reseat pressure (the pressure at which the valve closes) with the reseat pressure being lower than the cracking pressure. Hysteresis is one of the reasons why the typical LHV valve setting is approximately 30% higher than the load pressure to ensure that the reseating pressure will be high enough to maintain the load.

$$P_1 + P_3 R_v = P_M + P_2(R_v + 1) \tag{1}$$

where, P_1 is the load pressure, P_3 is the pilot pressure, $P_M = F_{spring}/A_1$ is the spring pressure, P_2 is the return pressure, and R_v is the pilot ratio. The pilot ratio is given by the ratio between the pilot pressure area and the relief area.

$$R_v = \frac{A_3}{A_1} \tag{2}$$

In Figure 2a, a typical application is shown. A directional valve is used in conjunction with an LHV to control the movement of an actuator. The pilot port of the LHV is connected to the opposite side of the actuator.

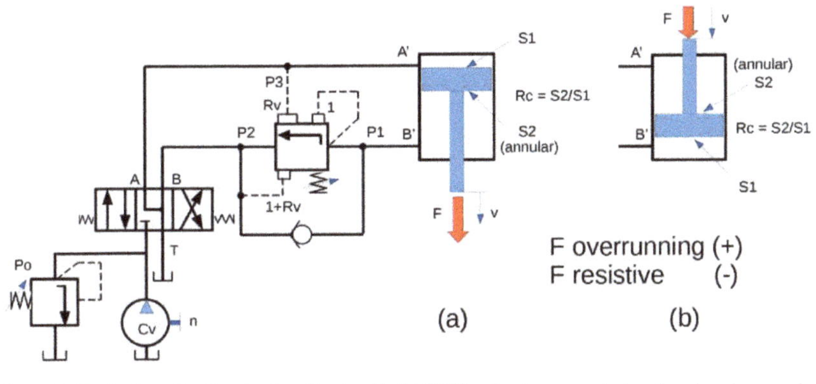

Figure 2. Typical application hydraulic circuit. (**a**) LHV valve is connected to the chamber on the rod side; (**b**) LHV valve is connected to the chamber on the piston side.

The force balance also makes it possible to determine the pressures in the hydraulic cylinder chambers (piston and rod side) as a function of a given load. This balance is established for two situations of the hydraulic cylinder, that is, when the LHV valve is connected to the chamber on the rod side (see Figure 2a) and when it is connected to the chamber on the piston side (see Figure 2b).

For case 2a

$$P_1 R_c - P_3 = P_{Load} \tag{3}$$

$$R_c = \frac{S_2}{S_1} \tag{4}$$

$$P_{Load} = \frac{F_{Load}}{S_1} \tag{5}$$

where, P_1 is the rod side chamber pressure, P_3 is the pilot pressure (equal to piston side chamber pressure), P_2 is the return pressure, and R_c is the hydraulic cylinder section ratio.

The sectional area ratio of the cylinder is given by the ratio between the effective sections of rod and piston side.

The load, F, can be considered resistive load (opposite direction for the actuator velocity "sign −"), or overruning load (same direction concerning the actuator speed "sign +"), as depicted in Figure 2.

Assuming that P_2 is very small and combining Equations (1) and (3), the pressures of the hydraulic cylinder chambers can arise from the pressures of the hydraulic cylinder chambers as a function of the spring pressure, P_M, and the equivalent load pressure, P_{Load}.

$$P_1 = \frac{1}{1 + R_c R_v} P_M + \frac{R_v}{1 + R_c R_v} P_{Load} \tag{6}$$

$$P_3 = \frac{R_c}{1 + R_c R_v} P_M - \frac{1}{1 + R_c R_v} P_{Load} \tag{7}$$

Analogously, for case 2b

$$P_1 - P_3 R_c = P_{Load} \tag{8}$$

where, P_1 is the piston side chamber pressure, P_3 is the pilot pressure (equal to the rod side chamber pressure), and P_2 is the return pressure.

In the same way, assuming that P_2 is very small and combining Equations (1) and (8), the pressures of the hydraulic cylinder chambers can be expressed as a function of the spring pressure, P_M, and the equivalent load pressure, P_{Load}.

$$P_1 = \frac{R_c}{R_c + R_v} P_M + \frac{R_v}{R_c + R_v} P_{Load} \tag{9}$$

$$P_3 = \frac{1}{R_c + R_v} P_M - \frac{1}{R_c + R_v} P_{Load} \tag{10}$$

Two types of operating curves can be found in the technical documents related to LHVs. On the one hand, the so-called "steady-state operating curve" expresses the load pressure as a function of the pilot pressure (P_3 vs. P_1). Figure 3 shows, for cases (2a) and (2b), the plots of the hydraulic actuator/LHV system based on Equations (1), (2) and (8), corresponding to the steady-state operating curves. The red lines are the characteristic curves of the LHV with different pilot ratios, while the solid black line corresponds to the unloaded hydraulic cylinder, and the dashed gray line corresponds to the hydraulic cylinder with an overrunnig load equivalent to 100 bar. Consequently, intersection point A corresponds to the LHV opening point and intersection point B to the LHV set pressure. This shows the effect of the setting pressure P_M and the pilot ratio R_v on the operation of the LHV. It is significant to note that this representation is used by some manufacturers to highlight differential aspects in the performance of the latest LHV versions, such as the two-stage valve, the multiple pilot ratio (adaptive) valve, and the setting self-adjusting (load match) valve.

Another variant of the operating curve for cases (2a) and (2b) is shown in Figure 4. These plots express the pressures P_1 and P_3 of the system hydraulic actuator/LHV in the cylinder chambers as a function of the equivalent load pressure, P_{Load}, which is the graphical representation of the two systems of linear Equations (6), (7), (9) and (10). The positive axis direction refers to overrunning loads, while the negative direction corresponds to resistive loads. The red line corresponds to the P_1 pressure and the black line to the pilot pressure, P_3. It can be seen that when the equivalent load pressure decreases, the pilot pressure increases.

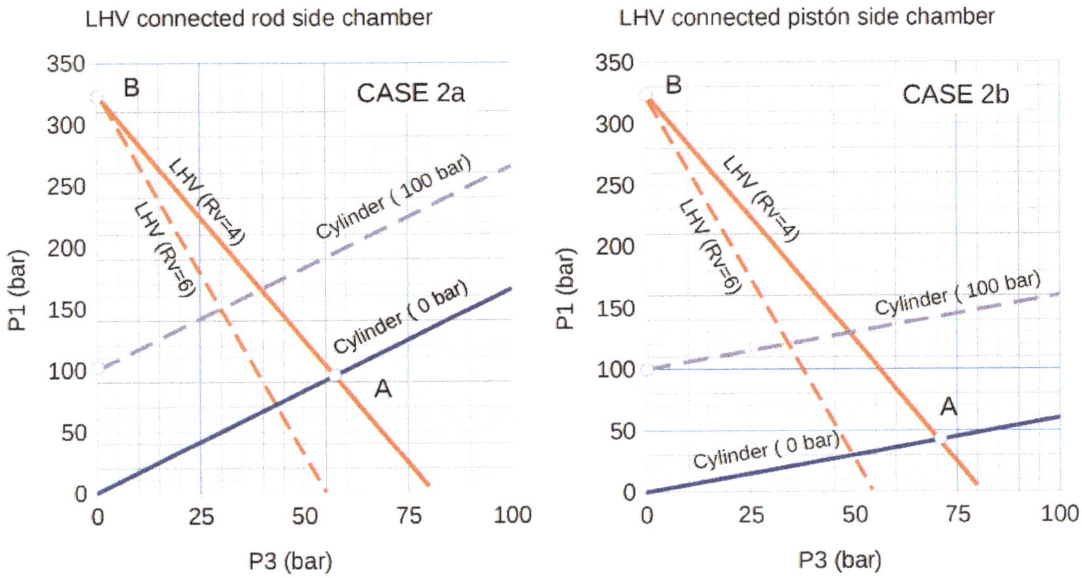

Figure 3. Steady-state operating curves $P_1 = f(P_3)$ for the two cases (2a) and (2b). Point A corresponds to the LHV opening point and point B to the LHV set pressure.

The plots in Figure 4 were first introduced by Professor Nicola Nervegna [1] and Professor Luca Zarotti [2]. Ritelli and Vacca used these graphs to evaluate energy aspects in [3], considering the hydraulic power consumption equal to the product of the pressure and the flow rate through the valve. If the flow rate is constant, the hydraulic power is proportional to the pressure in the actuator chamber P_1. Although these graphs have a high conceptual and qualitative value from the energetic point of view, the steady-state operating curves $P_1 = f(P_3)$ shown in Figure 3 can be used to evaluate the power requirements of the hydraulic system with LHV valves, as will be discussed in the following sections.

Table 1 shows the specifications of the essential hydraulic components that make up the hydraulic circuit used. These specifications were to draw Figures 3 and 4, while Table 2 shows the coordinates of the most significant points of the curves shown in Figure 4.

Table 1. The specifications of the primary hydraulic components of the hydraulic circuit imperative to depict graphs included in Figures 3 and 4.

Cylinder		LHV		System/Load	
Piston diameter	100 mm	Pilot ratio	4	Relief valve	250 bar
Rod diameter	63 mm	Setting pressure	325 bar	P_{Load} (min)	0 bar
Section ratio	0.603	Load max	270 bar	P_{Load} (example)	100 bar
		Nominal flow	90 L/min		

Table 2. List of critical points in the diagrams in Figure 4 in bar.

Case (2a)	x	y	x	y
A	P_{vlp}	P_{vlp}	250	250
B	$(1/R_v)\cdot P_M$	$1/R_v\ P_M$	81	81
C	$R_c\cdot P_M$	0	196	0
D	$R_c\ P_M$	P_M	196	325
E	$(1/R_v)\cdot P_M$	0	81	0
F	P_{vlp}	0	250	0
Case (2b)	x	y	x	y
A	$R_c\ P_{vlp}$	P_{vlp}	151	250
B	R_c/R_v	$1/R_v\ P_M$	49	81
C	P_M	0	325	0
D	P_M	P_M	325	325
E	$(R_c/R_v)\cdot P_M$	0	49	0
F	$R_c\ P_{vlp}$	0	151	0

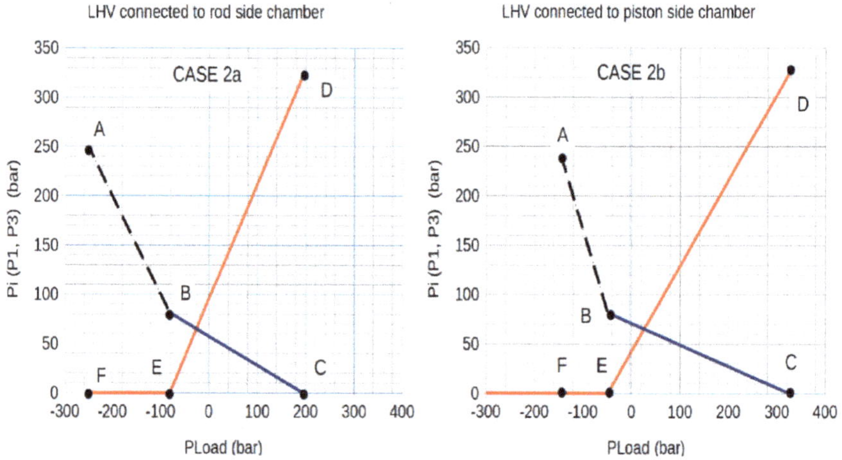

Figure 4. Steady-state operating curves $P_i(P_1,\ P_3) = f(P_{load})$ for the two cases (2a) and (2b). The meanings of points A–F are defined in Table 2.

3. The State of the Art

Figure 5 shows a historical compilation of publications published under the following designations: overcenter, counterbalance, and load-holding valves. This list is not exhaustive, but it is sufficient to show the evolution of the scientific and technological aspects. Two clear periods can be observed in this figure: a first period, extending from the beginning to the middle of the first decade of the 21st century, which is characterized by a scientific interest in finding out the causes or origins of instabilities, followed by a second stage focusing on finding designs and/or circuits capable of minimizing or eliminating their effects. In a second period, defined approximately by the last decade, new valve designs appear, modeling is carried out by means of concentrated parameter methods and/or distributed parameter methods, and studies related to energy consumption are found.

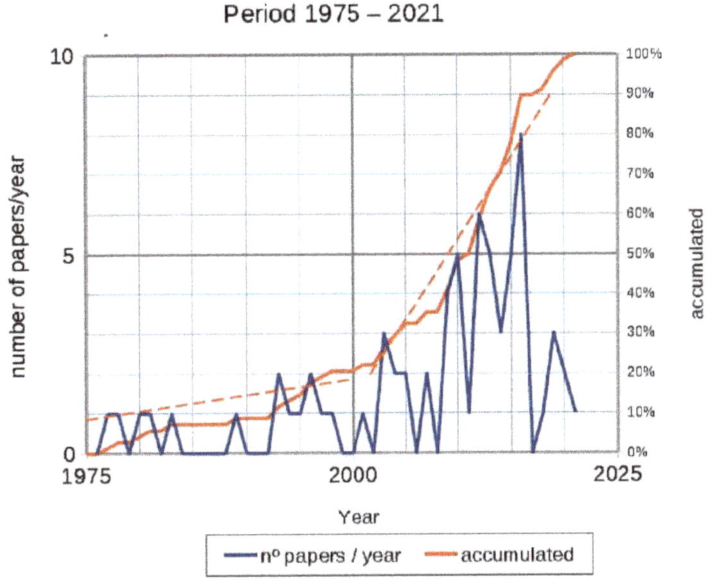

Figure 5. The number of papers/year (in blue) published related to load-holding valve issues shows an accumulated behavior (in red) with a tendency of 80% in the last 20 years.

3.1. About Instability

As mentioned, the severity of the oscillations is affected by a wide variety of parameters, some of which are hard to predict or change: external load on the actuator, the properties of the mechanical structure, the damping and hysteresis of the LHV, the operator input, as well as the volumes and restrictions in the hydraulic lines. All of these have been subjected to extensive investigations by, among others, refs. [4–14].

It is believed that the inherent stability problem of LHV is due to the fact that there is a phase lag between the pilot pressure and the outlet pressure of the actuator, and oscillations can be reduced by separating the pilot pressure and the actuator inlet pressure [15]. To improve the dynamic response of LHV, ref. [16] suggested that a pressure feedback system can indirectly eliminate oscillations and improve the stability of the hydraulic system. Another novel approach was developed by Sorensen et al. [17], using a low-pass filtered value of the secondary circuit load pressure for the pilot connection of the LHV. From the conceptual point of view, Groof [18] showed a simple criterion that can predict instability. The criterion is based on a third-order equation and indicates when a system is stable or unstable. It focuses only on the hydraulic system, without taking into account the mechanical stiffness of the complete system. When the LHV is mounted on the rod side of the cylinder (in Figure 2a), the criterion used for checking the stability of the system is defined by Equation (11), and when the LHV is mounted on the piston side of the cylinder (in Figure 2b) is defined by Equation (12)

$$\frac{G_{relief} C_{hB}}{R_c} > \frac{G_{pilot}}{C_{hA}} \quad (11)$$

$$\frac{G_{relief} C_{hB}}{C_{hA}} > \frac{G_{pilot}}{R_c} \quad (12)$$

where, G_{pilot}, gradient of pilot function, describes the opening characteristic as a function of the pilot pressure

$$G_{pilot} = \frac{dQ_{throughLHVvalve}}{dP_{pilot}} \tag{13}$$

G_{relief}, gradient of relief function, describes the opening behavior of the valve depended on the load pressure

$$G_{relief} = \frac{dQ_{throughLHVvalve}}{dP_{relief}} \tag{14}$$

C_h, hydraulic capacitance is given by the increase in effective stored fluid volume dV per change in pressure dP. The subscripts A and B indicate the hydraulic conduction and the volume of the cylinder hydraulic chamber on the corresponding side.

$$C_h = \frac{dV}{dP} \tag{15}$$

By way of illustration and for scenario (2b), the above criterion (Equation (12)) can be expressed by Equation (16).

$$\frac{V_2}{V_1} > R_v R_c \left(\frac{2P_1 A_1}{1 + 2P_1 + k \cdot x_{LHV} - k_f \cdot x_{LHV} \cdot P_1} \right) \tag{16}$$

where,

k, elastic constant of the spring of LHV
k_f, flow force constant
V_1, hydraulic cylinder piston side chamber volume
V_2, hydraulic cylinder rod side chamber volume
x_{LHV}, lifting height of the spool/poppet with respect to the seat surface of the LHV

According to the previous paragraphs, the most influential parameters of the stability are the setting pressure and the pilot ratio. A reduction of the pilot ratio results in a more stable operation, but at the cost of a higher pressure level and thus higher energy consumption, see ref. [3]. This is especially significant for small external loads. Adding damping when designing the hydraulic circuit is another proven approach. By increasing volumes, adding orifices, using logic valves, etc., more damping is introduced into the system. Special attention has been devoted to the pilot line. By manipulating the pressure in different ways on the LHV pilot port line (adding delays, creating a difference in the path back and forth), positive effects have been achieved. Sciancalepore and Vacca [19] proposed a solution by using LHVs to control the actuator speed while also reducing energy consumption. It consists of controlling the pilot port of the LHV through an external pressure source (adjustable pilot). Two different control strategies are presented: the "Smart LHV", where the LHV does not control the actuator velocity but it minimizes the system pressure; and the "Smart System" that uses the LHV to efficiently control the actuator velocity during overrunning load conditions.

3.2. About Modeling

Most of these investigations are related to modeling the steady-state and dynamic characteristics of load-holding valves in the time and frequency domains using a lumped parameter approach. However, most attempts to establish physical or semi-physical models of such valves have encountered many challenges, e.g., related to friction and resulting hysteresis, non-linear discharge area characteristics, varying discharge coefficients, and varying flow forces.

Today, there are several commercially available software packages such as MATLAB/Simulink™ (The MathWorks, Inc., Natick, Massachusetts, United States), AMESim™ (Siemens AG, Munich, Germany), Dymola™ (Dassault Systèmes, Paris, France), Maple/MapleSim™

(Waterloo Maple Inc., Waterloo, AB, Canada), and 20-Sim™ (Controllab Products B.V., Enschede, The Netherlands) that make work much more manageable. These packages may be classified into two different modeling approaches: a simple semi-physical model and a non-physical model (black box).

In the first approach, static modeling method based on force balance and the Bernoulli orifice pressure–flow equation is used to achieve the load velocity control ability of the valve. It does not take phenomena like friction and flow forces into account. In other cases, the dynamic modeling method is based on Newton's second law and fluid continuity equation. In both cases, experimental verification was required. In the second approach, the model uses two different pressure ratios to compute the flow through the valve together with a number of parameters that must be experimentally determined. Despite this, LHVs are rarely modeled accurately due to the effort required to obtain basic model parameters and the complexity involved in identifying equations for flow forces and friction.

Computational fluid dynamics (CFD) may be used to provide insight into some of these phenomena, but often experimental work and semi-physical or non-physical modeling approaches are required for time-domain simulation. Very interesting and to highlight are the works of refs. [14,20].

3.3. About Energetic Aspects

The energetic analysis of fluid power systems is becoming a requirement of machinery manufacturers, taking into account the trend towards the electrification of machines and the limited battery capacity. Many researchers are involved [21–25], and some are specific to studying the different hydraulic load retention systems in mobile and industrial applications [15,26,27].

Ritelli and Vacca [3] focused on the analysis of LHV energy savings. They quantified the energy consumption of LHVs with different pile ratios, showing that in a crane application, it is possible to achieve 59% energy savings with high pile ratios.

Regarding the energy efficiency of LHVs, there are some relevant previous efforts. LHVs with external and mixed control types were considered by ref. [13]. Significant examples are also the LoadMatch™ model, recently commercialized by Sun Hydraulics, which dynamically changes the valve setting as a function of the load pressure. In contrast, other works focus on introducing an external control to the LHV to influence its opening based on the instantaneous loading conditions. The most important contribution in this direction is the work in ref. [17], which proposed to use an auxiliary hydraulic circuit to control the pilot of the LHV, and ref. [28], which presented a method to control the pilot port of the CBV through an external pressure source (adjustable pilot). For both cases of "Smart CBV" and "Smart System", remarkable energy savings of 75% and up to 90%, respectively, were observed.

Despite scientific and technological advances reported and the wide variety of models and algorithms, it can be said that the use of this type of valve is far from easy. This wide range of tools is often not available to application engineers. In addition, the use of these tools still require a high level of application-specific knowledge. The impact outside academia remains limited. One reason may be that the design of hydraulic systems depends to some extent on the application.

In this paper, the authors highlighted the importance of understanding how LHVs work, not only in the critical opening condition but also outside of this condition, with the aim of helping to select the most suitable valve for the application according to its characteristics (extracted from the corresponding technical catalogs). Of course, experimentation is the best way to judge the system's power consumption but, the availability of a simple graphical method in the course to estimate the impact of the LHV on the overall power consumption may be very useful.

Of course, experimentation is the best way to judge the system's power consumption. However, the availability of a simple graphical method in the course to estimate the impact of the LHV on the overall power consumption may be very useful. In the next section, a brief background overview of the different types of performance curves is presented.

3.4. Performance Curves

Choosing the satisfactory LHV valve for a specific application is a challenging task. In order to apply a LHV correctly, it is crucial to understand how it works and to have sufficient technical information about its performance. After carefully analyzing the catalogs and other types of technical documents (release and technical papers, handbooks, user's manual, worksheets, prospectus, or brochures) the following comments can be summarized:

a They all provide a simple and generic description of how it works, symbols, valve cavity, specifications or technical data, and some performance graphs or characteristic curves. Typical technical data include maximum operating pressure, setting pressure (cracking pressure) interval, nominal flow, pilot ratio, internal leakage, and hysteresis, among others.

b Although graphical presentations of performance should help viewers quickly and easily understand the critical information, there is no unanimous agreement on which are the most appropriate. Of the different ways of expressing its performance, the next conditions may be pointed out:

   i Pressure drop in pilot operation condition (valve fully opened by the pilot pressure), (from port 1 to port 2, see Figure 1).

   ii Pressure drop in check valve operating condition (free flow) (from port 2 to port 1) and

   iii Pressure drop in pressure relief working condition (valve opened by the load). Unfortunately, some technical documents or specification sheets do not provide this minimum information, which is necessary for applying LHVs in a system.

c Generally, the curves shown at the end of Section 2 of this document and in Figures 3 and 4 are not usually included in the catalogs, but only in technical papers and to highlight the significant improvements in the performance of different valves.

4. Graphic Method to Estimate the Power Balance

The method is attractive because only the pressures in the three ports and the relief function curve of the valve are sufficient to evaluate the power consumption. The method has been inspired by the work done at the Agder University, Norway, led by Prof. M. Hansen [11]. It consists of a simple semi-physical model.

4.1. Description of the Four-Quadrant Diagram

For the description of the graphical method, the four-quadrant diagram of the Cartesian plane is a good description of the proposed method (Figure 6). The following variables are represented on the axes of this diagram:

x-axis (+): pilot pressure (P_3) in bar which corresponds to the pressure that prevails in the actuator rod chamber

x-axis (−): flow through the load holding valve (Q_1) in L/min, equal to the flow leaving the piston side actuator chamber

[-15]y-axis (+): load pressure (P_1) in bar, corresponding to the pressure in the actuator piston chamber

y-axis (−): flow rate entering the piston side actuator chamber (Q_S) in L/min according to the configuration of the hydraulic circuit shown in scenario (2b), Figure 2.

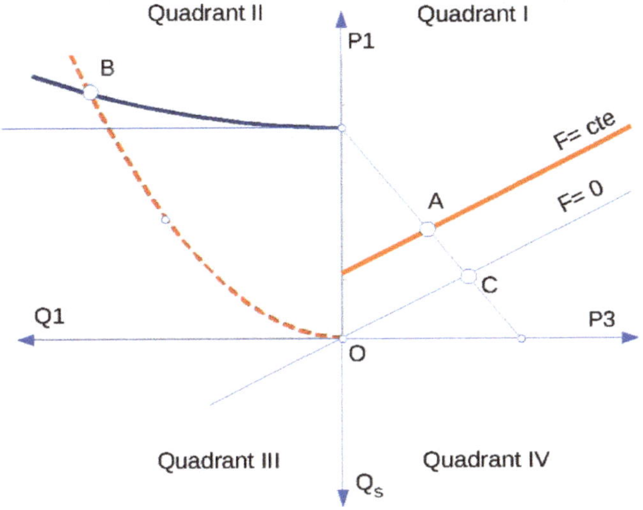

Figure 6. The cartesian plane is divided into four quadrants and characteristic curves of the hydraulic actuator and the load-holding valve. A represents the intersection between the curve of the cylinder function and that of the LHV when the force is constant; B is the intersection between the curve of the operating point of the LHV acting as a relief valve and the flow curve; C is the same as A but zero for the case of zero force.

The four-quadrant diagram allows us to plot the characteristic curves of an actuator and holding valve as follows:

- First quadrant "I": Upper right region. It is used to represent the characteristic curve of the hydraulic actuator and the characteristic curve $P_1 = f(P_3)$ of the load-holding valve.
- Second quadrant "II": Upper left region. It is used to show the characteristic curve of the load holding valve acting as a pressure-limiting valve (relief function).
- Third quadrant "III": Lower left region. It is used as an auxiliary two-dimensional space.
- Fourth quadrant "IV": Lower right region. It is used as an auxiliary two-dimensional space.

Alternatively, it can be used to depict the characteristic curve of the load-holding valve that expresses the pilot pressure (P_3) as a function of the flow rate (Q_1) through the load-holding valve for a predetermined load pressure value (P_1)

4.1.1. Steady-State Operating Curve and Hydraulic Cylinder Characteristic Curve

Equations (1) and (3) have been plotted in quadrant I. The intersection between the two curves gives the operating point of the hydraulic system (actuator/LHV valve) when the cylinder is subjected to a load, overrunning or resistive, as in Figure 3.

4.1.2. Characteristic Curve $P_1 = f(Q)$ of an LHV Valve Acting as a Pressure Limiting Valve (Relief Function)

Considering the simplified diagram of the holding valve poppet shown in Figure 7 and assuming that P_3 is not working and the flow is in a steady-state and incompressible fluid (ρ is constant), the application of the principle of conservation of momentum to the control volume allows us to write:

$$I_2 - I_1 = P_1 A_1 - P_2 A_2 - F_{solid-fluid} \tag{17}$$

$$F_{fluid-solid} = -F_{solid-fluid} \tag{18}$$

where: P_1A_1 and P_2A_2 denote the net force due to the pressure distribution in the inlet and outlet sections, respectively, I_1 and I_2 are the momentum fluxes of the flow entering and leaving the control surface,

$$I_1 = \frac{\rho Q^2}{A_1} \quad (19)$$

$$I_2 = \frac{\rho Q^2}{A_2(x)} \cos(\alpha) \quad (20)$$

where

$$A_2(x) = \pi \cdot d \cdot sen(\alpha) \cdot x = K_2 \cdot x \quad (21)$$

$$K_2 = \pi \cdot d \cdot sen(\alpha) \quad (22)$$

and $F_{fluid\text{-}solid}$ is the force on the poppet exerted by the fluid.

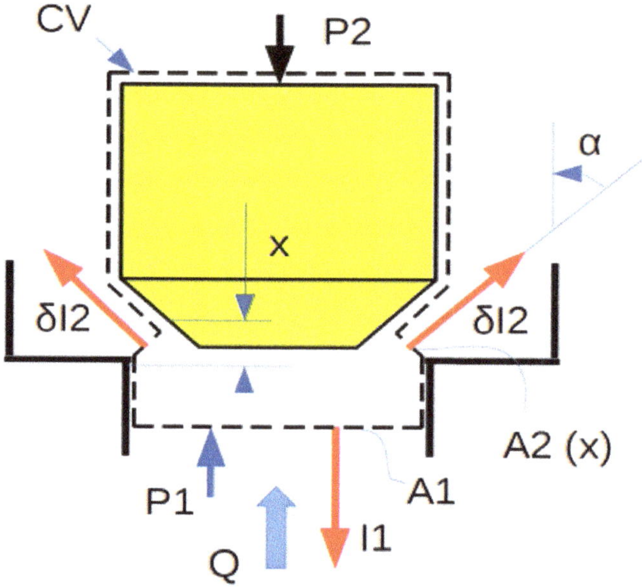

Figure 7. Control volume of the poppet/seat surface system of an LHV valve.

Applying Newton's second law of motion to the valve poppet (considered a free solid)

$$F_{fluid-solid} - (F_{spring} + M_{solid}\, g) = M_{solid}\dot{v} \quad (23)$$

Accepting an equilibrium position defined by the distance, x, considering that M_{solid} is negligible then:

$$F_{fluid-solid} = F_{spring} \quad (24)$$

$$F_{spring} = K_1(x + x_0) \quad (25)$$

where, K_1 is the constant of the spring

$$F_{precompresion} = K_1 x_0 \quad (26)$$

$$P_M = \frac{F_{precompresion}}{A_1} \quad (27)$$

Combining the previous Equations (17)–(27), it follows that the relief function curve

$$P_1 = P_M + \frac{K_{spring}}{A_1}x + K_2Q^2 \tag{28}$$

with

$$K_2 = \frac{\rho}{A_1^2} - \frac{\rho}{A_1\cdot(\pi d)\cdot x\cdot \tan\alpha} \tag{29}$$

On the other hand, for a predefined distance between poppet and seat surface, the flow rate can be calculated by the equation

$$Q_{out}(y,P) = K_2\cdot x\cdot C_d\cdot\sqrt{2\frac{P_1}{\rho}} \tag{30}$$

or alternatively

$$P_1 = K_3Q^2 \tag{31}$$

with

$$K_3 = \left(\frac{1}{K_2\cdot x\cdot C_d\cdot\sqrt{\frac{2}{\rho}}}\right)^2 \tag{32}$$

In quadrant II, see Figure 8, the relief function curve (Equation (28)) and the flow rate curve (Equation (31)) are represented (not scaled). The point of intersection between the two curves (point B) indicates the operating point of the LHV acting as a relief valve; that is, the valve opens at pressure P and allows fluid flow Q.

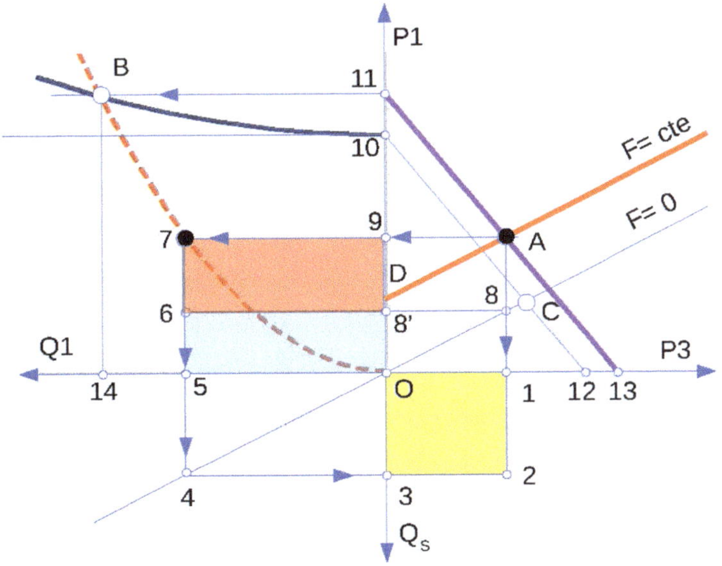

Figure 8. Scheme to display the described graphics approach. $N_{LHV} = A_{Blue} + A_{Red} = A_{O579}$, $N_S = A_{Yellow} = A_{Blue} = A_{O123} = A_{0568'}$ and $N_M = A_{Red} = A_{6798'}$.

4.1.3. Power Balance Applied to the Actuator/LHV Hydraulic System

Before explaining how the power balance is proposed, it is helpful to explain the different curves shown in the four-quadrant diagram (Figure 8), as well as the points that are considered most representative:

- Curve 10–12: characteristic curve $P_1 = f(P_3)$ of the LHV valve in the static balance position (closing condition). It corresponds to the representation of Equation (1). Load force $F_{load} = 0$
- Curve 11–13: characteristic curve $P_1 = f(P_3)$ of the LHV valve in the permanent regime when a flow rate Q_1 flows through the valve, as a consequence of the opening of the obturator by the action of pressure P_1 and pilot pressure P_3. Load force constant $F_1 > 0$
- Curve O–C: characteristic curve $P_1 = f(P_3)$ of the hydraulic actuator when it is subjected to zero force. It corresponds to the representation of Equation (3).
- Curve D–A: characteristic curve $P_1 = f(P_3)$ of the hydraulic actuator when it is subjected to an overrunning force, F = constant.
- Curve 10–B: characteristic curve $P = f(Q)$ of the LHV valve acting as a pressure limiting valve (relief function). See Equation (28).
- Curve O–7–B: LHV poppet pressure drop in the open position as a result of the shutter opening due to the action of the pressure P_1 and the pilot pressure P_3. See Equation (31).

Assuming that the operating point of the hydraulic system (hydraulic actuator/LHV valve) is defined by point A, that is, the intersection of curves (11–13) and (D–A), this point corresponds to the operating point of the LHV when a flow rate Q_1 flows as a result of the opening of the obturator under the action of pressure P_1 and pilot pressure P_3, thus defined by the coordinates (P_3, P_1).

Other points of interest are:

- point 7: This point is defined by the coordinates (Q_1, P_1), intersection of the curve (0–B), with the line of constant pressure equal to P_1.
- point 5: defined by the coordinates (Q_1, 0)
- point 1: defined by the coordinates (P_3, 0)
- point 3: defined by the coordinates (0, Q_S)

Based on the hydraulic circuit (e.g., scenario 2b) outlined in Figure 2, the following powers involved in movements are defined:

N_M power due to the movement of the actuator as a consequence of being subjected to the overruning force F.

$$N_M = F_{overrunning} \cdot v_{actuator} \tag{33}$$

N_S, power supplied by the pump feeding the hydraulic actuator chamber on the rod side

$$N_S = P_S Q_S = P_{S(:point\ 1:)} Q_{S(:point\ 3:)} \tag{34}$$

N_{LHV}, power related to the fluid flow through the LHV valve as a result of the combined action of pressure P_1 and pilot pressure P_3

$$N_{LHV} = P_1 Q_1 = P_{1(:point\ 9:)} Q_{1(:point\ 5:)} \tag{35}$$

A simple power balance allows us to establish the following equation:

$$N_{LHV} = N_S + N_M \tag{36}$$

N_S power is defined by the rectangle (O123) (yellow area), where side O3 has magnitude Q_S and side O1 has magnitude equal to P_3. Note that the area (O123) is equal to the area (O568′). Using the hydraulic actuator curve (for condition $F = 0$) extended from Quadrant I to Quadrant III, represented by the "40C" line, the following relationships are established:

$$P_{1(:point\ 8':)} = R_c P_{3(:point\ 1:)} \tag{37}$$

$$Q_{1(:point\ 5:)} = \frac{Q_{S(:point\ 3:)}}{R_c} \tag{38}$$

multiplying Equations (37) and (38), it turns out

$$P_{1(:point\ 8':)}Q_{1(:point\ 5:)} = P_{3(:point\ 1:)}Q_{S(:point\ 3:)} \quad (39)$$

Equation (39) highlights the aforementioned equality, that is the area (O568′) = area (O123). On the other hand, note that the area of the rectangle (O579) (A_{Blue} + A_{Red}), where side O5 has magnitude Q_1 and side O9 is equal to P_1, represents the power N_{LHV}, i.e., the power related to the fluid flow through the LHV valve. Then, by virtue of Equation (36), it can be deduced that the area of the rectangle (6798′) (A_{Red}), represents the power due to the movement of the actuator as a consequence of being subjected to the overrunning force F.

Based on these equations, the ratio ϕ is defined between the power provided by the pump and the power related to the flow through the LHV valve, see Equation (40). This power ratio is an index that indicates the percentage of the pump power needed to open the LHV valve and, therefore, an index of the energetic goodness of the LHV valve.

$$\phi = \frac{N_S}{N_{LHV}} = \frac{area(O568')}{area(O579)} = \frac{Area_{Blue}}{(Area_{Blue} + Area_{Red})} \quad (40)$$

5. Experimental Validation

5.1. Lab Testing

To validate the proposed methodology, first of all, the performance of an LHV valve was tested. Specifically, the reference valve tested at the LABSON Laboratory (Terrassa-Barcelona) was the LHV-4 valve (note: there was no particular reason for the choice, only its availability in the laboratory). Figure 9 shows the diagram of the hydraulic circuit used.

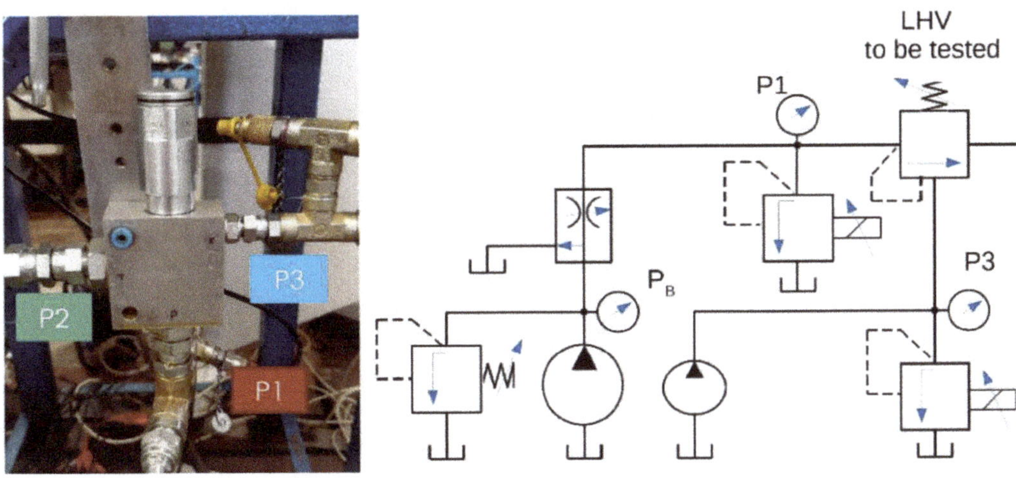

Figure 9. LHV-4 picture. Diagram of the hydraulic circuit used where: P_1: upstream water pressure; P_2: downstream pressure; P_3: pilot pressure.

Figure 10 presents the results obtained. Figure 10a shows the steady-state operating curve of LHV valve, while Figure 10b shows of relief function curve. On the one hand, the relief functions with setting pressure of 150 and 250 bar are shown. These curves were evaluated based on the adjustment curves shown in Figure 10a. On the other hand, the relief curve with setting pressure 100 bar, obtained by varying the pressure P_1 with the help of a proportional pressure relief valve, showed its hysteresis phenomenon.

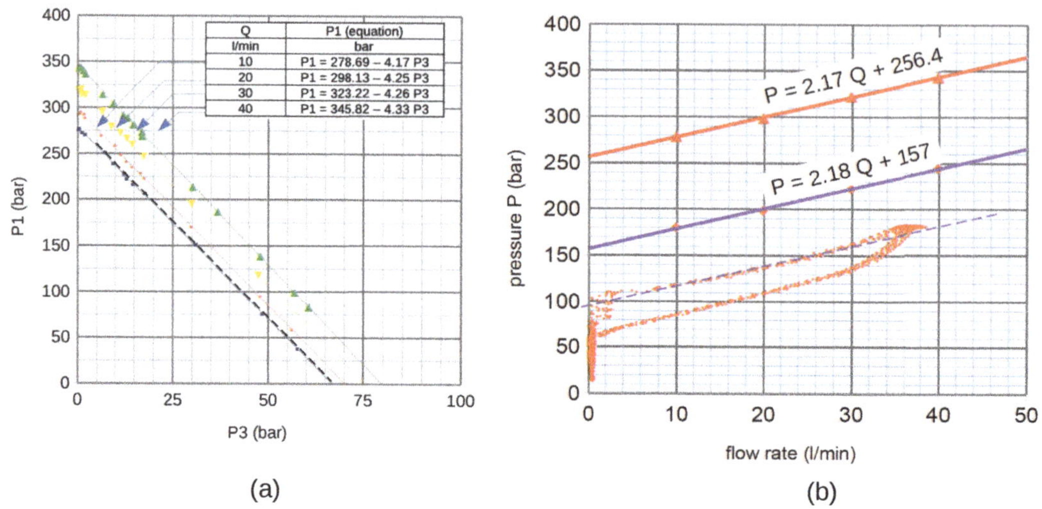

Figure 10. LHV experimental curves: (**a**) steady-state operating curves and (**b**) relief function curve.

5.2. Field Testing

This section describes the experimental tests performed to validate the methodology proposed in the previous section. Several tests were performed on the hydraulic lifting cylinder of a telescopic handler (called M_1). For the first series of tests, the M_1 telehandler was equipped with the LHV4-M_1 valve, the characteristics of which are well known (Figure 10), replacing the original valves of the commercial version.

Figure 11 shows panoramic images of the experimental tests carried out. The tests were carried out with different load conditions of the telehander.

Figure 11. The telehandler machine conveniently instrumented.

The hydraulic cylinder was actuated by means of a manual directional control valve. The field test was equipped with multiple pressure sensors, a flow sensor, and a cylinder position sensor. Displacement and angle sensors were also fitted to the boom.

Sensor data acquisition and electronic actuation were carried out with the help of a multi-axis motion controller RMC200, equipped with input and output modules (Delta Computer Systems). It is easy to use, and the software provided with it has powerful plotting capabilities.

Sensors and their main characteristics are listed in Table 3. Calibration values were obtained from the manufacturer's information and from comparison with a standard when the manufacturer's information was not available. All values were checked in preliminary tests. Sensor values were measured and recorded at a frequency of 1 kHz. Higher rates were not considered necessary for the energy analysis. Digital filtering at 100 Hz was required for the calculation of the speeds, especially with the angle sensor. The data were processed with commercial software (DIAdem, Excel) to calculate the total energy consumption.

Table 3. List of main characteristics of used sensors.

Magnitude	Manufacturer	Model	Range	Accuracy
Pressure	WIKA	MH2	0–250/0–400 bar	>0.5% span
Flow	HYDAC	EVS3199TF	6–60 L/min	>2% act. val.
Position	Micro Epsilon	WDS-1500_PS60-SR-U	0–1500 mm	>+/−1.5 mm
Angle (tilt)	SICK	TMM 56E-PMH045	+/−45°	+/−0.3°
Temperature	Omega	PT100	10–100°	>0.36 °C

Figure 12 shows the results obtained with machine M_1 in which a 90 × 50 hydraulic cylinder and the LHV4 valve were installed to move the telescopic arm. Figure 12a shows the temporal history of the load pressure P_1, the return pressure P_2, and pilot pressure P_3, together with the hydraulic cylinder position x_1 and pump flow Q_S companion. It is observed that there is a time interval in which the values P_1, P_2, P_3, and Q_1 remained essentially constant, which corresponds to the downward cylinder movement, x_1, and, consequently, to the time interval in which the valve is working. The companion Figure 12b shows the working pressure space of rod side P_1 versus the pilot P_3 pressures, respectively. The average values are represented by a small circle in Figure 12a,b.

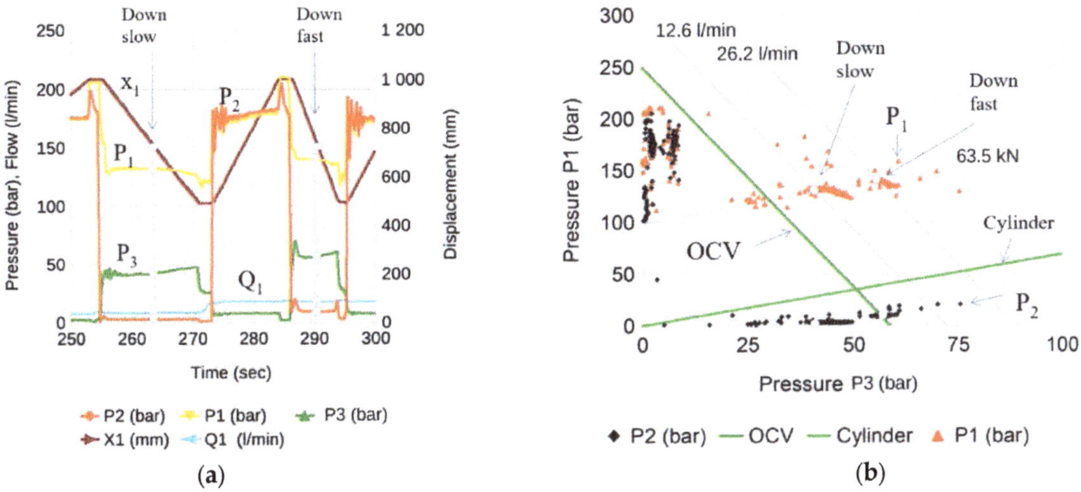

Figure 12. Example of experimental results obtained on M_1 machine using LHV$_4$-M_1 valve where: (**a**) shows the temporal evolution of the experimentally measured variables. (**b**) experimental results plotted on steady-state operating curves for $P_1 = f(P_3)$.

Procedure

The method is presented over the description set out in Figure 13 with real values (and Figure 8 as reference).

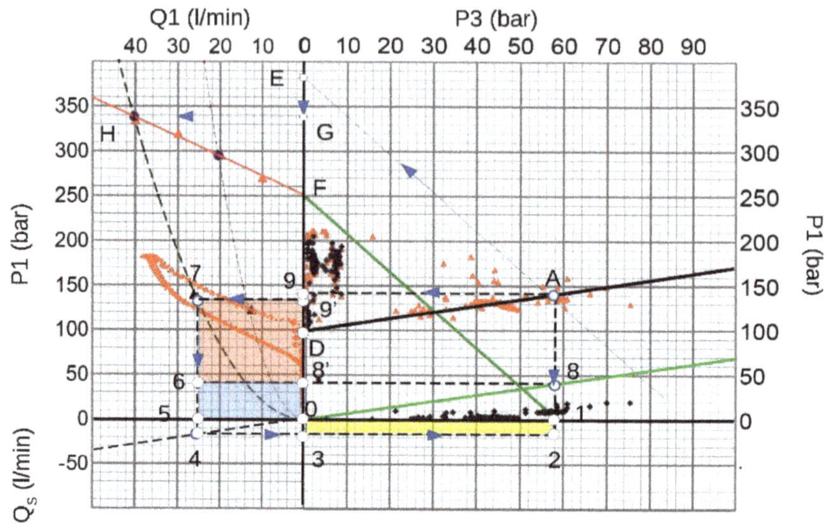

Figure 13. Graphical representation of the power balance for a real case.

Actions to be carried out over the first quadrant:

1. Plot the working point A (e.g., P_1 = 142 bar and P_3 = 58 bar);
2. Plot LHV curve based on its setting point P_m = 250 bar and the pilot ratio 4.25 as it is shown in Table 1 (e.g., straight-line 1F);
3. Plot an auxiliary line parallel to 1F at the A point projecting it to the pressure axis (e.g., E point);
4. Calculate the effect of the return pressure using the equation $P_2 \cdot (R_v + 1)$, which is represented by the segment EG. G point represents P_M pressure shown in Equation (1);
5. Trace a horizontal line crossing at A point to obtain the working pressure (e.g., point 9);
6. Obtain the differential pressure P_1-P_2 by plotting 9′ position (e.g., segment length between 0–9′). Actions in the second quadrant:
7. Plot the relief LHV function curve taking the setting pressure as 250 bar (e.g., the experimental curve shown in Figure 10b);
8. Draw a horizontal line starting at G and crossing the LHV function curve (e.g., H point is obtained). Therefore, the K_3 constant can be calculated from Equation (3) obtained using H coordinated, $K_3 = \frac{P_H}{Q_H^2}$;
9. The parabolic performance following Equation (31);
10. Trace a horizontal line across 9′ to obtain the coordinates of intersection with the parabolic curve OH to bring point 7, which corresponds to the orifice working conditions;
11. Draw a vertical segment from point 7 to the flow axis to obtain the flow Q_1 through the valve (point 5);
12. Extend the cylinder actuator characteristic curve from the first quadrant to the third quadrant (segment "04", as an extension of segment "08");
13. The pump flow rate Q_S is represented by point 3, which is calculated from point 4 by crossing the vertical axis;
14. Extend line "4–3" to the intersection with the vertical line through point 1, obtaining point 2;
15. The graphics method provides flow rate (represented by points 3 and 5 compared with experimental ones, as shown in Table 4).

Table 4. Working values.

			Go down Slow	Go down Fast
load pressure	P_1	bar	132	142
return pressure	P_2	bar	3	9.21
pilot pressure	P_3	bar	42	58
supply flow	Q_S	l/min	9	18
valve flow	Q_1	l/min	13	26
setting pressure	P_m	bar	250	250

Power ratios ϕ can be obtained applying Equation (40).

- "(0123)" area represents the power, N_S (yellow area)
- "(0568′)" area has an identical area of "(0123)" (blue area)
- "(679′8′)" area is equivalent to the power load, F·v (red area)
- "(0579′)" area is the power dissipated in the valve, $N_{LHV} = P_1Q_1$, which is equal to the sum of "(0568′)" and "(679′8′)"

The yellow area, identical to the blue area, shows that using LHV valves in a hydraulic system introduces additional energy consumption.

To conclude this research work, it is essential to apply and validate the proposed methodology in more cases. For this purpose, two models of telehandlers (M_1 and M_2) were equipped with seven different LHV valves. The mixed sets of valves and machines were distinguished with LHV_i-M_j where i = {1 ... 7} and j = {1,2}. The relevant data, setting pressure, and relief characteristic curve were obtained from the commercial catalogs of the valves.

Figure 14 shows the high correlation between the flows estimated by applying the proposed method and the experimental tests carried out in the field with the two machines presented and the set of seven valves used. The high correlation measured through the coefficient $R_2 \approx 0.99$ is notable. Taking into account that the proposed method allows estimating the flows and powers involved in the operation of the hydraulic cylinder/holding valve system, it seems reasonable to compare the calculated values with the actual values measured in the experimental tests in order to obtain the accuracy of the method. The estimated and measured flow rates for machine M_1 are shown in Figure 14a, where the dashed lines define an error range of ±7.5%. On the other hand, the values for the machine M_2 are shown in Figure 14b, where the dashed lines define an error range of +8%/−12%.

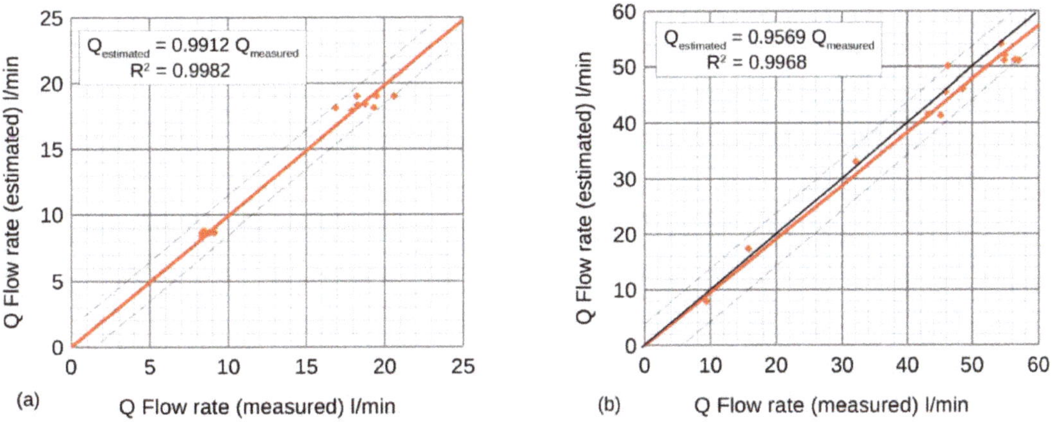

Figure 14. Comparative graphs between measured and estimated flows (Q_S) for the two machines: 14 (**a**) machine M_1 and 14 (**b**) machine M_2.

Figure 15 shows the power ratio, ϕ, as a function of the net force to which the lifting cylinder is subjected for machines M_1, Figure 14a, and M_2, Figure 14b.

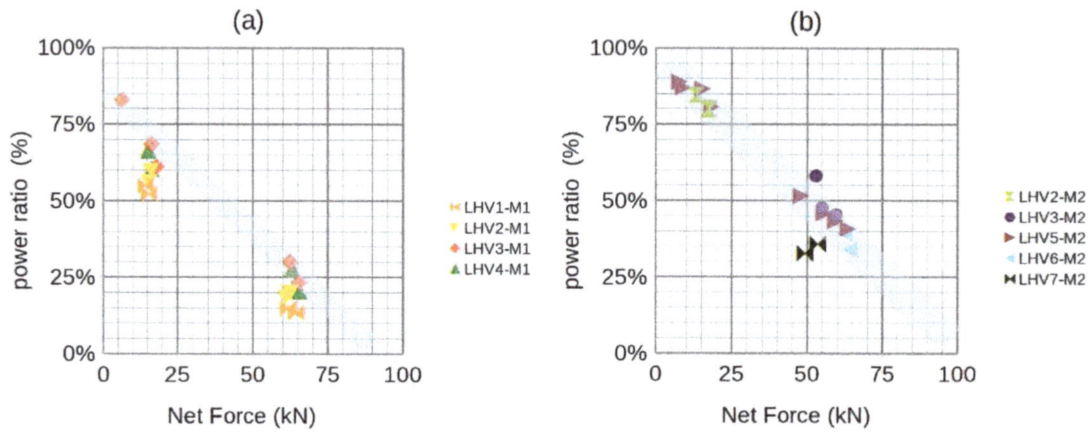

Figure 15. Graph power ratio as a function of the net force (for all cases). (**a**) M_1 machine; (**b**) M_2 machine.

Figure 16 shows the order of magnitude of the errors made in the estimation of power according to the following definition:

$$error(\%) = \frac{N_{supply}(estimated) - N_{supply}(mesured)}{N_{supply}(mesured)} \quad (41)$$

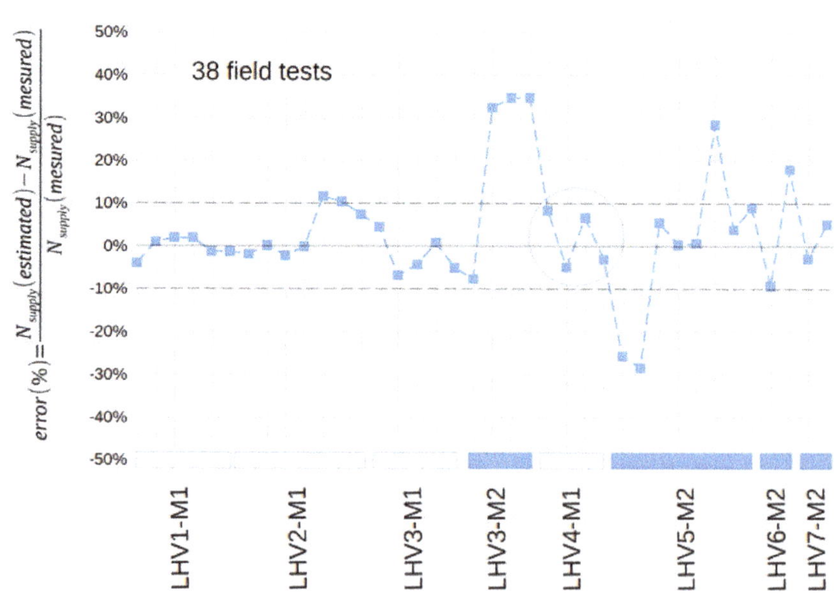

Figure 16. Bounds of the power estimation errors of the applied method.

6. Conclusions

In this paper, the authors highlighted the importance of understanding how LHVs work in the critical opening condition. Outside of this state, it helps select the most suitable valve for the application according to its characteristics.

Going deeper into the knowledge of these valves requires the use of tools (software and models). Still, it involves a high level of application-specific knowledge (mainly done in academia). On the contrary, the methodology presented in this paper is an ad hoc methodology ready to use for an end-user.

The primary method's advantage is that the information provided by a catalog and the pressure readings in the three ports of the valve is sufficient to estimate the power ratio. Despite using DAQ and high-level instrumentation to prove the goodness of this methodology, the end-user does not need them to evaluate power. The end-user can use simple gauge-pressure sensors to carry out the proposed method and obtain good results.

First, the graphic method estimates the flow rate that passes through the valve and the pump's flow rate with great precision. Therefore, the powers involved are calculated. Secondly, the functional relationship between the power ratio and the load's net force is established. The less load, the more significant the power provided. Figure 15 summarizes the performance differences found between the several valves tested. Some of the contrasted result from the design innovations incorporated in recent years to improve performance and especially achieve more significant energy savings.

Figure 16 shows the method's validity considering the proposed simplifications and hypotheses. The method generally estimates the power ratios within the error range of $\pm 10\%$, which can be considered very acceptable.

Author Contributions: The investigation was led and supervised by L.J.B., E.C. and P.R. Experimental works, models, data processing and illustrations were completed by L.J.B., G.R. and E.C. The manuscript was finalized by L.J.B., P.-J.G.-M., G.R. and E.C. All authors have read and agreed to the published version of the manuscript.

Funding: This research received no external funding.

Institutional Review Board Statement: Not applicable.

Informed Consent Statement: Not applicable.

Data Availability Statement: Not applicable.

Acknowledgments: Our gratitude to the entities that have lent their machines to evaluate their behavior, during our R&D activities, especially AUSA (Manresa-Spain). Also, Roquet Group and IHBER, who provided important hydraulic devices used in the research.

Conflicts of Interest: The authors declare no conflict of interest. The graphs in Figure 14, which show very significant differences in performance between the different types of LHV valves. The authors consider it inappropriate to identify the types of valves used to avoid possible interpretations that could directly or indirectly affect manufacturers.

Nomenclature

Symbol	Description	Unit
A	area	m^2
C_d	orifice coefficient	-
C_{ha}	hydraulic capacitance	m^3/bar
F	force	N
G_{pilot}	gradient pilot function	
G_{relief}	gradient relief function	
K_1	spring constant	N/m
$K_{2,3}$	generic constant	-
M	mass	kg
N	power	W
P	load pressure	bar
P_{pilot}	pilot pressure	bar
P_{relief}	relief pressure	bar

Q	flow rate	l/min
R	ratio	-
S	cylinder area	m^2
V	volume	m^3
v	velocity	m/s
x	position	m
subscripts		
1	rod side	
2	return	
3	pilot	
c	cylinder	
LHV	Load Holding Valve	
s	supply	
v	pilot	
greek		
ρ	fluid density	kg/m^3
ϕ	energetic goodness' ratio	-

References

1. Nervegna, N. *Oleodinamica e Pneumatica*; Politeko, T., Ed.; Politeko: Arlington, VA, USA, 2003.
2. Zarotti, G.L.; Nervegna, N. *Rational Design of Mobile Hydraulics by Digital Computer*; SAE Ed.: Warrendale, PA, USA, 1979.
3. Ritelli, G.F.; Vacca, A. Energetic and Dynamic Impact of Counterbalance Valves in Fluid Power Machines. *Energy Convers. Manag.* **2013**, *76*, 701–711. [CrossRef]
4. Overdiek, G. Design and Characteristics of Hydraulic Winch Controls by Counterbalance Valves. In *European Conference on Hydrostatic Transmissions for Vehicle Applications*; Mechanical Engineering Publications Limited for the Institution of Mechanical Engineers: Aachen, Germany, 1981.
5. Persson, T.; Krus, P.; Palmberg, J.-O. The Dynamic Propertles of Over-Center Valves in Mobile Systems. In Proceedings of the 2nd International Conference on Fluid Power Transmission and ControI, Hangzhou, China, 20–22 March 1989; Zhejiang University: Hangzhou, China, 1989.
6. Miyakawa, S. Stability of a Hydraulic Circuit with a Counter-Balance Valve. *Bull. JSME* **1978**, *21*, 1750–1756. [CrossRef]
7. Handroos, H.; Halme, J.; Vilenius, M. Steady State and Dynamic Properties of Counterbalance Valves. In Proceedings of the 3rd Scandinavian International Conference on Fluid Power (SICFP'93), Linköping, Sweden, 25–26 May 1993.
8. Chapple, P.J.; Tilley, D.G. Evaluation Techniques for the Selection of Counterbalance Valves. In Proceedings of the Expo and Technical Conference for Electrohydraulic and Electropneumatic Motion Control Technology, Anaheim, CA, USA, 23–24 March 1994.
9. Rahman, M.M.; Porteiro, J.L.F.; Weber, S.T. Numerical Simulation and Animation of Oscillating Turbulent Flow in a Counterbalance Valve. In Proceedings of the IECEC-97 Thirty-Second Intersociety Energy Conversion Engineering Conference (Cat. No.97CH6203), Honolulu, HI, USA, 27 July–1 August 1997; Volume 2, pp. 1525–1530.
10. Lisowski, E.; Stecki, J.S. Stability of a Hydraulic Counterbalancing System of a Hydraulic Winch. In Proceedings of the Sixth Scandinavian Intl. Conf. on Fluid Power, Tampere, Finland, 26–28 May 1999; pp. 921–933.
11. Andersen, T.O.; Hansen, M.R.; Pedersen, P.; Conrad, F. The Influence of Flow Force Characteristics on the Performance of over Centre Valves. In Proceedings of the 9th Scandinavian International Conference on Fluid Power, Linköping, Sweden, 1–3 June 2005.
12. Liu, X.; Liu, X.; Wang, L.; Chen, J. The Dynamic Analysis and Experimental Research of Counter Balance Valve Used in Truck Crane. In Proceedings of the 2010 International Conference on Electrical and Control Engineering, Wuhan, China, 25–27 June 2010; College of Mechanical Science and Engineering: Tampere, Finland, 2010.
13. Zähe, B.; Anders, P.; Ströbel, S. A New Energy Saving Load Adaptive Counterbalance Valve. In Proceedings of the 10th International Conference on Fluid Power, Dresden, Germany, 8–10 March 2016; pp. 425–436.
14. Jakobsen, J.H.; Hansen, M.R. CFD Assisted Steady-State Modelling of Restrictive Counterbalance Valves. *Int. J. Fluid Power* **2020**, *21*, 119–146. [CrossRef]
15. Andersson, B.R. Energy Efficient Load Holding Valve. In Proceedings of the 11th Scandinavian International Conference on Fluid Power SICFP, Linköping, Sweden, 2–4 June 2009; Volume 9.
16. Kjelland, M.B.; Hansen, M.R. Numerical and Experiential Study of Motion Control Using Pressure Feedback. In Proceedings of the 13th Scandinavian International Conference on Fluid Power, SICFP2013, Linköping, Sweden, 3–5 June 2003.
17. Sørensen, J.K.; Hansen, M.R.; Ebbesen, M.K. Novel Concept for Stabilising a Hydraulic Circuit Containing Counterbalance Valve and Pressure Compensated Flow Supply. *Int. J. Fluid Power* **2016**, *17*, 153–162. [CrossRef]
18. de Groof, A.G. *Instability of Counterbalance Valves*; Hogeschool: Rotterdam, The Netherlands, 2014.
19. Sciancalepore, A.; Vacca, A.; Weber, S. An Energy-Efficient Method for Controlling Hydraulic Actuators Using Counterbalance Valves with Adjustable Pilot. *J. Dyn. Syst. Meas. Control* **2021**, *143*, 111007. [CrossRef]

20. Christensen, M. Design Af Hydraulisk Lastholdeventil. Design Af Mekaniske Systemer. Master's Thesis, Aalborg Universitet, Aalborg, Denmark, 2007.
21. Vukovic, M.; Leifeld, R.; Murrenhoff, H. Reducing Fuel Consumption in Hydraulic Excavators—A Comprehensive Analysis. *Energies* **2017**, *10*, 687. [CrossRef]
22. Mahato, A.C.; Ghoshal, S.K. Energy-Saving Strategies on Power Hydraulic System: An Overview. *Proc. Inst. Mech. Eng. Part J. Syst. Control Eng.* **2021**, *235*, 147–169. [CrossRef]
23. Roquet, P.; Raush, G.; Berne, L.J.; Gamez-Montero, P.-J.; Codina, E. Energy Key Performance Indicators for Mobile Machinery. *Energies* **2022**, *15*, 1364. [CrossRef]
24. Berne, L.J.; Raush, G.; Gamez-Montero, P.J.; Roquet, P.; Codina, E. Multi-Point-of-View Energy Loss Analysis in a Refuse Truck Hydraulic System. *Energies* **2021**, *14*, 2707. [CrossRef]
25. Rydberg, K.-E. *Energy Efficient Hydraulics—System Solutions for Minimizing Losses*; National Conference on Fluid Power, "Hydraulikdagar'15"; Linköping University: Linköping, Sweden, 2015.
26. Lin, T.; Chen, Q.; Ren, H.; Huang, W.; Chen, Q.; Fu, S. Review of Boom Potential Energy Regeneration Technology for Hydraulic Construction Machinery. *Renew. Sustain. Energy Rev.* **2017**, *79*, 358–371. [CrossRef]
27. Axin, M. Fluid Power Systems for Mobile Applications with a Focus on Energy Efficiency and Dynamic Characteristics. Ph.D. Thesis, Fluid and Mechatronic Systems Department of Management and Engineering, Linköping University, Linköping, Sweden, 2013.
28. Sciancalepore, A.; Vacca, A.; Pena, O.; Weber, S.T. Lumped Parameter Modeling of Counterbalance Valves Considering the Effect of Flow Forces. In Proceedings of the ASME/BATH 2019 Symposium on Fluid Power and Motion Control, FPMC 2019, Longboat Key, FL, USA, 7–9 October 2019.

energies

MDPI

Article

Theoretical Analysis of Active Flow Ripple Control in Positive Displacement Pumps

Paolo Casoli [1,*], Carlo Maria Vescovini [1], Fabio Scolari [1] and Massimo Rundo [2]

[1] Department of Engineering and Architecture, University of Parma, 43124 Parma, Italy; carlomaria.vescovini@unipr.it (C.M.V.); fabio.scolari@unipr.it (F.S.)
[2] Department of Energy, Politecnico di Torino, C.so Duca degli Abruzzi 24, 10129 Turin, Italy; massimo.rundo@polito.it
* Correspondence: paolo.casoli@unipr.it

Abstract: Positive displacement machines present a well-known major drawback that is the oscillation in delivered flow rate. This paper presents two active solutions for reducing the flow ripple generated by a pump with an external device actuated by means of a piezo-stack actuator. The work is focused on a theoretical analysis, with the aim of collecting information about the performance of the solutions proposed and their main advantages and drawbacks. The active methods proposed involve a cylindrical actuator connected to the delivery line of the pump. The piston could be actuated directly by a piezo-stack actuator or by a differential pressure modulated by a proportional piezo actuated valve. The actuators were modelled and a control algorithm based on Least Mean Square algorithm was used to achieve the adaptability for both systems at different operating conditions. The developed mathematical model permits to define the great potential of these solutions that can drastically reduce the flow ripple. The first architecture presented resulted as the best solution, while the second one allowed reduction of the production cost.

Keywords: positive displacement pump; active control; flow ripple; pressure ripple

Citation: Casoli, P.; Vescovini, C.M.; Scolari, F.; Rundo, M. Theoretical Analysis of Active Flow Ripple Control in Positive Displacement Pumps. *Energies* **2022**, *15*, 4703. https://doi.org/10.3390/en15134703

Academic Editor: Kamel Hooman

Received: 9 June 2022
Accepted: 24 June 2022
Published: 27 June 2022

1. Introduction

Positive displacement machines are commonly utilized in hydraulic circuits since they provide high power density together with robustness and reliability. However, one of the major drawbacks is the oscillation of the delivered flow rate, which in turn generates a pressure ripple and finally produces noise and vibrations that can cause stress to the components along the hydraulic circuit. This phenomenon can negatively affect durability and functionality of the hydraulic components [1]; furthermore, vibrations can affect the human being and environment [2].

Researchers dealing with the reduction in the flow and pressure ripple generated by positive displacement pumps have focused on different solutions, many of which contemplate the geometric optimization of components. In axial piston pumps, for example, a careful design of the port plate allows a smooth transition of the fluid pressure between suction and delivery phase and vice versa [3–5]. In gear pumps instead, side bushings are designed with particular grooves to improve the machine performance from this point of view [6–11].

Other methods involve devices external to the pump and could be classified as passive or active methods.

Passive methods contemplate fitting external devices, such as resonators, outside the pump. One major drawback of passive devices connected on a pump delivery line is that they usually are designed and set to operate in particular conditions, being effective only at certain pressures or against particular frequencies [12–14]. This means, for example, that if the pump shaft speed changes, these systems lose their effectiveness, with the pressure ripple frequency being determined directly by the pump rotational speed. In [13], a passive solution is proposed with the aim of reducing the pressure ripple both on the suction and on the delivery side; a particular flow ripple attenuator is presented and connected to suction and delivery lines.

To overcome passive method limitations, the solutions based on an active control can adapt themselves to the operating conditions in terms of fluid pressure and machine shaft speed. Researchers have also considered the combination of passive–active methods as reported in [12,14], where passive hoses are dedicated to the highest frequency reduction, while an active actuator acts on the main lower frequencies.

In particular, in [12,14] the presented solution is based on an actuator with a variable volume installed on the delivery side of the pump; the actuator acts by means of a piezo-electric actuator that is controlled with an adaptive filter with a feedforward algorithm.

Active methods are mostly based on generating a flow signal that superimposed to the pump delivery flow rate allows reduction of its oscillation: these solutions should generate a signal with a correct amplitude and phase to cancel the original one. The main drawback of active systems is that they are sophisticated devices that must be installed outside the pump, creating with it a relatively complex and bulky unit compared to the pump alone.

In [15], an active fluid borne noise attenuation has been applied to an aircraft hydraulic pump, the attenuator is an in-house rotary valve design, the pump and valve are shaft coupled so the effect of the speed is considered.

Focusing on axial piston pumps, the literature reports studies that aimed to reduce the amount of the structure-borne noise acting on the vibrations of the swash plate, that is controlled by a servo valve [16], while in [17,18] a similar solution is focused to compensate the uneven flowrate normally produced.

A similar solution is proposed in [19] where the swash plate is controlled to make adjustment to the instantaneous displacement. In [19], a second method is presented that uses a piezo stack to actuate a cylinder connected to the pump outlet.

In this paper, an active method to achieve flow ripple reduction in a positive displacement pump is presented. This method is based on an actuated piston placed on the delivery side of the pump that acts in counter phase with respect to the main flow ripple and it is able to significantly reduce the main harmonic of the flow oscillation.

The aim of this work is to move the first step of the research, so only a theoretical analysis is presented. The models of the two solutions were developed in Simcenter Amesim™ environment. Two solutions for reducing the flow ripple are proposed, both based on the presence of a piston that, oscillating, generates a suitable flow. The difference between the two solutions is in the way the piston is actuated. This paper analytically presents the two solutions pointing out their advantages and disadvantages.

With the aim of making the two solutions able to adapt to changing pump working conditions, an adaptive control logic based on the LMS (Least Mean Square) algorithm was implemented to drive both active systems. The control logic was developed in Simulink™ and cosimulation with the Amesim physical model was performed.

2. Description of the Active Methods

2.1. Piston Actuated by Piezo-Stack Actuator (1st Solution)

This active method of reducing the flow ripple, whose schematic is reported in Figure 1, involves a cylindrical actuator connected to the delivery line of the pump.

The piston is actuated by a piezo-stack actuator powered by an appropriate voltage amplifier with a feedforward logic signal based on the LMS algorithm. Thanks to this configuration, the entire system behaves like an adaptive filter that modulates its response according to the different operating conditions of the volumetric machine, adapting to variations in delivery pressure, flow rate, and rotation speed of the pump.

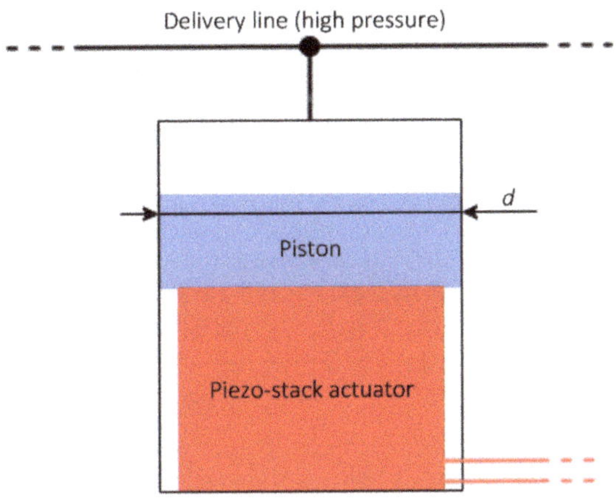

Figure 1. Schematic of piezo-stack actuated piston active method.

2.2. Hydraulically Actuated Piston (2nd Solution)

The second proposed active method is similar to the previous one, but a different solution was implemented to actuate the piston as shown in the schematic of Figure 2. In this solution, the piston is actuated by the fluid whose pressure is modulated by a three-way proportional valve, actuated by means of a piezo-stack actuator. The piston presents two cylindrical sections of different diameters.

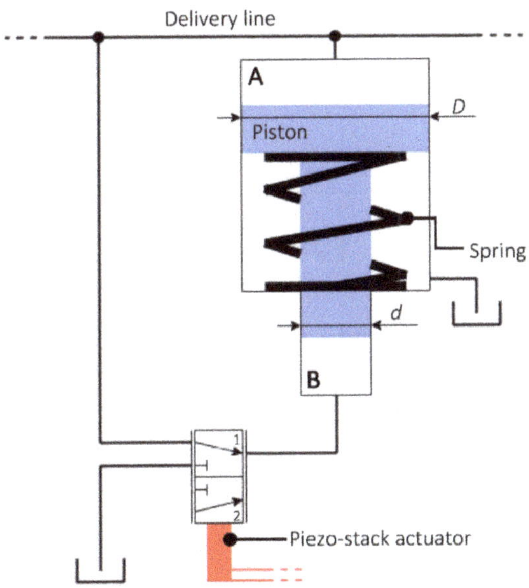

Figure 2. Schematic of hydraulically actuated piston active method.

The piston is housed in such a way to be in contact with a fluid at different pressure. The face of the piston with the larger diameter D is in contact with the volume A, directly connected to the delivery line of the hydraulic circuit. A spring is inserted to oppose the force that the fluid in A at delivery pressure generates on the piston. The chamber in which the spring is housed is connected to the tank and therefore no hydrostatic forces are generated on surfaces in contact with this chamber. The face with the smaller diameter d

is in contact with volume B. Such a volume can alternatively be connected to the delivery or tank line by means of a three-way proportional valve. The valve is operated by means of a piezo-stack actuator powered by a suitable voltage amplifier. As for the first system, the control signal used to power the piezo actuator uses an LMS algorithm to implement a feedforward logic.

The active system operates as follows. When the piezo-stack actuator keeps the valve in position 1, as in the figure, the volume B is connected with delivery line and high-pressure fluid can flow into it pushing the piston upwards. With the upper side of the piston being of greater diameter than the lower side, the total contribution of this phase is a release of fluid towards the delivery line. Then, when the piezo stack actuator brings the valve in position 2, chamber B is connected to low pressure tank line where the fluid can flow while the piston comes back down, pushed by the delivery pressure that acts on the greater side of the piston.

The piezo-stack actuator used in this method to operate the valve can be smaller than the one used in the first method, because it does not have to act directly against the delivery pressure and so it has to withstand a definitely lower force.

In the operation of this active system, the drainage of a certain amount of fluid is expected. In fact, as it has been explained, at each switching cycle of the piezo-actuated valve a small volume of pressurized fluid is diverted from the pump delivery into volume B to allow the piston to be actuated and then released towards the tank to depressurize the volume. The drainage obtained is, however, modest as found thanks to the developed model.

3. Sizing and Definition of Active System Operating Parameters

3.1. Piston Actuated by Piezo-Actuator (1st Solution)

The piston and the piezo-actuator were sized starting from the analysis of the flow ripple generated by the pump. By integrating the flow rate oscillating component in the time domain, it is possible to obtain the fluid volume that the piston should move at each pulsation of the piezo-actuator to completely cancel the flow ripple. As shown in Figure 3, the aforementioned volume V is obtained as the peak-to-peak value of the integral of flow ripple.

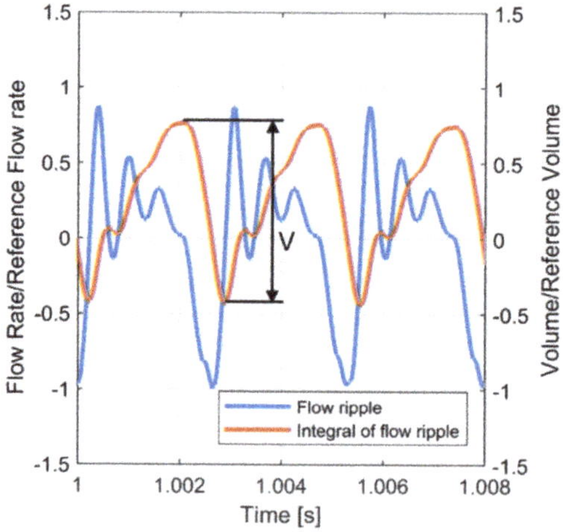

Figure 3. Simulated pump delivery flow ripple and its integral, 250 bar, 2500 rpm.

The volume V was identified starting from the flow ripple generated by the pump at its most difficult working conditions. This is to ensure that the active system is effective in the whole working range of the pump, even in the most severe conditions.

Naming d the piston diameter and s the maximum displacement of the piezo-stack actuator, which corresponds to the stroke of the piston, it is possible to relate these two parameters thanks to the fluid volume V that must be handled by the piston:

$$V = s \cdot \pi \frac{d^2}{4} \tag{1}$$

from which d can be written as a function of s:

$$d = \sqrt{\frac{4V}{\pi s}} \tag{2}$$

The diameter d depends the load F on the piston and the actuator at a fixed pressure p. When the pressure is at its maximum, p_{max}, the load is also at its maximum F_{max}, and writing d as a function of s as in the (2) it is possible to obtain

$$F_{max} = p_{max} \cdot \frac{V}{s} \tag{3}$$

The piezo-stack actuator used to actuate the piston should then be able to withstand such a load.

3.2. Hydraulically Actuated Piston

For the active method with a hydraulically actuated piston, the sizing and definition of the operating parameters are more difficult due to the complex interaction between the different elements that make up the system. For this reason, the sizing was performed through successive optimization steps, carried out using the optimization tools available in the Amesim software. The parameters subject to optimization are the upper diameter of the piston D, the lower diameter of the piston d, and the stiffness of the spring coupled to the piston. The goal of the optimization is to maximize the reduction of the first harmonic of the flow ripple, containing at the same time the drainage of the fluid that occurs, as explained above, during the operation of this device.

The Amesism tool used to perform this optimization relies on the NLPQL (Non-Linear Programming by Quadratic Lagrangian) optimization algorithm.

4. Mathematical Model

4.1. Fluid Model

A simple fluid model was used to represent the hydraulic working fluid, gas/vapor release is not taken into account since the pressure at the delivery side of the pump always stays well above the cavitation threshold. The temperature was considered constant and equal to T = 50 °C. Density was calculated as a function of pressure variation considering the liquid bulk modulus β:

$$\rho = \rho_0 \cdot [1 + 1/\beta \cdot (p - p_0)] \tag{4}$$

4.2. Pump Model

With the aim of generating a flow ripple to test the two active solutions, a pump model was used. The pump mathematical model adopted in this study is based on the axial piston pump Amesim submodel. This already available model permitted the authors to focus the research only on the active solutions for reducing the flow ripple. A brief description of the model is presented. The pump is treated as a series of interconnected open volumes according to the Filling and Emptying concept.

The pressure in every chamber is obtained by integrating the pressure time derivative considering the volumes as capacitive elements:

$$\frac{dp}{dt} = \frac{\beta}{\rho} \frac{1}{V(t)} \left(\sum \dot{m} - \rho \frac{dV(t)}{dt} \right) \tag{5}$$

where ρ is the fluid density, β is the bulk modulus, $V(t)$ the instantaneous volume of the chamber, $\dot{m}$ is the flow rates from the inlet port and to the outlet port, respectively. Flow rates going through inlet and outlet ports for every volume chamber are computed with the orifice law:

$$\dot{m} = c_q A(t) \rho \sqrt{\frac{2|\Delta p|}{\rho}} \tag{6}$$

where c_q is the discharge coefficient and $A(t)$ is the flow area.

The geometrical features of the simulated pump were used to define all geometrical parameters needed by the Amesim axial piston pump mathematical submodel.

The target of the pump model is only to generate a typical flow ripple of a positive displacement pump, in this particular case axial piston pumps were considered. Figures 4 and 5 report the output flow ripple of the model and its Fast Fourier Transform (FFT), respectively.

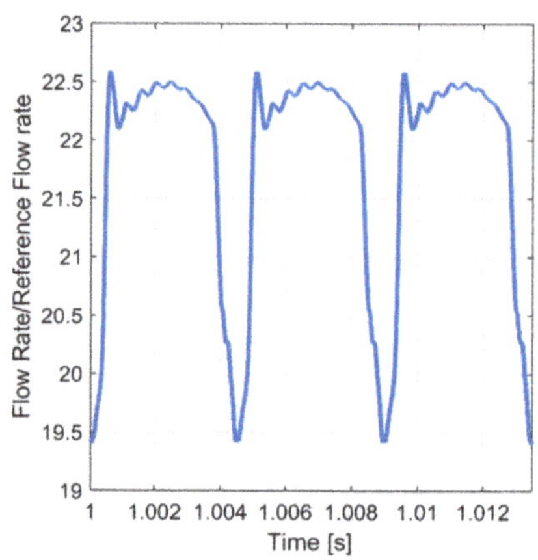

Figure 4. Flow rate at 1500 r/min, 250 bar.

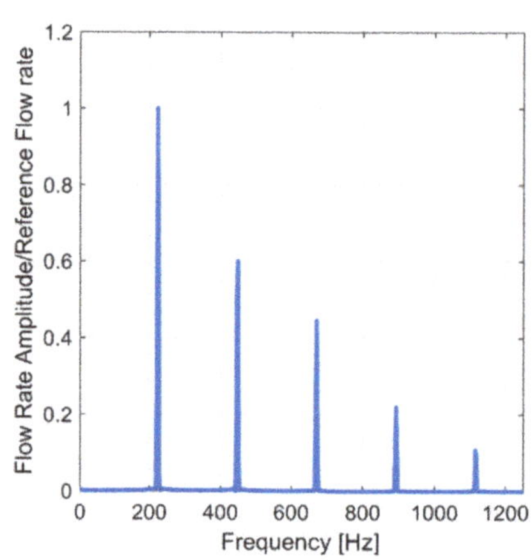

Figure 5. FFT of the flow rate at 1500 r/min, 250 bar.

4.3. Piezo-Stack Actuator Model

The piezoelectric actuator model was simulated with the Amesim linear piezoelectric stack actuator submodel.

In this model, the force F and the voltage U applied to the piezo-stack actuator are related to the displacement dZ and the electric charge dQ by the following relation:

$$\begin{pmatrix} F \\ U \end{pmatrix} = \begin{pmatrix} K_{11} & K_{12} \\ K_{21} & K_{22} \end{pmatrix} \cdot \begin{pmatrix} dZ \\ dQ \end{pmatrix} \quad (7)$$

Thanks to typical the piezo-stack actuator specifics (electrical capacitance, closed circuit stiffness, blocking load), it is possible to obtain the parameters $K11$, $K12$, $K21$, and $K22$ that determine the functioning.

The described model is simplified in comparison to the complexity of a real piezo-stack actuator. In fact, the model does not take into account either hysteresis effects or the dependence of the piezo-stack electrical capacitance on the temperature and operating frequency. If hysteresis effects can be neglected for this study, the variation in electrical capacitance must be considered. In fact, an increase in electrical capacitance is the cause of a decrease in the piezo-stack actuator ability to respond to high frequency control signals, as illustrated in Section 4.4. For this reason, in the model, an electrical capacitance overestimated by a safety factor is used.

4.4. Piezo-Stack Actuator Power Supply

The high performance piezo-stack actuators considered for this study are characterized by an operating voltage assumed to be in the range of 0–1000 V. To achieve this voltage, a high-power amplifier must be used.

Assuming that the power amplifier has a maximum output current, therefore, current must be limited to stay below this threshold, the current limitation is the cause of delay and reduction in the power amplifier response to the control signal. In fact, having the piezo-stack actuator at a considerable electric capacitance, a certain amount of time is required for the voltage to build up at its ends in case of a limited charge current. The electrical circuit in which the piezo-stack actuator is plugged in behaves, therefore, like an RC circuit, as shown in Figure 6.

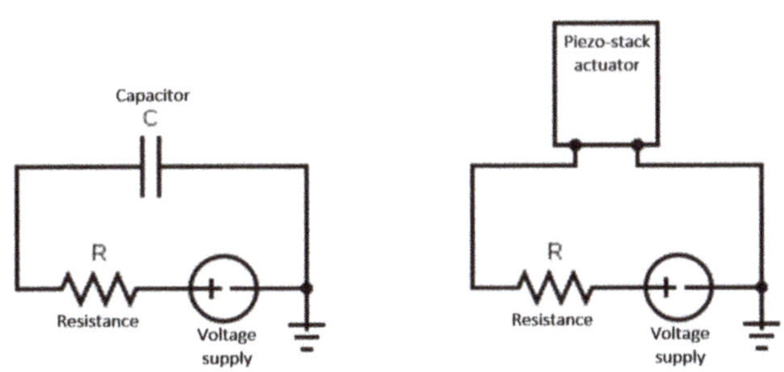

Figure 6. Analogy between RC circuit and circuit with piezo-stack actuator.

To ensure that the current stays below the maximum threshold imposed by the power amplifier, an electrical resistance is needed, connected in series with the piezo-stack actuator. The resistance must be sized to be as small as possible because an excessive reduction in the power amplifier output current brings a greater response delay and a high frequency performance loss.

In Figures 7 and 8, it is possible to see the influence of the electrical resistance on the power amplifier current and on the voltage imposed to the piezo-stack actuator when a 0–1000 V step is used as the control signal on the system.

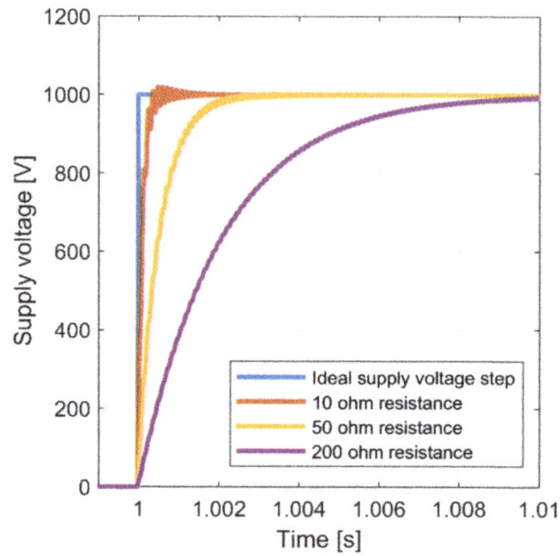

Figure 7. The 0–1000 V step response of the circuit with a piezo-stack actuator and different resistance values.

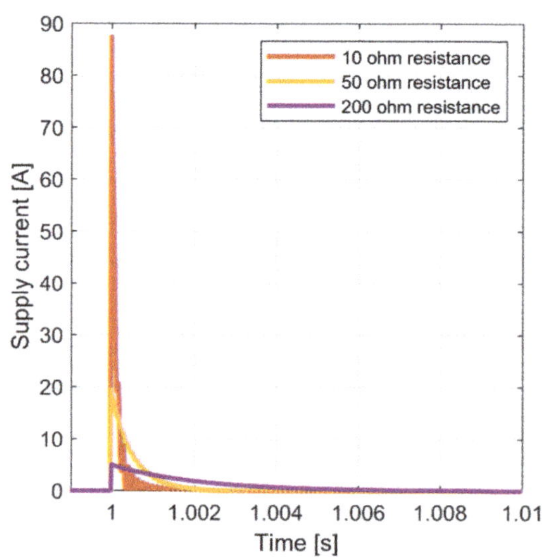

Figure 8. Electrical current in the piezo-stack actuator circuit, 0–1000 V step input for different resistance values.

5. Control Algorithm Model

The active systems are controlled to behave like adaptive filters, which are FIR (Finite Impulse Response)-type filters whose parameters are constantly updated through a specific algorithm. In this case, the algorithm used is the LMS (least Mean Square) algorithm. The scheme of a generic adaptive filter is reported in Figure 9.

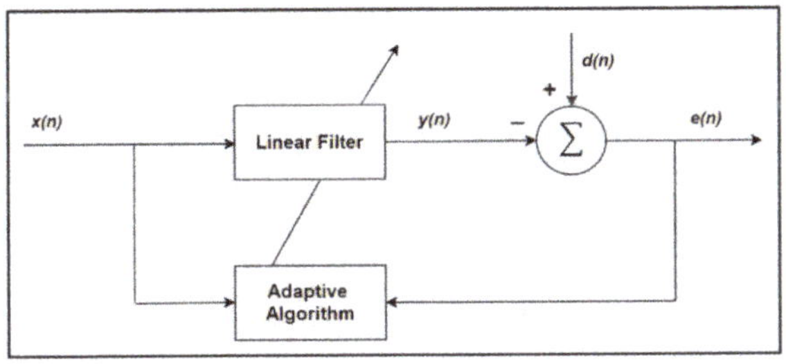

Figure 9. Scheme of a generic adaptive filter.

The adaptive algorithm is fed with an input signal $x(n)$, which is a sinusoid having the same frequency of the error signal $e(n)$ pulsation that must be reduced. The adaptive algorithm computes the parameters used by the linear filter to minimize the error signal.

The aforementioned linear filter is fed with $x(n)$ input signal and gives a cancellation signal $y(n)$ as output. $y(n)$ is subtracted from $d(n)$ which is the actual pump flow rate containing the desired average value and the unwanted oscillating component.

The error signal $e(n)$ used to feed the adaptive filter is the pressure ripple measured at pump delivery and cleaned of the average value by means of a high-pass filter.

The LMS algorithm was implemented in the Simulink environment using two blocks: the LMS update block, which outputs the weights used as input in the Linear Filter block, as shown in Figure 10. The FIR linear filter length, N, was set to 32, which means that it is composed of 32 filter weights $w(n)$ that are constantly updated thanks to the LMS block.

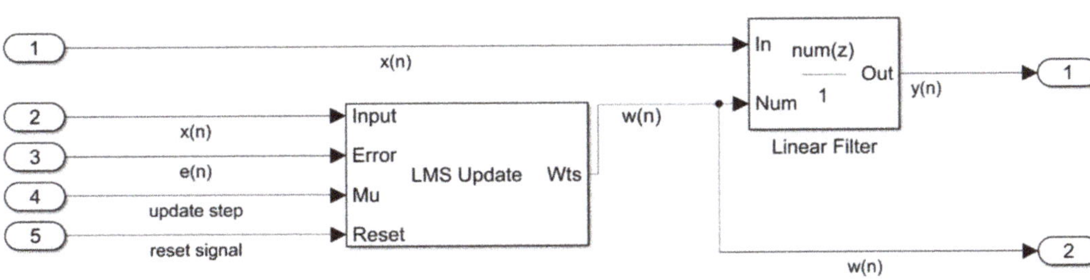

Figure 10. Scheme of the Simulink control algorithm model.

In LMS algorithm, the updating of the N filter weights $w(n)$ happens as follows:

$$w(n+1)_k = w(n)_K + \mu e(n)x(n-K), \;\; K = 0, 1, \ldots, N-1 \tag{8}$$

where K is the index of the filter weight and μ is the update step size that must be large enough to provide fast convergence of the filter weights and at the same time small enough to ensure algorithm stability.

6. Simulations Results

Simulations of the operation of the two studied active flow ripple cancellation systems were carried out with the aim of verifying their effectiveness for different working conditions of the positive displacement pump. In particular, the target was to evaluate the capacity to reduce the flow ripple and the ability to adapt to different conditions of the delivery pressure, delivered flow rate, and rotation speed of the volumetric pump.

Simulations consist of imposing different working cycles to the pump during which operating conditions change. Precisely, two different cycles were tested, each composed of three steady sections characterized by different pressure, pump speed, and swash plate angle.

6.1. Piston Actuated by Piezo-Actuator (1st Solution)

In the first run, the pump working cycle was set as shown in Table 1.

Table 1. Pump working conditions in first simulation cycle.

Cycle Section	1	2	3
Average delivery pressure [bar]	250	135	30
Pump rotation speed [rpm]	2500	2500	2500
Swash plate angle/max. angle [%]	100	100	50
Average flow rate [L/min]	200	200	100

Between every steady-state section, a short transition section was also inserted to allow the working conditions to change smoothly.

In Figures 11–13, the FFT of pump delivery flow rate in the three steady-state conditions is reported. It is possible to see the effect of flow ripple reduction produced by the active system comparing the main harmonic amplitude with and without active system intervention.

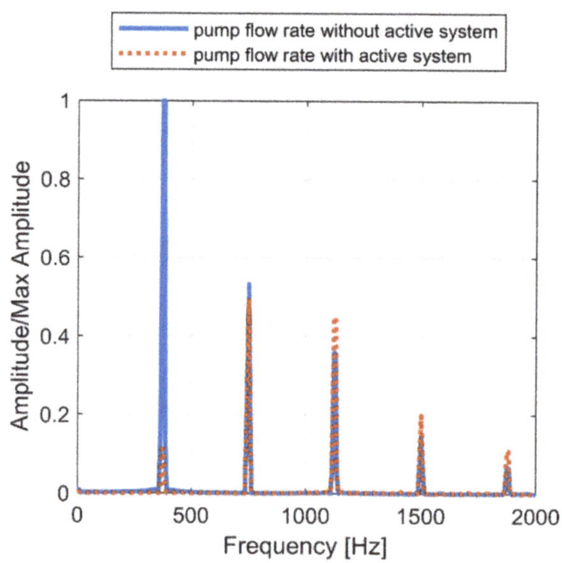

Figure 11. Simulation cycle 1, section 1: flow ripple FFT at 250 bar, 2500 rpm, 100% swash plate inclination.

In each one of the three cycle sections, the first order harmonic of the flow ripple is clearly reduced. The reduction amounts to 89% for cycle section 1, to 92% for section 2, and to 95% for section 3.

It is possible to note that higher order harmonics may show a small increase or decrease even if the active system is meant to act only on the first main one. This is attributable to the introduction of the active system that brings with it small changes in the delivery line characteristics, like fluid volume for example, modifying the frequency response of the hydraulic system. This phenomenon can be seen in all the other simulated conditions, for which the aforementioned explanation remains valid.

The active method also shows good performances in autonomously updating the control system parameters to adapt to pump working condition changes. In Figure 14, 4 of the 32 weights of the FIR filter updated with the LMS algorithm during simulation are shown. Only 4 weights are reported for simplicity. As it can be seen from the graph, each weight converges rapidly to a new value after the working condition changes. Convergence time is in the order of tenths of a second.

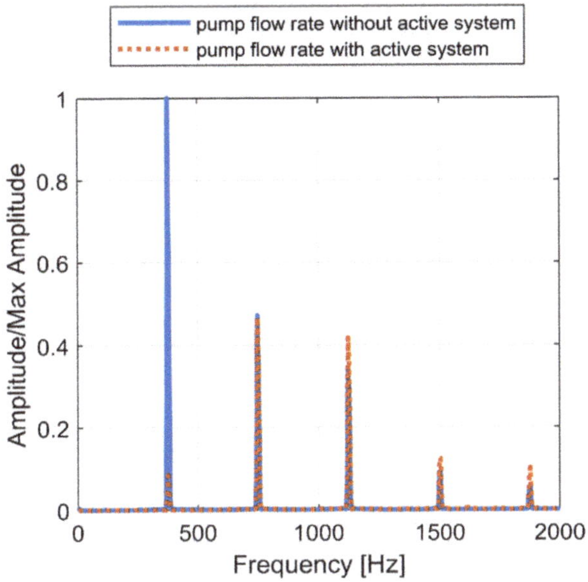

Figure 12. Simulation cycle 1, section 2: flow ripple FFT at 135 bar, 2500 rpm, 100% swash plate inclination.

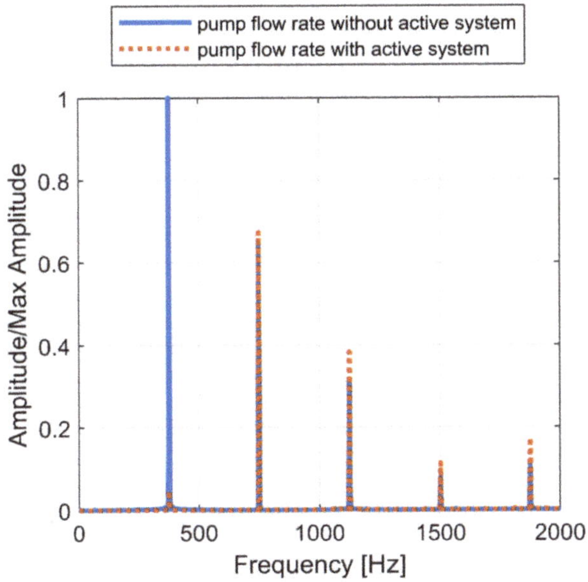

Figure 13. Simulation cycle 1, section 3: flow ripple FFT at 30 bar, 2500 rpm, 50% swash plate inclination.

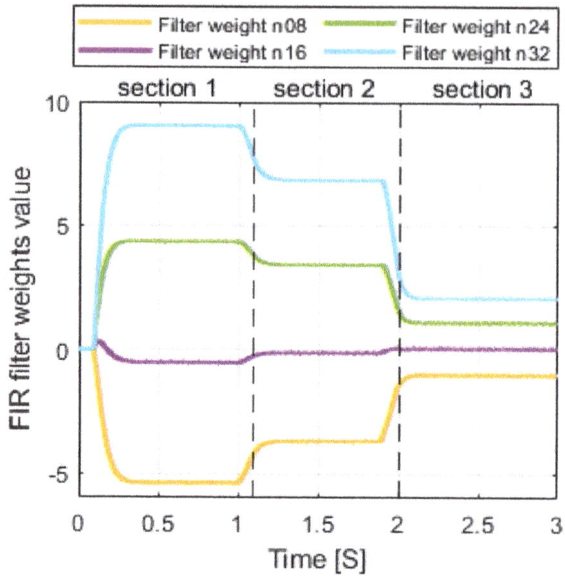

Figure 14. Simulation cycle 1: weight trend in FIR discrete filter.

Since in the first run cycle only constant speed and decreasing pressure situations were tested, the second run cycle contemplates a decrease in pump speed between sections 1 and 2 and an increase in pressure between sections 2 and 3, as it can be seen in Table 2.

Table 2. Pump working conditions in second simulation cycle.

Cycle Section	1	2	3
Average delivery pressure [bar]	250	90	220
Pump rotation speed [rpm]	2500	1500	1500
Swash plate angle/max. angle [%]	100	100	100
Average flow rate [L/min]	200	120	120

In Figures 15–17, results referring to the simulated cycle indicated in Table 2 are reported. Results regarding section 1 are not reported because the section 1 of the second cycle has the same parameters of the section 1 of the first cycle. From the results, it is possible to state that also in conditions of variable pump speed and increasing delivery pressure the control algorithm can allow the active system to adapt to varying pump working conditions. Flow ripple first harmonic reduction for sections 1–3 amounts, respectively, to 89%, 82%, and 74%.

During functioning of the active system, the supply current of the piezo stack actuator reaches very high values of over 8 A. The characteristic power supply maximum current of a piezo-actuator is usually below this value; for this reason, it was decided to limit the supply current by appropriately increasing the electrical resistance placed in series with the piezo-actuator, in order to maintain peak current below the 6 A threshold.

A simulation was run with pump working cycle identical to the first run (Table 1) to compare effectiveness with and without electric current limitation. Results are reported in Figures 18–21.

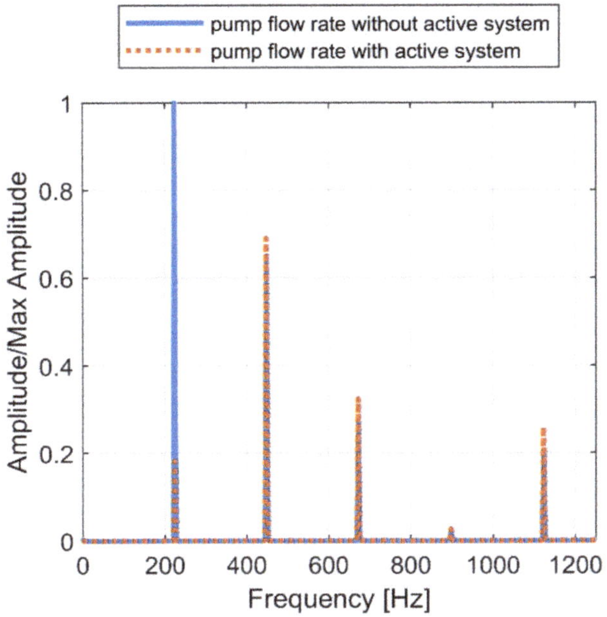

Figure 15. Simulation cycle 2, section 2: flow ripple FFT at 90 bar, 1500 rpm, 100% swash plate inclination.

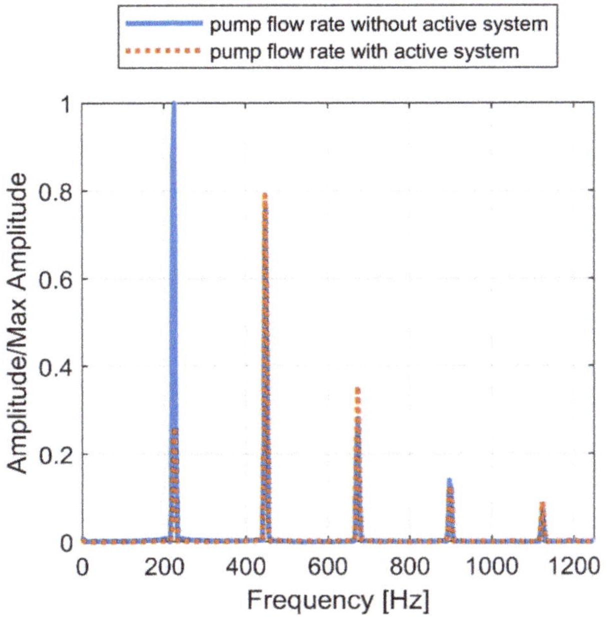

Figure 16. Simulation cycle 2, section 3: flow ripple FFT at 220 bar, 1500 rpm, 50% swash plate inclination.

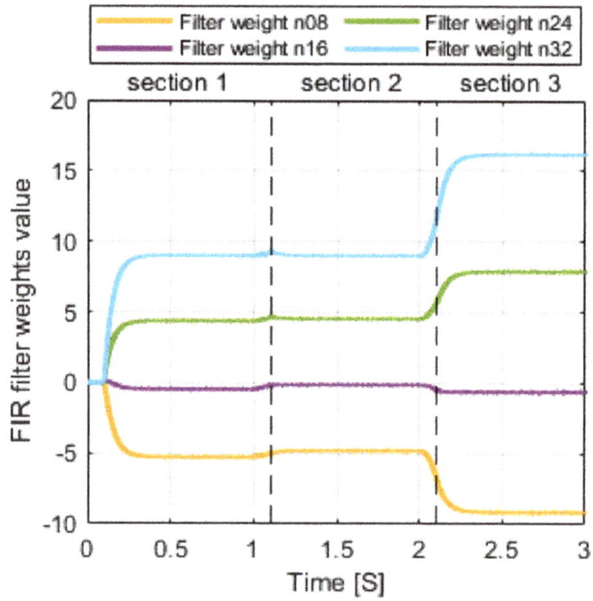

Figure 17. Simulation cycle 2: weight trend in FIR discrete filter.

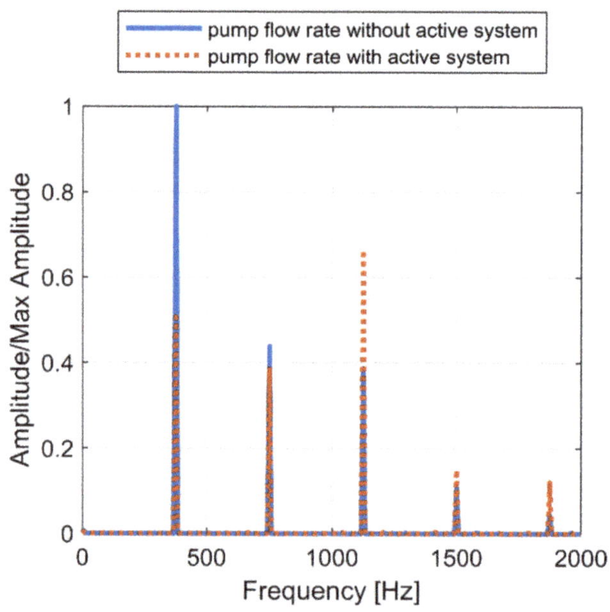

Figure 18. Limited current simulation, section 1: flow ripple FFT at 250 bar, 2500 rpm, 100% swash plate inclination.

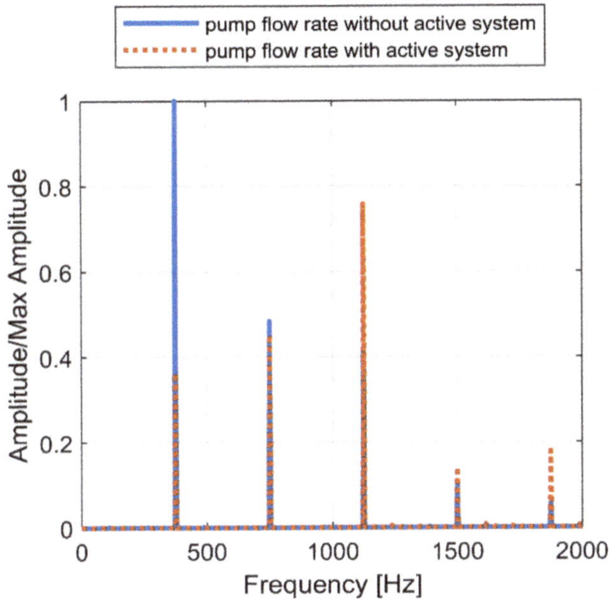

Figure 19. Limited current simulation, section 2: flow ripple FFT at 135 bar, 2500 rpm, 100% swash plate inclination.

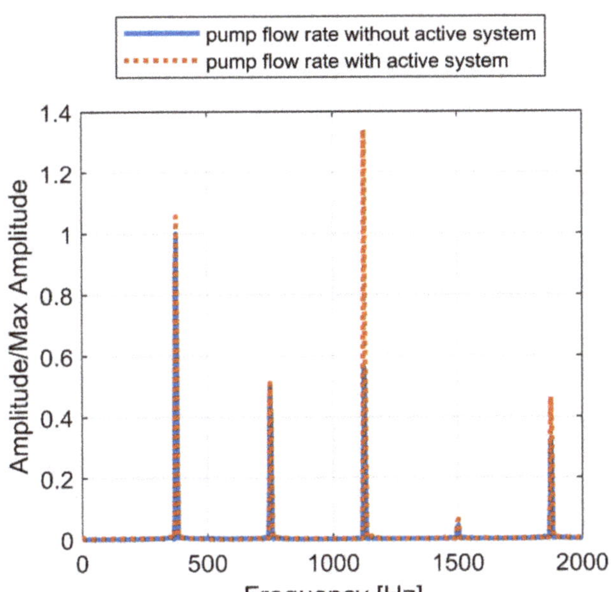

Figure 20. Limited current simulation, section 3: flow ripple FFT at 30 bar, 2500 rpm, 50% swash plate inclination.

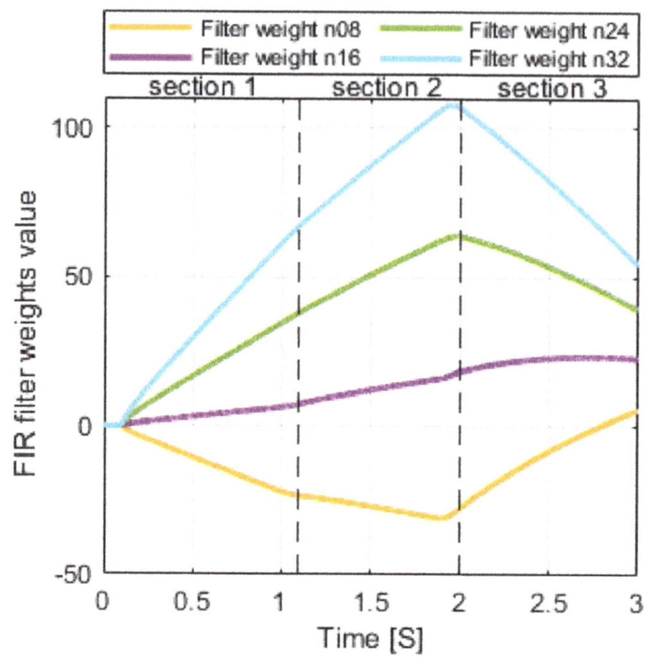

Figure 21. Limited current simulation: weight trend in FIR discrete filter.

With the supply current limited to 6 A, the effectiveness of the active system, as currently dimensioned, is significantly reduced with respect to the case with higher current. In addition, the adaptive algorithm is no more capable of successfully updating the FIR filter weights that now tend to diverge, explaining why flow ripple reduction cannot be achieved.

A possible solution to this problem is represented by using a greater number of piston–actuator pairs, to use smaller piezo-stack actuators which are provided with a higher bandwidth requiring a smaller supply current. Another possible solution is to design an ad hoc power supply, capable of providing peak currents suitable for making the application effective. However, both solutions would be very expensive, making the system grow in complexity and costs.

6.2. Piston Actuated Hydraulically via Piezo-Valve (2nd Solution)

The same working cycles used for the 1st solution were applied to the 2nd solution which relies on a piezo-valve to hydraulically actuate the piston.

In Figures 22–25, results referring to the first cycle stage (Table 1) are reported. The reduction in the first harmonic of flow ripple is clearly effective, reaching values of 97% in section 1 of the cycle, 96% in section 2, and 93% in section 3. However, convergence in FIR filter weights appears to be slower if compared to the 1st piezo-actuated active system case. In particular, in section 3 of the cycle, filter weight stable values were not reached in 1 s time, and therefore it is possible to see that they are still updating when the simulation stops, with their trend still sloping.

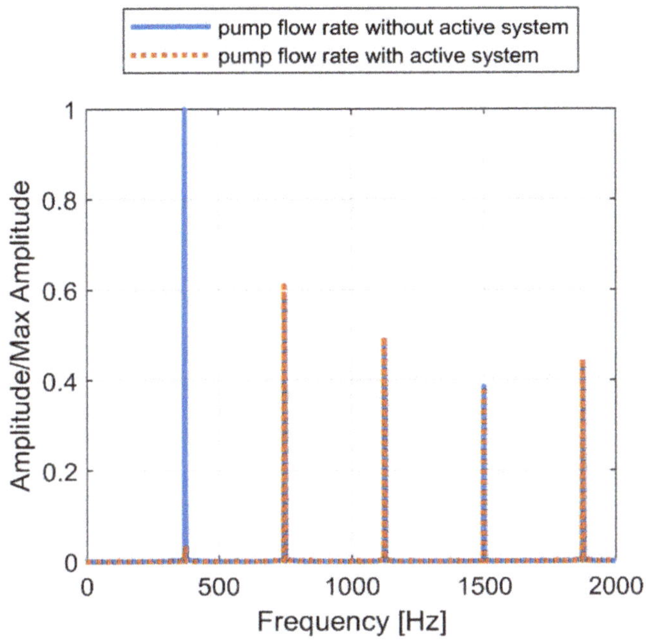

Figure 22. Simulation cycle 1, section 1: flow ripple FFT at 250 bar, 2500 rpm, 100% swash plate inclination.

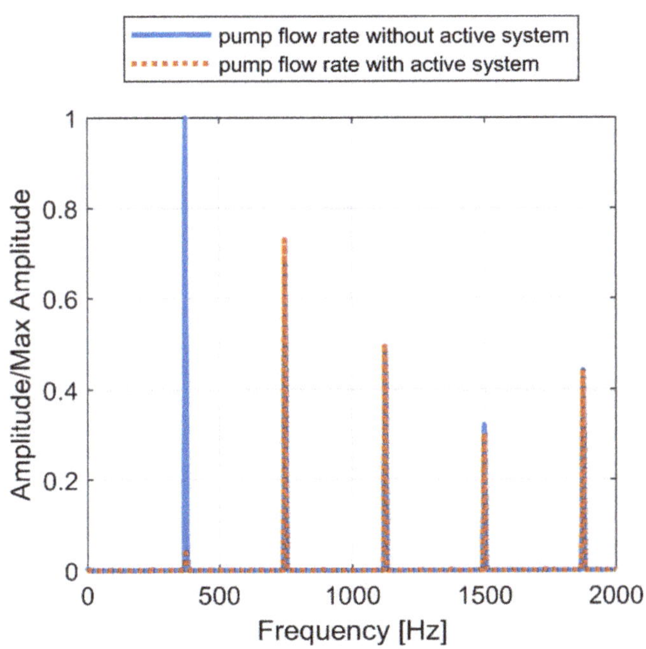

Figure 23. Simulation cycle 1, section 2: flow ripple FFT at 135 bar, 2500 rpm, 100% swash plate inclination.

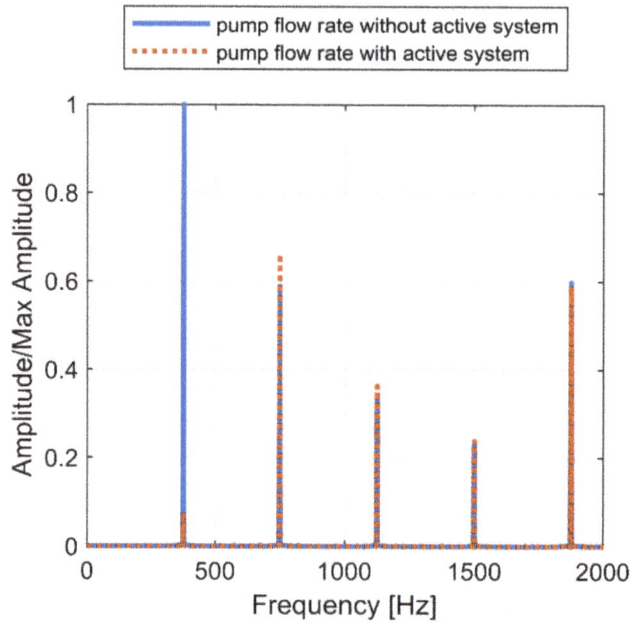

Figure 24. Simulation cycle 1, section 3: flow ripple FFT at 30 bar, 2500 rpm, 50% swash plate inclination.

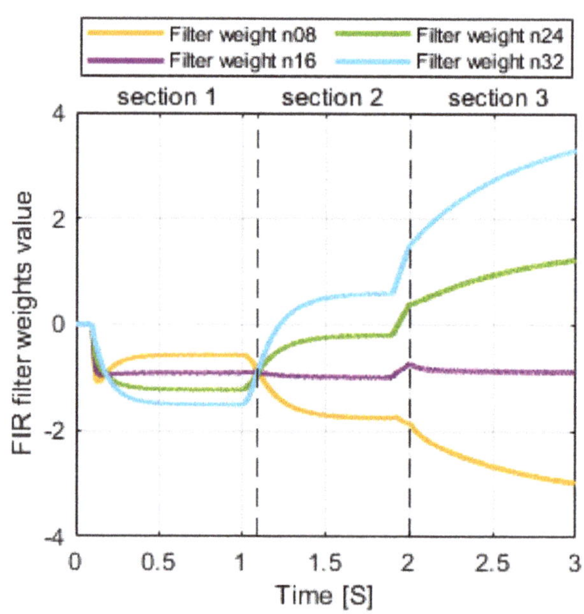

Figure 25. Simulation cycle 1: weight trend in FIR discrete filter.

In Figures 26–28, results referring to the second cycle pattern (Table 2) are reported. Results regarding section 1 are not reported because the section 1 of the second cycle has the same parameters of the section 1 of the first cycle. In this case, it is possible to see by the trend of the FIR filter weights that the control algorithm is not able to keep up with changing working conditions of the pump. In particular, after being able to converge in the first section of the cycle, filter weights then rapidly diverge starting from the section number 2, compromising the effectiveness of the active system.

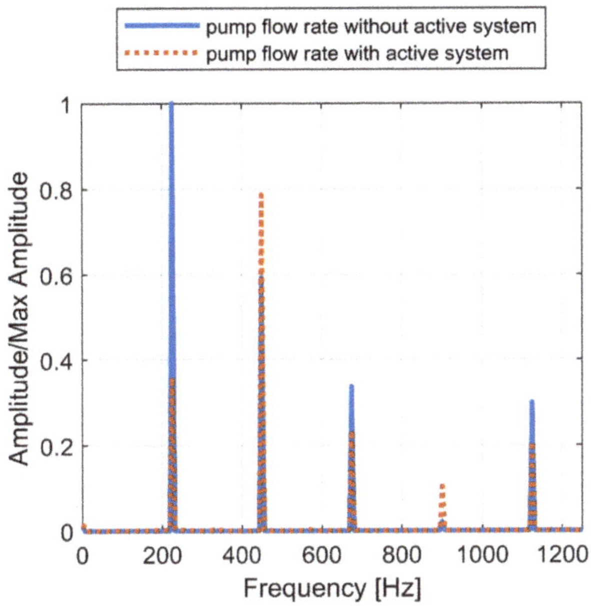

Figure 26. Simulation cycle 2, section 2: flow ripple FFT at 90 bar, 1500 rpm, 100% swash plate inclination.

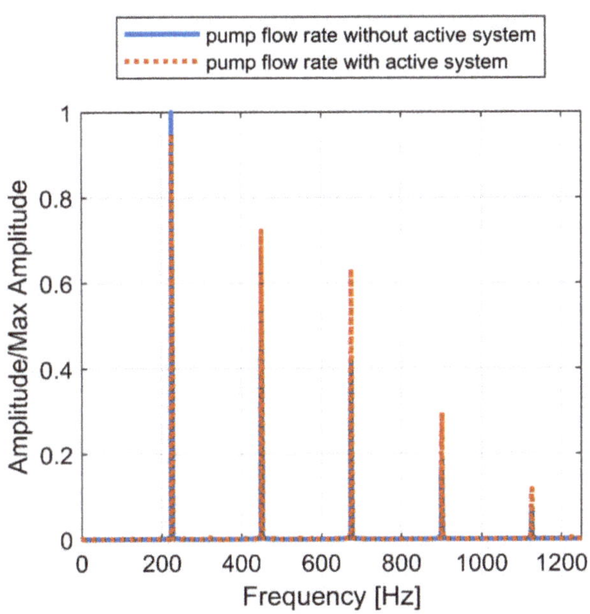

Figure 27. Simulation cycle 2, section 3: flow ripple FFT at 220 bar, 1500 rpm, 50% swash plate inclination.

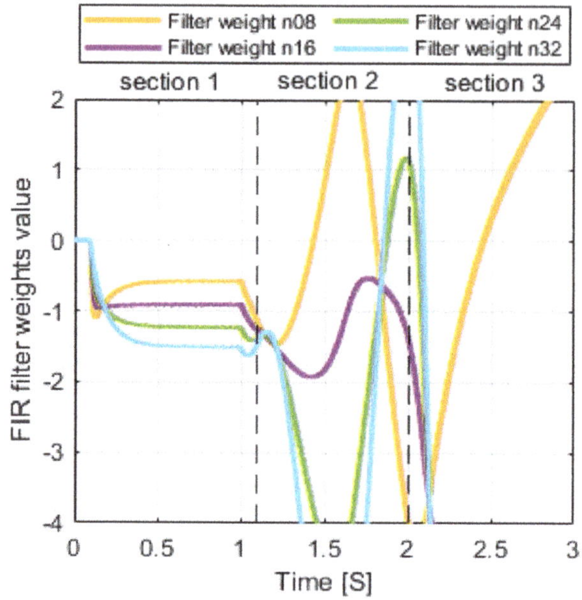

Figure 28. Simulation cycle 2: weight trend in FIR discrete filter.

One possible explanation for the unwanted behavior of the 2nd active system control algorithm is the fact that the control signal generated by the FIR block and the piston actual movement is not correlated as strongly as in the 1st active system presented. In the 2nd system, the control signal operates a piezo-valve which has an influence on the piston oscillation, but the correlation is much smaller than in a control signal directly operating the piston through a piezo-stack actuator as in the 1st active system.

The LMS algorithm needs a strong correlation between the generated control signal and the effective cancellation signal to function properly. In this second active system, the oscillation of the piston, which actually produces the flow ripple cancellation signal, depends strongly on the average delivery pressure and on the first harmonic of the flow ripple. As a consequence, the algorithm is not able to adapt itself promptly when the pump working condition changes.

The quantity of fluid drained in the operation of this active system is acceptably small, never exceeding 1.5% of the pump delivery flow rate in any of the simulated operating conditions.

7. Conclusions

In this paper, two active methods to achieve pump flow ripple reduction were proposed. Both methods rely on pistons producing flow ripple cancellation signals, one involving a direct piezo-actuated piston and the second involving a piston actuated hydraulically via a piezo-valve. Mathematical models of the two methods were presented, including a control algorithm used to achieve adaptability for both systems.

Both methods can be applied to any kind of positive displacement pump, even if for this work an axial piston pump was used for simulation purposes.

Results of numerical simulations were discussed, showing the potentialities and problems of the two proposed methods.

In particular, it was demonstrated how the first active solution can heavily reduce the flow ripple and effectively adapt to changing working conditions, as long as a sufficient supply current is available. Thanks to this active system, in fact, the reduction of 80–90% of the flow ripple main harmonic was achieved.

The second active method permits to avoid some of the first method disadvantages. In fact, a smaller piezo-stack actuator is needed to operate the valve reducing the cost and dimensions and preventing high supply current related problems. However, with the method being more complex, the control algorithm has to be improved to allow the active system to reliably adapt to pump working condition changes. However, this active system was proved to be promising, achieving reduction of the flow ripple main harmonic of over 90% when correctly functioning, implying only a small fluid drainage of 1.5% of pump delivery flow.

Regarding future developments on the research presented in this paper, the next step will be the design and optimization of more stable and reliable control algorithms, working on the basis represented by the one used for this research. Then, prototypes of the two solutions presented will be designed and an experimental campaign will be performed to validate simulation results.

Author Contributions: Conceptualization, P.C. and C.M.V.; methodology, P.C. and C.M.V.; software, C.M.V.; validation, P.C., C.M.V. and M.R.; writing—original draft preparation, P.C., F.S. and C.M.V.; writing—review and editing, P.C., F.S., C.M.V. and M.R.; supervision, P.C.; project administration, P.C. All authors have read and agreed to the published version of the manuscript.

Funding: This research received no external funding.

Institutional Review Board Statement: Not applicable.

Informed Consent Statement: Not applicable.

Data Availability Statement: Not applicable.

Acknowledgments: The authors would like to acknowledge the active support of this research by Casappa S.p.A., Parma, Italy.

Conflicts of Interest: The authors declare no conflict of interest.

References

1. Karpenko, M.; Prentkovskis, O.; Šukevičius, Š. Research on high-pressure hose with repairing fitting and influence on energy parameter of the hydraulic drive. In *Eksploatacja i Niezawodnosc—Maintenance and Reliability*; Polish Maintenance Society: Lublin, Poland, 2022; Volume 24, pp. 25–32, ISSN 1507-2711.
2. Stosiak, M. The impact of hydraulic systems on the human being and the environment. *J. Theor. Appl. Mech.* **2015**, *53*, 409–420. [CrossRef]
3. Ye, S.; Zhang, J.; Xu, B. Noise Reduction of an Axial Piston Pump by Valve Plate Optimization. *Chin. J. Mech. Eng.* **2018**, *31*, 57. [CrossRef]
4. Johansson, A.; Olvander, J.; Palmberg, J.-O. Experimental verification of cross-angle for noise reduction in hydraulic piston pumps. *Proc. Inst. Mech. Eng. Part I J. Syst. Control Eng.* **2007**, *221*, 321. [CrossRef]
5. Harrison, A.M.; Edge, K. Reduction of axial piston pump pressure ripple. *Proc. Inst. Mech. Eng.* **2000**, *214*, 53–64.
6. Borghi, M.; Zardin, B. Axial Balance of External Gear Pumps and Motors: Modelling and Discussing the Influence of Elastohydrodynamic Lubrication in the Axial Gap. In Proceedings of the ASME International Mechanical Engineering Congress and Exposition, Houston, TX, USA, 13–19 November 2015. [CrossRef]
7. Zhao, X.; Vacca, A. Theoretical investigation into the ripple source of external gear pumps. *Energies* **2019**, *12*, 535. [CrossRef]
8. Zhao, X.; Vacca, A. Numerical analysis of theoretical flow in external gear machines. *Mech. Mach. Theory* **2017**, *108*, 41–56. [CrossRef]
9. Zhou, J.; Vacca, A.; Casoli, P. A Novel Approach for Predicting the Operation of External Gear Pumps Under Cavitating Conditions. *Simul. Model. Pract. Theory* **2014**, *45*, 35–49. [CrossRef]
10. Rundo, M.; Altare, G.; Casoli, P. Simulation of the filling capability in vane pumps. *Energies* **2019**, *12*, 283. [CrossRef]
11. Corvaglia, A.; Ferrari, A.; Rundo, M.; Vento, O. Three-dimensional model of an external gear pump with an experimental evaluation of the flow ripple. *Proc. Inst. Mech. Eng. Part C J. Mech. Eng. Sci.* **2021**, *235*, 1097–1105. [CrossRef]
12. Pan, M.; Ding, B.; Yuan, C.; Zou, J.; Yang, H. Novel Integrated Control of Fluid-Borne Noise in Hydraulic Systems. In Proceedings of the BATH/ASME 2018 Symposium on Fluid Power and Motion Control FPMC 2018, Bath, UK, 12–14 September 2018.
13. Shang, Y.; Tang, H.; Sun, H.; Guan, C.; Wu, S.; Xu, Y.; Jiao, Z. A novel hydraulic pulsation reduction component based on discharge and suction self-oscillation: Principle, design and experiment. *Proc. Inst. Mech. Eng. Part I J. Syst. Control Eng.* **2020**, *234*, 433–445. [CrossRef]
14. Pan, M.; Johnston, N.; Plummer, A. Hybrid Fluid-borne Noise Control in Fluid-filled Pipelines. *J. Phys. Conf. Ser.* **2016**, *744*, 012016. [CrossRef]

15. Waitschat, A.; Thielecke, F.; Behr, R.M.; Heise, U. Active Fluid Borne Noise Reduction for Aviation Hydraulic Pumps. In Proceedings of the 10th International Fluid Power Conference, Dresden, Germany, 8–10 March 2016.
16. Kim, T.; Ivantysynova, M. Active Vibration/Noise Control of Axial Piston Machine Using Swash Plate Control. In Proceedings of the ASME/BATH 2017 Symposium on Fluid Power and Motion Control, Sarasota, FL, USA, 16–19 October 2017; Paper No. FPMC2017-4304.; p. V001T01A053. [CrossRef]
17. Casoli, P.; Pastori, M.; Scolari, F.; Rundo, M. Active pressure ripple control in axial piston pumps through high-frequency swash plate oscillations—A theoretical analysis. *Energies* **2019**, *12*, 1377. [CrossRef]
18. Casoli, P.; Pastori, M.; Scolari, F. Swash plate design for pressure ripple reduction—A theoretical analysis. In Proceedings of the AIP Conference, Modena, Italy, 11 September 2019. [CrossRef]
19. Hagstrom, N.; Harens, M.; Chatterjee, A.; Creswick, M. Piezoelectric actuation to reduce pump flow ripple. In Proceedings of the ASME/BATH 2019 Symposium on Fluid Power and Motion Control 2019 FPMC, Sarasota, FL, USA, 7–9 October 2019.

energies

MDPI

Article

Multi-Chamber Actuator Mode Selection through Reinforcement Learning–Simulations and Experiments

Henrique Raduenz [1,*], Liselott Ericson [1], Victor J. De Negri [2] and Petter Krus [1]

[1] Division of Fluid and Mechatronics Systems, Linköping University, 581 83 Linköping, Sweden; liselott.ericson@liu.se (L.E.); petter.krus@liu.se (P.K.)
[2] Laboratory of Hydraulic and Pneumatic Systems, Federal University of Santa Catarina, Florianópolis 88040-900, Brazil; victor.de.negri@ufsc.br
* Correspondence: henrique.raduenz@liu.se

Abstract: This paper presents the development and implementation of a reinforcement learning agent as the mode selector for a multi-chamber actuator in a load-sensing architecture. The agent selects the mode of the actuator to minimise system energy losses. The agent was trained in a simulated environment and afterwards deployed to the real system. Simulation results indicated the capability of the agent to reduce energy consumption, while maintaining the actuation performance. Experimental results showed the capability of the agent to learn via simulation and to control the real system.

Keywords: reinforcement learning; multi-chamber actuator; mode selection

Citation: Raduenz, H.; Ericson, L.; De Negri, V.J.; Krus, P. Multi-Chamber Actuator Mode Selection through Reinforcement Learning–Simulations and Experiments. *Energies* **2022**, *15*, 5117. https://doi.org/10.3390/en15145117

Academic Editors: Paolo Casoli and Massimo Rundo

Received: 20 June 2022
Accepted: 12 July 2022
Published: 13 July 2022

1. Introduction

In a multiple hydraulic actuator system, the load on each actuator results in different pressure levels. If the actuators are controlled by a load-sensing architecture with pressure-compensation valves and a single pump, the supply pressure is controlled according to the highest required pressure. This causes a mismatch between the pressures in the actuators and the supply that is compensated for by the throttling on the pressure-compensation valves. These resistive control losses are one of the major sources of energy losses in such hydraulic systems [1,2].

One way of reducing such losses is to use a multi-chamber actuator in one of the loads. A hydraulic diagram of a pressure-compensated valve-controlled load-sensing architecture, where Load 2 is driven by a multi-chamber actuator, is shown in Figure 1.

The set of on/off valves between the proportional valve and the actuator allows for the combinations of different chambers that define the possible actuator modes to be used. The capability to select different actuator modes allows the modulation of the resultant pressure from the load to make $p_{LS,2}$ similar to $p_{LS,1}$. This is shown in the illustrative flow-pressure diagrams in Figure 2.

The possibility to modulate the pressure, Figure 2b, enables the reduction in resistive control losses. A conventional actuator, Figure 2a, cannot perform this modulation and, thus, results in higher resistive control losses. A deeper discussion on this system architecture is provided in [3].

This architecture has another degree of freedom to be controlled, which is the mode selection. The mode that minimises the losses is a function of the speed and forces of all the actuators, but because these vary along with the system operation, this is not a trivial problem to solve.

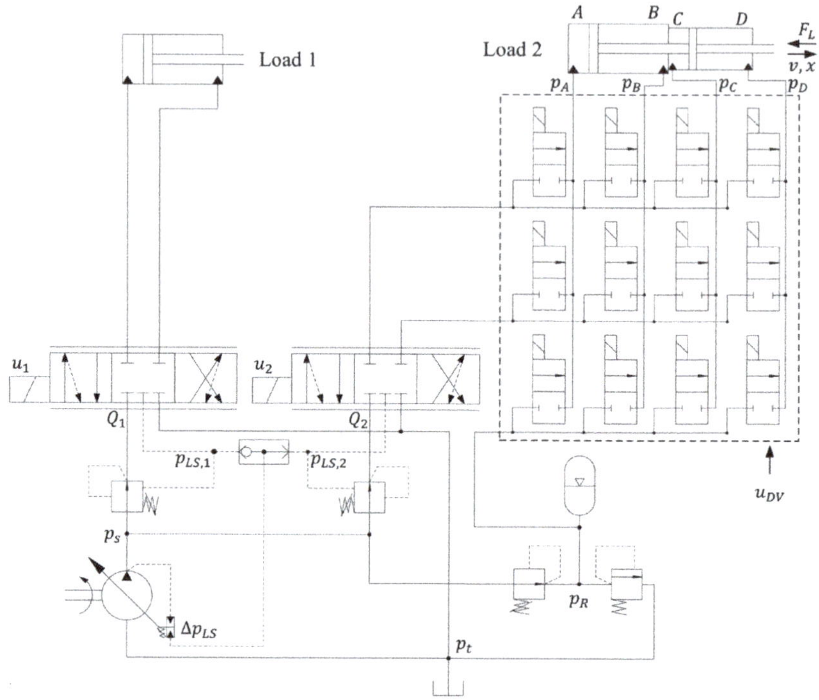

Figure 1. Hydraulic-system architecture.

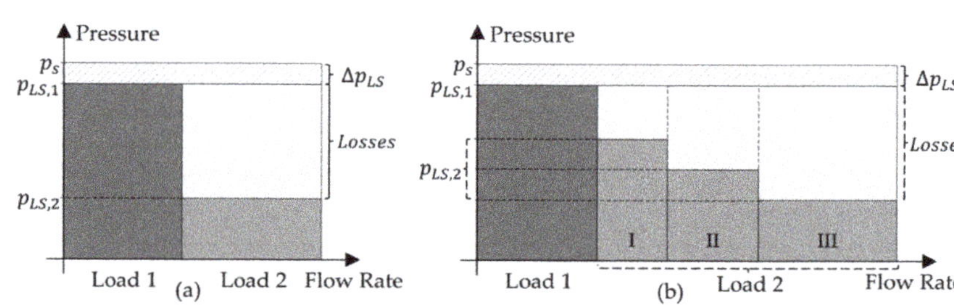

Figure 2. Flow-pressure diagram showing the effect of mode selection on resistive control losses. (**a**) With a conventional actuator; (**b**) with a multi-chamber actuator.

Multi-chamber actuators usually belong to architectures where the selection of modes is responsible for the control end goal, such as speed or position. The speed or position error is translated to a reference force that the multi-chamber actuator must exert. In [4], this force reference is compared to the available forces for each mode, and the mode that results in the smallest force error is applied to the system. In [5], a mode selection based on the minimisation of the force error is implemented, as in [4]; however, there is no mention of avoiding frequent switching. In [6], the same mode-selection strategy as in [4] is compared to one that penalizes high-amplitude pressure changes between different modes. In [7], the authors describe the operation of a controller that selects the mode that minimises the energy consumption, while still being able to drive the current load. In [8], the mode selection is also based on a predefined range of force capacity for each mode. The implemented mode is the one that can exert the requested force, and the difference in force is controlled by the proportional pressure control in one of the chambers.

Model predictive control (MPC) is studied in [9] to solve the problems related to force spikes when in mode switching. The mode selection is still performed based on minimising the force difference between the reference and the available modes and on minimising the energy losses associated with switching. The force difference is handled by means of throttle control, to adjust the pressure in each chamber to the calculated pressure reference. MPC is focused on minimising the force transients when switching. A similar study also using MPC is presented in [10], where the results also indicate an advantage of MPC over simpler mode-selection strategies, as presented in [4], for example.

Although not used for a multi-chamber actuator, the selection of modes in [2] is performed based on the calculation of the capacity of each mode to overcome the current force over the actuator, while, apparently, minimising the pump pressure and flow rate.

The authors in [11] describe a state machine to handle the mode selection based on, for example, the load pressure and pressure levels defined for each mode. The authors emphasise possible instability and non-smooth operation of the actuator during mode switching and suggest that these problems could be minimised by acting on the proportional valve control, which to some extent follows the ideas presented in [8,9]. In [12], the mode selection of the independent metering valve system is also performed based on the current load force and the force capability of each mode.

The control problem here is not the same as the ones studied by the works presented above. Here, the mode selection acts as an enabler for the control end goal, which is still performed by the proportional valve. Therefore, the control goal for the multi-chamber actuator is to select a mode that enables the requested motion to be completed, while minimising the resistive control losses. From another point of view, the mode selector can be interpreted as performing energy management because the goal of using the multi-chamber actuator is to reduce energy losses. Instead of designing a controller for the mode selection based on the analytical equations of the energy losses in the system, in this work a controller for the mode selection based on RL was studied.

Reinforcement learning (RL), with neural networks as the actor and critic function approximators, has proven to be a powerful and successful control-development methodology for complex problems, such as playing Atari [13]. The main advantages are automatic learning by interacting with the environment and finding optimised control solutions that are inherently difficult to be solved analytically. Additionally, machine learning algorithms also perform well in scenarios that they are not exactly trained for, which is likely to occur in the operation of mobile machines, due to their vast field of application.

In the control of mobile working machines, Ref. [14] presents an approach based on RL, with the Q-learning algorithm for the control of hybrid system power to minimise fuel consumption. What the authors also show is the possibility of first training the controller with simulation models of the system and then applying the controller to the real system. An RL-based energy-management strategy is proposed in [15] for hybrid construction machines. The authors use a combination of Dyna learning and Q-learning, but the study is limited to simulation results.

Other applications of RL for mobile machines include: Ref. [16], where it is used, based on cameras, lidar, and motion and force sensors, to perform bucket loading of fragmented rock with a multi-objective target, including maximisation of the bucket loading; Ref. [17], where it is trained for the motion control of a forestry crane while minimising energy consumption; and Ref. [18], where it is used for the trajectory tracking control of an excavator arm, with the controller generating the valve-control signals directly.

It is observed in these papers that the development of RL controllers usually starts with a pre-training of the agent in a simulation environment. Advantages of this approach are: it avoids undesirable real-world consequences; it reduces costs associated with obtaining real experience; and the simulations typically run faster than real time [19]. The reviewed papers also show the controller's capacity to find and implement solutions for complex tasks.

This paper demonstrates, through simulation and experimental results, the training and implementation of a RL-based controller for the mode selection of a multi-chamber hydraulic actuator. The actuator is part of a multi-actuator load-sensing architecture driven by a single pump. The selection of different modes allows for the reduction in resistive losses in the control valves, due to a better match of pressures between actuators. A Deep Q-Learning (DQN) agent was created and trained to learn how to select the modes to minimise the system energy losses.

Paper Contribution and Objectives

Research on the topic of this paper has been presented in previous publications by our group. In [3], the system architecture is described along with the potential efficiency improvements. The first implementation of the RL-based approach to control this system was studied and presented in [20]. The present work builds on those previous publications, where the main difference from [20] is the use of a load-sensing pump instead a constant pressure supply, which makes the learning task significantly harder and closer to a real, mobile system application.

The objective and, consequently, the contribution of this paper are to show experimental results demonstrating that RL can also be used to control complex hydraulic systems. In this case, it is about the selection of modes of multi-chamber actuators aiming to reduce resistive control losses. In particular, it is shown that RL finds an optimised control solution with reduced need for manually designing the controller. However, the study is limited to training the agent in simulation and not on continuing the learning after being deployed to the real system.

2. Available Modes, Model Description, and Control Structure

The multi-chamber actuator used in this study has four chambers and is connected to three supply lines (A/B/R). This gives a total of 87 possible modes. However, most modes are ruled out due to the reasons presented in [3]. The modes used in this study are given in Table 1. Mode 1 is the only mode that is not an agent's decision, because it is implemented as a rule to ensure safe operation. The steady state force displayed in Table 1 is calculated considering a pressure of 100 bar on port A and 15 bar on ports B and R. This is done to show which modes can exert higher force.

Table 1. Available modes and connected chambers.

Mode	A	B	R	Force [kN]
1	-	-	-	-
2	AC	BD	-	77
3	ABC	D	-	70
4	A	BD	C	55
5	AB	D	C	48
6	C	BD	A	12

The model of the physical system was developed in the multi-domain system simulation tool HOPSAN [21] and describes the motion of the boom arm as a function of the motion of the actuators. It also models the dynamic behaviour of the pressures, flow rates, internal leakage, and closing and opening of the valves. The model was validated against experimental results, with a detailed description presented in [20].

The controller and training algorithm for the DQN agent were implemented in MATLAB/Simulink [21]. During the training in simulation, the agent learns by interacting with the system. After training, the agent controls the real system.

The actual controller of the system is not only composed of the trained agent selecting the optimal mode, it also contains other control and safety rules. An overview of the controller structure is shown in Figure 3. The reward branch of the agent is only used during the training phase.

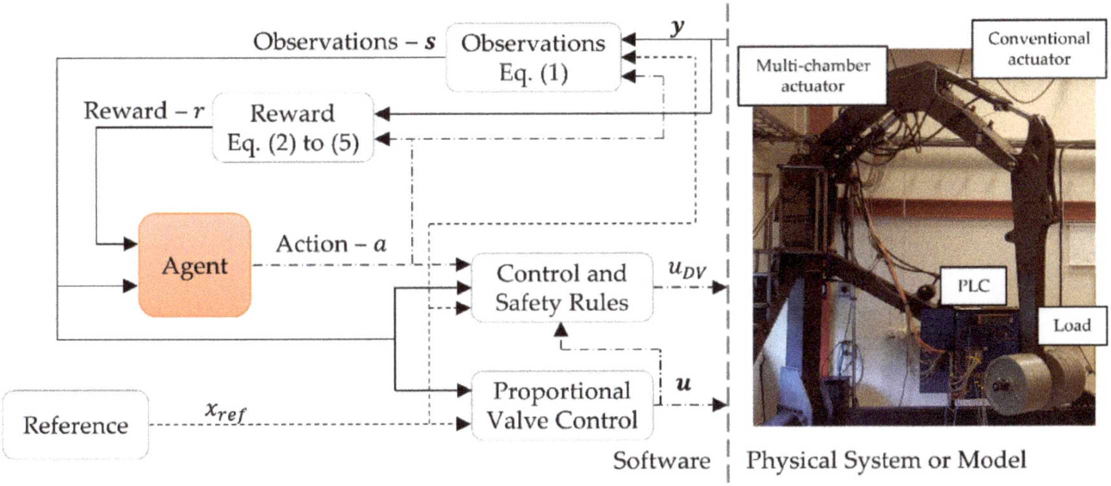

Figure 3. Structure of the controller and picture of the test bench.

In Figure 3, it is also seen how the interaction of the agent with the environment takes place during the learning phase and then in the controlling phase. Paraphrasing [13] for the present control problem: *The agent interacts with the environment in a sequence of actions, observations, and rewards. At each agent time step* (t)*, the agent observes the observations* $s(t)$*, selects an action* $a(t)$ *from the set of modes (*Table 1*), and applies it to the environment. It observes the new observations* $s(t+1)$ *and the reward* $r(t)$*. It tries to maximise the reward.*

From the RL perspective, everything that is not the agent is considered as the environment it is interacting with. It is important to make the environment in the simulation as close to the real environment as possible. For the control of hydraulic systems in mobile applications, along with having a sufficiently accurate model of the physical system, there is the need to include additional safety rules. Therefore, the environment is the combination of the system model and the additional rules. The added control and safety rules for this system are:

- Apply mode 1 if $|u_2| < 0.002$ m;
- Limit the available modes according to position and external load;
- Compensate u_2 due to the difference in areas in each mode; and
- Use mode 4 for the lowering motion ($u_2 < -0.002$ m).

The first rule is used because the valve has an overlap of 2 mm, so the on/off valves can be closed. The second rule prevents the agent from choosing a mode that cannot drive the load with the available maximum pump pressure, otherwise the load could fall. For the same operator control input signal to the proportional valve, the third rule ensures approximately the same actuator speed for the different modes, see [3,20] for details.

The fourth rule is implemented because the return motion, due to the much larger area of the actuator connected to the return than the supply, results in excessive throttling on the meter-out edge of the valve. This causes the pump to operate at maximum pressure due to a perceived high load, which causes the pressure-compensation valves to operate fully open and, thus, the compensation losses are close to zero. Thus, the selection of modes, based on minimising the compensation losses, does not work for the returning motion. Therefore, due to this design constraint, the return motion is not an agent's decision.

While the RL agent is responsible for the mode selection, another P controller 'mimics' the machine operator controlling the proportional valves. The P controller is implemented inside the proportional valves control block in Figure 3.

3. Learning Setup for the Agent

The application has a continuous observation space (pressures, speed, etc.) and a discrete action space (modes to be selected). A DQN agent [13,22] is suitable for this type of observation and action space, and it mainly consists of a neural network calculating the value for taking a certain action given the current observations. The value is an estimation of the sum of the reward that the agent can collect over a future time horizon, by following a certain sequence of control actions. In this case, the network at each agent's time step predicts the value of taking each action. A greedy function selects and implements in the system the action that has the highest estimated value, which is, in other words, the action that would lead to the best performance, according to the reward function. This agent is, thus, a non-linear map between the system states (observations of pressures, speed, position, …) and the optimised control action (modes). The reader is referred to [13,22] for a description of the type of agent and training algorithm.

Each agent action corresponds to one mode (Table 1). Inside the 'Control and Safety Rules' block in Figure 3, a lookup table maps the action to the corresponding vector u_{DV} of the open/closed digital valves that implements the mode in the system.

The structure and parameters of the network are presented in Table 2. No sensitivity analysis was performed to evaluate the size of the network on the performance of the task. However, the size of the network was chosen to be small.

Table 2. Structure of the agent network.

Layer	Size
Feature inputs (observations)	9 input features
Fully connected with Rectified Linear Unit activation function	70 neurons
Fully connected with Rectified Linear Unit activation function	35 neurons
Fully connected output (actions)	5 neurons

The observations s used as input features are

$$s(t) = [a, u_2, p_A, p_B, p_C, p_s, p_{LS,1}, v, x], \tag{1}$$

where a is the previous action, u_2 is the proportional valve control signal, p_{A-B-C} are the chambers' pressures, p_s is the supply pressure, $p_{LS,1}$ is the load-sensing pressure for the conventional actuator, v is the actuator speed, and x is the position.

The reward function r to be maximised is composed of three terms,

$$r = K_1 r_{Velocity} + K_2 r_{Power} + K_3 r_{Switch}, \tag{2}$$

$$r_{Velocity} = -1 \; if \; |v| < 0.001 \text{ m/s}, \tag{3}$$

$$r_{Power} = -(|Q_2(p_s - p_{LS,2})| + |\, Q_1(p_s - p_{LS,1})|)/P_{Norm}, \tag{4}$$

$$r_{Switch} = \begin{cases} r_{Switch} & if \; a_{t+1} = a_t \\ r_{Switch} - 1 & if \; a_{t+1} \neq a_t \end{cases}. \tag{5}$$

The velocity term ($r_{Velocity}$) penalizes the agent if a motion is requested ($|u_2| > 0.002$), and the multi-chamber actuator does not move with a minimum velocity. This encourages the agent to learn to meet a minimum control-performance requirement. The power loss term (r_{Power}) is a penalty based on the hydraulic system control losses, to encourage the agent to find a mode that results in smaller energy losses due to pressure compensation. The switch term (r_{Switch}) penalizes the agent for frequent mode switching.

Table 3 presents the parameters and load cases used during training. To increase the agent's robustness, noise is added to the measurements used as observations in the simulations. R is a uniformly distributed random variable with a maximum amplitude of 1.

Table 3. Agent and task parameters.

Parameter	Value
Initial exploration factor	0.99
Minimum exploration factor	0.15
Exploration factor decay	7.5×10^{-6}
Target smooth factor	2.0×10^{-3}
Learning rate	7.5×10^{-4}
Mini-batch size	128
Look ahead steps	4
Experience buffer length	4.0×10^{4}
Agents sample time [s]	0.80
Discount factor	0.99
Gradient decay factor	0.90
Maximum number of episodes	1.8×10^{4}
Load Cases and Task	**Value**
Initial position [m]	$0.10 + 0.02R$
Final position [m]	$0.40 + 0.05R$
External load [kg]	[40 80 120 160 200 240]
Load 1 load sensing pressure $p_{LS,1}$ [bar]	[60 100] $\pm$ $3R$
Load 1 flow rate Q_1 [lpm]	10

In a real system, the load pressure from the conventional actuator ($p_{LS,1}$) varies according to the load. However, a simplification is made by setting it to constant values to allow for easier interpretation of the results both in the simulation and in the experiments. In the experiments, this signal is emulated with an additional hydraulic circuit. This means that the agent is exposed to 12 different scenarios of external load and pressure on the conventional actuator, and the task is to lift the load from an initial position to a final position while maximising the reward function.

Dynamic systems have oscillations, and these must be taken into consideration when setting the agent-control time step, defining the reward function, and considering what type of information is stored as experience. This is represented in Figure 4, by showing a dynamic response of the system state (s) due to an action a_t.

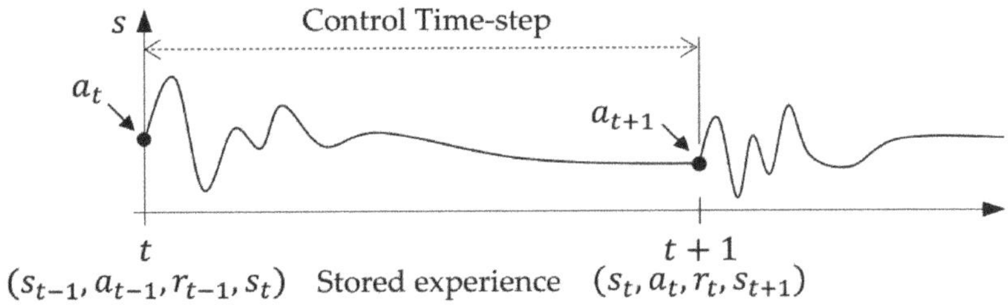

Figure 4. Agent time step versus system dynamics.

If the time step is too small, the stored experience will contain dynamic effects rather than steady-state conditions. This might affect the agent's ability to learn because the reward function could give misinformation. However, the reward from the action at t is only observed at $t+1$ because it is usually a function of the final state and the transition between the states. Therefore, the size of the control time step can be adjusted to avoid these effects. For the current system, the mode selection does not need to occur frequently to avoid unnecessary switching. Thus, the agent time step is set to avoid these oscillations.

4. Results

The agent was trained to lift a load from a certain initial position to a final position, with 12 load cases that including 2 pressure levels on $p_{LS,1}$ and 6 external loads. Figure 5 presents the agent's training progress, where Q_0 is the estimated value.

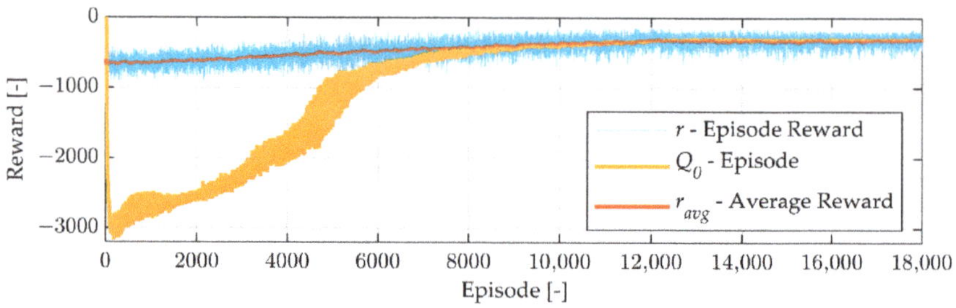

Figure 5. Agent's training progress.

After training, the agent is tested in simulation for all 12 load cases, and its capability to complete the task in an efficient way is evaluated. The energy loss (E_{loss}) is calculated with Equations (6) and (7), with the variables extracted from the simulation.

$$P_{loss} = P_{loss,1} + P_{loss,2} = |Q_1(p_s - p_{LS,1})| + |Q_2(p_s - p_{LS,2})| \quad (6)$$

$$E_{loss} = E_{loss,1} + E_{loss,2} = \int_{t_i}^{t_f} (P_{loss,1} + P_{loss,2})dt \quad (7)$$

4.1. Simulation Results

An estimation of what mode results in the lowest power losses is made by evaluating Equation (6) for each mode and the steady-state conditions of speed (v) and position (x) of the multi-chamber actuator. $p_{LS,2}$ is calculated based on the external load and the position of the actuator, given the kinematics of the boom arm. The mode that results in the lowest power loss is selected as the best. This gives an indication of what mode selection to expect from the agent's learning. The solution for the two load cases, with 120 kg external load and the two pressure levels on $p_{LS,1}$, are shown in Figure 6. Moreover, the safety boundary condition limiting the number of modes that can be applied to the system is shown, which was implemented as a rule. The resultant pressure ($p_{LS,2}$) for the different modes is also shown, where it is seen that, to some extent and as expected, the best mode leads to a pressure level close to $p_{LS,1}$.

What Figure 6a,b show is that modes that can exert lower forces are better when the actuator is retracted, while modes that can exert higher forces are better when it is extended. This results in a closer match of $p_{LS,1}$ and $p_{LS,2}$, as shown in Figure 6c,d. Due to the geometry of the boom, the further the actuator is extended, the higher the load force on it. However, the mode that results in the lowest power losses is also dependent on the flow rate, Equation (6), and this is the reason why weaker modes might be better (Figure 6b) at higher speeds.

Figure 6 indicates what to expect from the agent's learning when it comes to the minimisation of energy losses. However, as shown in Equation (2), this is not the only optimisation objective. Figures 7 and 8 present simulation results for the trained agent performing the mode selection, for the two load cases. What the figures also show is the performance on the same tests when a conventional actuator with areas equivalent to mode 2 was used instead of the multi-chamber actuator. The abbreviations Conv and RL are used in the figures for the conventional actuator case and the multi-chamber actuator with mode selection, respectively. Power results are normalised with respect to the maximum power of each load case.

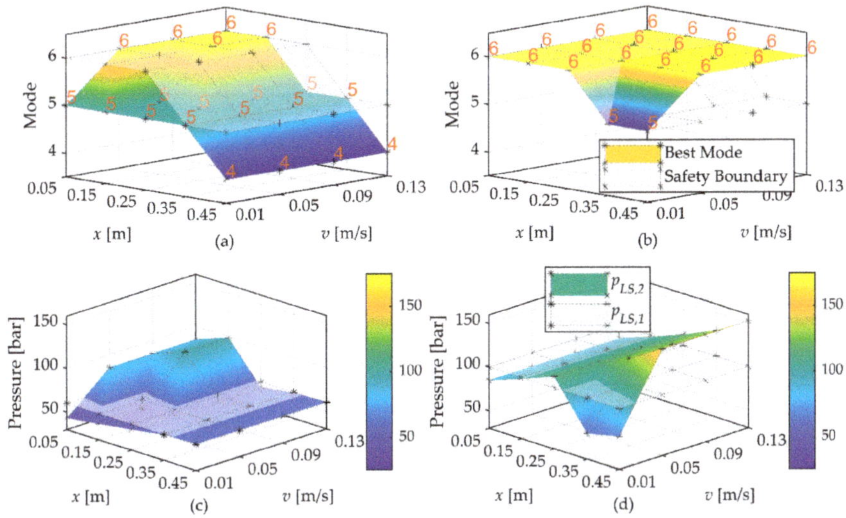

Figure 6. Best modes for 120 kg load. (**a**) For $p_{LS,1} = 60$ bar; (**b**) for $p_{LS,1} = 100$ bar; (**c**) resultant $p_{LS,2}$ for $p_{LS,1} = 60$ bar; (**d**) resultant $p_{LS,2}$ for $p_{LS,1} = 100$ bar.

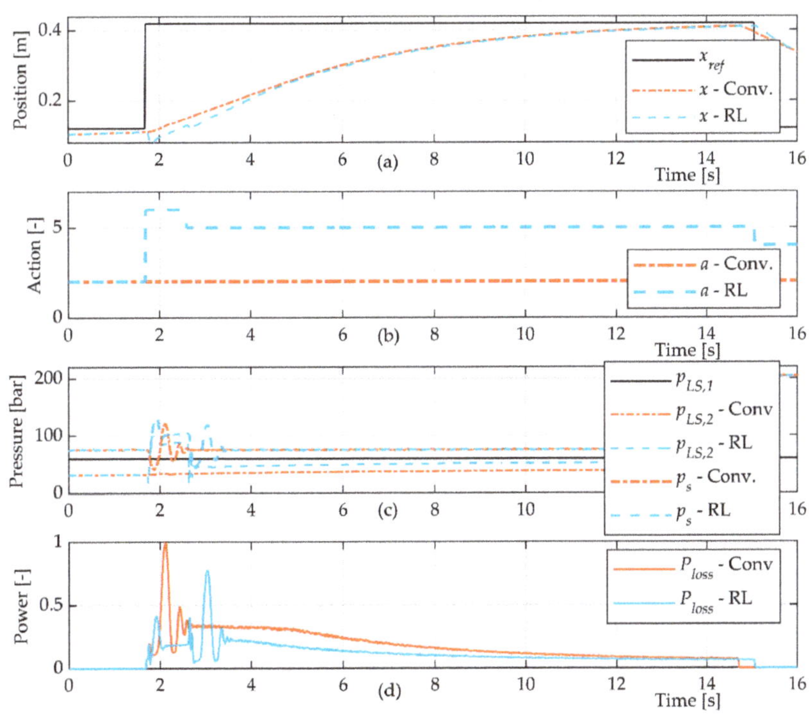

Figure 7. Simulation result, 40 kg load and $p_{LS,1} = 60$ bar. (**a**) Position; (**b**) action; (**c**) system and load-sensing pressures; (**d**) power loss.

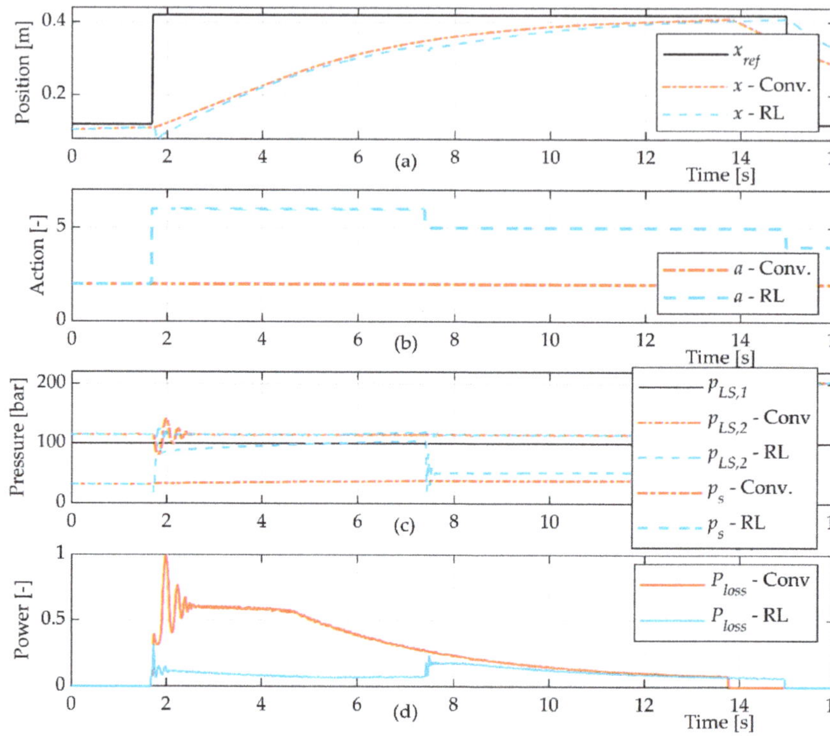

Figure 8. Simulation result, 40 kg load and $p_{LS,1} = 100$ bar. (**a**) Position; (**b**) action; (**c**) system and load-sensing pressures; (**d**) power loss.

Comparing Figure 7b with Figure 7d shows that the agent does learn to select modes that reduce the energy losses. This results in a better match of pressures than the conventional case, as shown in Figure 7c. By choosing a mode with smaller areas and resulting in a better match of $p_{LS,1}$ and $p_{LS,2}$, the power losses are reduced significantly. A similar learning and decision-making capacity is observed for the other test case, as shown in Figure 8.

Comparing Figures 8b and 8d shows that the agent learns to choose the better mode based on the similar principles of reducing the mismatch of pressure (Figure 8c) and choosing the modes with a smaller area.

All load scenarios that the agent was trained for were simulated, and Equation (7) is used to compare the energy losses. The results, normalised by the highest total energy losses for all test cases of each level of $p_{LS,1}$, are presented in Figures 9 and 10.

Results from the energy analysis show that the agent learned to select modes that result in lower energy losses for almost all the load scenarios it was trained for. For some load cases, the chosen solution is close to or equal to the conventional case. This happens in the load scenarios presented in Figure 9d–f. This means that in the worst case the agent runs the system like a conventional actuator, but it does not choose something worse than that, for example, a mode that cannot drive the load. Although it must be remembered that there is a safety function preventing a weak mode from being applied for high loads, the results showed that most of the time the agent chooses modes within the safety boundary.

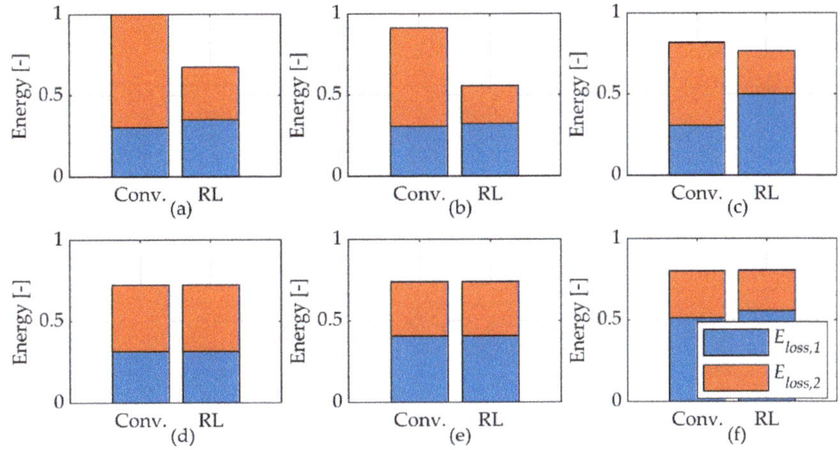

Figure 9. Energy comparison, $p_{LS,1} = 60$ bar: (**a**) 40 kg; (**b**) 80 kg; (**c**) 120 kg; (**d**) 160 kg; (**e**) 200 kg; (**f**) 240 kg.

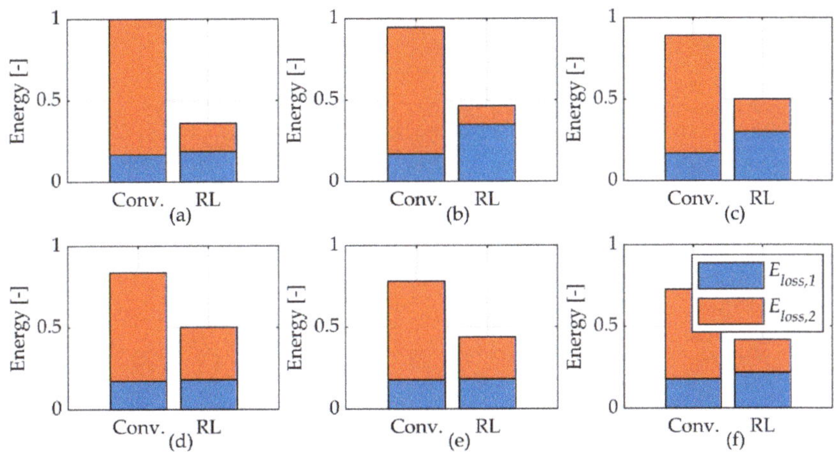

Figure 10. Energy comparison, $p_{LS,1} = 100$ bar: (**a**) 40 kg; (**b**) 80 kg; (**c**) 120 kg; (**d**) 160 kg; (**e**) 200 kg; (**f**) 240 kg.

4.2. Experimental Tests

The trained agent was deployed to the real system and evaluated on the same load scenarios used in the training. The experimental test results, for the same load cases of Figures 7 and 8, are shown in Figures 11 and 12. Simulation results are plotted in the same figures. However, in the experiments, the agent observes different conditions than in simulation, which can lead to different decisions.

The agent is able to apply in practice the same modes it learned from the optimisation (Figure 11b), even though there are differences, for example, in the actuator position (Figure 11a) and system pressure (Figure 11c).

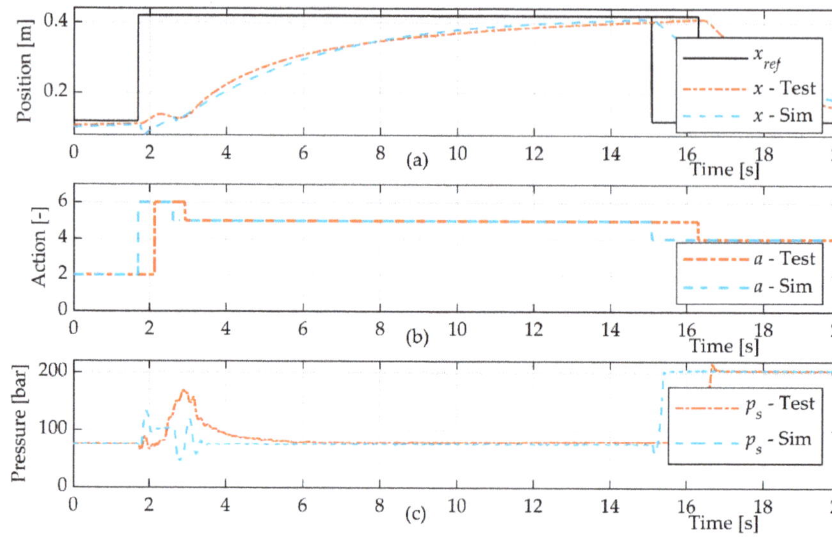

Figure 11. Experimental (Test) and simulation (Sim) results, 40 kg load and $p_{LS,1} = 60$ bar. (**a**) position and reference; (**b**) action; (**c**) system pressure.

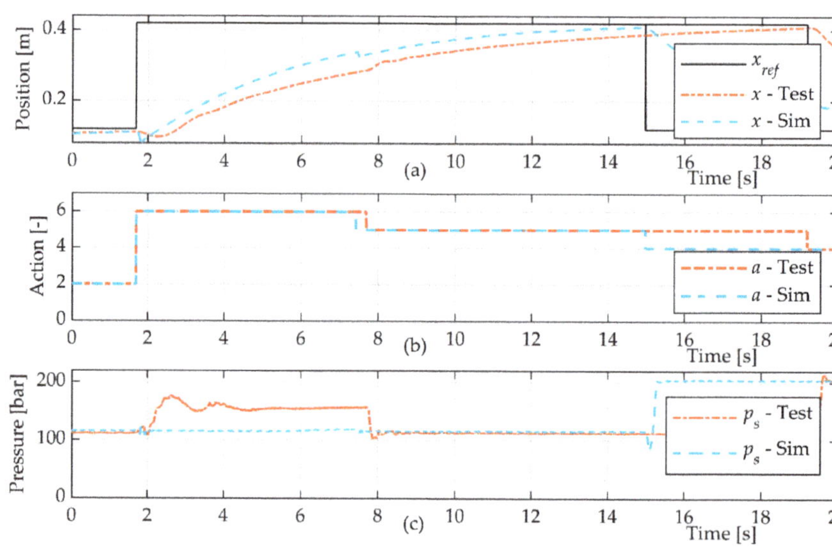

Figure 12. Experimental (Test) and simulation (Sim) results, 40 kg load and $p_{LS,1} = 100$ bar. (**a**) Position and reference; (**b**) action; (**c**) system pressure.

As for the previous load case, the agent is able to apply in practice the same modes it learned from the optimisation.

As said, all load cases were tested experimentally, and comparisons of selected modes are shown in Figures 13 and 14, where Sim and Test refer to the simulation and experiment, respectively.

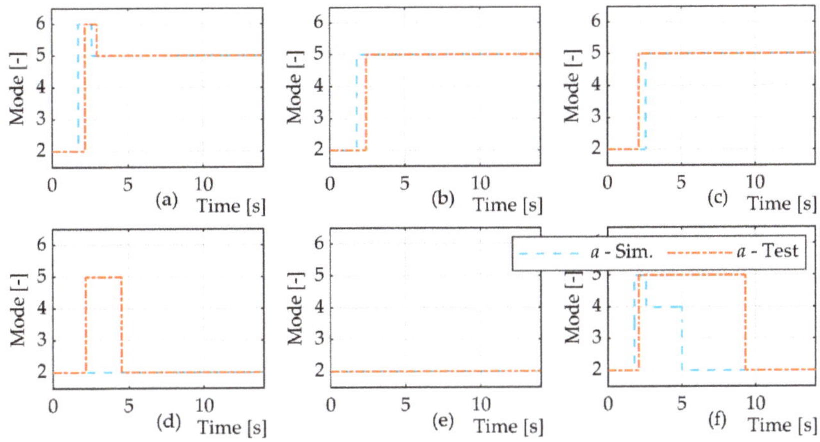

Figure 13. All test cases, $p_{LS,1}$ = 60 bar: (**a**) 40 kg; (**b**) 80 kg; (**c**) 120 kg; (**d**) 160 kg; (**e**) 200 kg; (**f**) 240 kg.

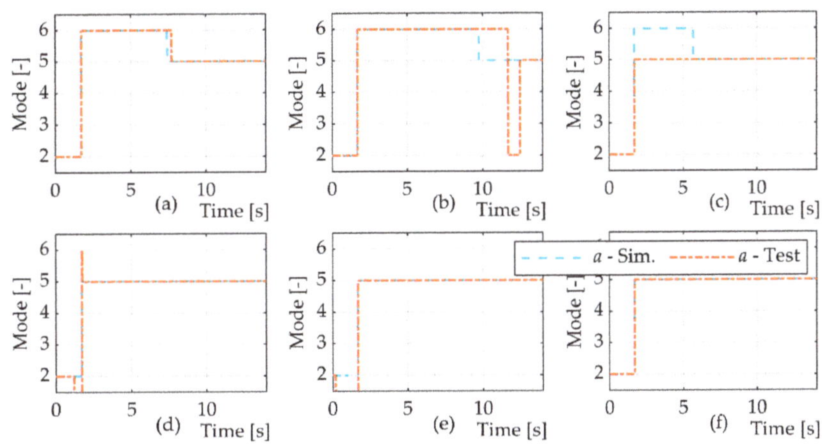

Figure 14. All test cases, $p_{LS,1}$ = 100 bar: (**a**) 40 kg; (**b**) 80 kg; (**c**) 120 kg; (**d**) 160 kg; (**e**) 200 kg; (**f**) 240 kg.

The experimental results show that in most of the cases the agent was able to implement in the real system the decisions that were also taken in the simulation. The load cases that failed are shown in Figure 13d,g and Figure 14b,c. Even though these cases show some deviation from what was achieved in the simulation, they are not wrong for the complete test. This means that only a few decisions were wrong in these tests. However, they still followed the logic of choosing weaker modes at the start of the motion and stronger modes at the end of the motion, and, therefore, the agent was still able to complete the motion.

Relatively high sensitivity to the variation in measured variables was observed, and this resulted in low repeatability in some of the load cases. The worst situation was for the load case of 120 kg load and $p_{LS,1}$ 60 bar, an example of which is shown in Figure 15.

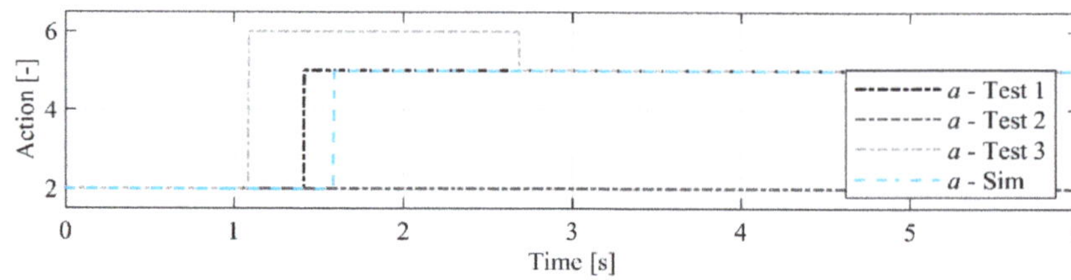

Figure 15. Issues with repeatability for 120 kg load and $p_{LS,1} = 60$ bar.

Figure 15 shows that such controllers, as expected, might underperform when operating under different conditions than they were trained for. However, for the trained agent presented in this study, the decisions, despite being different, still allowed the agent to complete the task in an acceptable way.

5. Discussion

The energy analysis based on the simulation results showed the reduction in energy losses caused by the selection of suitable modes by the agent. However, this does not mean that the training of the agent converged to the global optimal solution that results in the lowest energy losses possible. The optimisation objective has other terms, and, therefore, the results presented in Figure 6 are only an estimation of what are the best modes. Future studies should aim at finding the optimal solution and comparing this with what the agent finds during the training.

The model of the system was judged to be sufficiently accurate to describe the main characteristics and behaviours of the system. Still, there were deviations from the actual behaviour of the system. Although sufficient to pre-train the agent, it is likely that the performance of the agent could be further improved by letting it, in a safe manner, interact with the real system and train from those experiences in order to correct model deviations.

Machine learning-based controllers are a black-box type of controller. There is an inherent risk associated with them when deployed to control real systems, usually related to the controller operating outside the training domain. Therefore, it must be ensured that the control actions applied to the system do not lead to any unsafe and/or unstable behaviours. In this study, this was ensured by implementing safety rules that would overwrite the agent's decisions when necessary.

One limitation of this study was that the agent was evaluated under approximately the same load conditions and tasks that it was trained for. In the simulation, random noise was introduced to all measurements from the system in order to increase its robustness in the experimental tests. However, there are still questions to be answered and an evaluation to be made regarding the robustness and generalisation capability of the agent. This is especially the case for more realistic load cases, such as digging and variable $p_{LS,1}$ pressure.

6. Conclusions

This paper presents simulation and experimental results that demonstrate the capability of a controller, developed under a reinforcement learning framework, to learn and control the mode selection of a hydraulic system. The controller was trained in simulation and afterwards was deployed to control the real system. The main control objective was to optimise the selection of modes of the multi-chamber actuator so as to reduce the resistive control losses on a valve-controlled load-sensing system. The controller was able to automatically find optimised control decisions and to apply them in the real system. The controller was also shown to be robust to differences between the training domain and the application domain, thus extrapolating the knowledge to unseen situations. Although able to control the real system, the trained agent must be accompanied by safety rules to ensure the safe operation of the system. Future work should be dedicated to evaluating how close the solution is to the global optimum and to assessing its robustness in application, because a high sensitivity to variations in the input features was observed for some of the test cases.

Author Contributions: Conceptualisation, H.R., L.E., V.J.D.N. and P.K.; methodology, H.R.; validation, H.R.; formal analysis, H.R.; investigation, H.R.; data curation, H.R.; writing—original draft preparation, H.R.; writing—review and editing, H.R., L.E., V.J.D.N. and P.K; supervision, L.E., V.J.D.N. and P.K. All authors have read and agreed to the published version of the manuscript.

Funding: This research was supported by the Brazilian Coordination for the Improvement of Higher Education Personnel (CAPES), the Brazilian National Council for Scientific and Technological Development (CNPq), and the Swedish Energy Agency (Energimyndigheten) Grant No. P49119-1.

Institutional Review Board Statement: Not applicable.

Informed Consent Statement: Not applicable.

Data Availability Statement: Not applicable.

Conflicts of Interest: The authors declare no conflict of interest.

Nomenclature

Variable	Denotation	Value	Unit
x	Multi-chamber actuator position	-	m
x_{ref}	Position reference for Load 2	-	m
v	Multi-chamber actuator speed	-	m/s
F	Multi-chamber actuator load	-	N
m	Multi-chamber actuator mode	-	-
$u = [u_1,u_2]$	Proportional valve control signals (target spool position)	-	m
u_{DV}	On/off valves control signals	-	A
Q_1, Q_2	Load 1 and 2 flow rates	-	m^3/s
$p_{LS,1}, p_{LS,2}$	Load 1 and 2 load sensing pressures	-	Pa
p_s	Supply pressure	-	Pa
Δp_{LS}	Load sensing delta pressure	15×10^5	Pa
$[A_A, A_B, A_C, A_D]$	Areas of the multi-chamber actuator	$[27\ 3\ 9\ 1]A_D$	m^2
p_A, p_B, p_C, p_D	Multi-chamber actuator chambers pressures	-	Pa
p_t	Tank pressure	0	Pa
p_R	Return pressure	15×10^5	Pa
a	Agent action	-	-
r	Reward	-	-
$r_{Velocity}$	Velocity reward term	-	-
r_{Power}	Power reward term	-	-
r_{Switch}	Switch reward term	-	-
P_{Norm}	Power reward normalisation factor	10×10^4	W
K_1, K_2, K_3	Gains of the reward function	$2, 800, 2$	-
s	Agent observations	-	-
y	Measurements from the system	-	-
R	Random variable	-	-
P_{loss}	Power loss		W
$E_{loss,1}, E_{loss,2}$	Energy losses on the control valves	-	J

References

1. Vukovic, M.; Leifeld, R.; Murrenhoff, H. Reducing fuel consumption in hydraulic excavators—A comprehensive analysis. *Energies* **2017**, *10*, 687. [CrossRef]
2. Ketonen, M.; Linjama, M. Simulation study of a digital hydraulic independent metering valve system for an excavator. In Proceedings of the 15th Scandinavian International Conference on Fluid Power, SICFP'17, Linköping, Sweden, 7–9 June 2017.
3. Raduenz, H.; Ericson, L.; Heybroek, K.; de Negri, V.J.; Krus, P. Extended analysis of a valve-controlled system with multi-chamber actuator. *Int. J. Fluid Power* **2021**, *23*, 79–108. [CrossRef]

4. Linjama, M.; Vihtanen, H.; Sipola, A.; Vilenius, M. Secondary controlled multi-chamber hydraulic actuator. In Proceedings of the 11th Scandinavian International Conference on Fluid Power, SICFP09, Linköping, Sweden, 2–4 June 2009.
5. Belan, H.C.; Locateli, C.C.; Lantto, B.; Krus, P.; de Negri, V.J. Digital secondary control architecture for aircraft application. In Proceedings of the Seventh Workshop on Digital Fluid Power, Linz, Austria, 26–27 February 2015.
6. Dell'Amico, A.; Carlsson, M.; Norlin, E.; Sethson, M. Investigation of a digital hydraulic actuation system on an excavator arm. In Proceedings of the 13th Scandinavian International Conference on Fluid Power SICFP2013, Linköping, Sweden, 3–5 June 2013; pp. 505–511. [CrossRef]
7. Huova, M.; Laamanen, A.; Linjama, M. Energy efficiency of three-chamber cylinder with digital valve system. *Int. J. Fluid Power* **2010**, *11*, 15–22. [CrossRef]
8. Heemskerk, E.; Eisengießer, Z. Control of a semi-binary hydraulic four-chamber cylinder. In Proceedings of the Fourteenth Scandinavian International Conference on Fluid Power, Tampere, Finland, 20–22 May 2015.
9. Heybroek, K.; Sjöberg, J. Model predictive control of a hydraulic multichamber Actuator: A feasibility study. *IEEE/ASME Trans. Mechatron.* **2018**, *23*, 1393–1403. [CrossRef]
10. Donkov, V.H.; Andersen, T.O.; Pedersen, H.C.; Ebbesen, M.K. Application of model predictive control in discrete displacement cylinders to drive a knuckle boom crane. In Proceedings of the 2018 Global Fluid Power Society PhD Symposium (GFPS), Samara, Russia, 18–20 July2018; pp. 408–413. [CrossRef]
11. Yuan, H.; Shang, Y.; Vukovic, M.; Wu, S.; Murrenhoff, H.; Jiao, Z. Characteristics of energy efficient switched hydraulic systems. *JFPS Int. J. Fluid Power Syst.* **2014**, *8*, 90–98. [CrossRef]
12. Vukovic, M.; Murrenhoff, H. Single edge meter out control for mobile machinery. In Proceedings of the ASME/Bath Symposium on Fluid Power & Motion Control, FPMC2014, Bath, UK, 10–12 September 2014.
13. Mnih, V.; Kavukcuoglu, K.; Silver, D.; Graves, A.; Antonoglou, I.; Wiersta, D.; Riedmiller, M. Playing Atari with deep reinforcement learning. *arXiv* **2013**, arXiv:1312.5602v1.
14. Zhu, Q.; Wang, Q. Real-time energy management controller design for a hybrid excavator using reinforcement learning. *J. Zheijang Univ.-Sci. A (Appl. Phys. Eng.)* **2017**, *18*, 855–870. [CrossRef]
15. Zhang, W.; Wang, J.; Liu, Y.; Gao, G.; Liang, S.; Ma, H. Reinforcement learning-based intelligent energy management architecture for hybrid construction machinery. *Appl. Energy* **2020**, *275*, 115401. [CrossRef]
16. Backman, S.; Lindmark, D.; Bodin, K.; Servin, M.; Mörk, J.; Löfgren, H. Continuous control of an underground loader using deep reinforcement learning. *Machines* **2021**, *9*, 216. [CrossRef]
17. Andersson, J.; Bodin, K.; Lindmark, D.; Servin, M.; Wallin, E. Reinforcement learning control of a forestry crane manipulator. In Proceedings of the 2021 IEEE/RSJ International Conference on Intelligent Robots and Systems (IROS), Prague, Czech Republic, 27 September–1 October 2021.
18. Egli, P.; Hutter, M. A general approach for the automation of hydraulic excavator arms using reinforcement learning. *IEEE Robot. Autom. Lett.* **2022**, *7*, 5679–5686. [CrossRef]
19. Sutton, R.S.; Barto, A.G. *Reinforcement Learning—An Introduction*, 2nd ed.; MIT Press: Cambridge, MA, USA, 2018; ISBN 9780262039246.
20. Berglund, D.; Larsson, N. Controlling a Hydraulic System Using Reinforcement Learning: Implementation and Validation of a DQN-Agent on a Hydraulic Multi-Chamber Cylinder System. Master's Thesis, Linköping University, Linköping, Sweden, 2021.
21. HOPSAN Multi-Domain System Simulation Tool, Division of Fluid and Mechatronic System, Department of Management and Engineering, Linköping University, Linköping, Sweden. Available online: https://liu.se/en/research/hopsan (accessed on 1 June 2022).
22. DQN Agent, Mathworks Reinforcement Learning Toolbox. Available online: https://mathworks.com/help/reinforcement-learning/ug/dqn-agents.html (accessed on 1 June 2022).

energies

Article

A Variable Pressure Multi-Pressure Rail System Design for Agricultural Applications

Xiaofan Guo *, Jacob Lengacher and Andrea Vacca *

Maha Fluid Power Research Center, Mechanical Engineering, Purdue University, West Lafayette, IN 47907, USA
* Correspondence: guo92@purdue.edu (X.G.); avacca@purdue.edu (A.V.); Tel.: +1-765-496-0154 (X.G.)

Abstract: This paper presents a solution for reducing energy loss in the hydraulic control system of agricultural tractors and their implements. The solution is referred to as a multi-pressure rail (MPR) and provides power to the hydraulic functions following a pressure control logic, as opposed to the traditional flow control logic typical of hydraulic systems used in off-road vehicles. The proposed hydraulic control system allows for elimination of redundant flow control valves in the state-of-the-art system, which cause excessive throttling losses leading to poor overall energy efficiency. Related work on MPR technology targets construction vehicles, where the MPR solution can allow energy recovery during overrunning loads and better engine management. This paper alternatively addresses the case of agricultural applications where functions mostly operate under resistive load conditions with slow dynamics, which offers an opportunity to target throttle losses. For this purpose, the paper introduces a variable pressure control strategy to handle the instantaneous pressure at each rail. To develop both the controller and the hydraulic system architecture, a stationary test rig is conceived and used to validate a numerical simulation model of the MPR system and its control strategy. Particular focus is given to the dynamic behavior of the system during the switches of a function between different pressure rails, which needs to ensure reduced oscillations of the flow provided to each hydraulic function. Once validated, the simulation model is used to predict the energy savings of the MPR solution in an actual application: a 435 hp hydraulic tractor powering a 16-row planter, for which operating features during typical drive cycles were available to the authors. The results show up to 59% total power reduction at the pump shaft, corresponding to 89.8% system efficiency gain.

Keywords: agricultural tractors and implements; high pressure control system; multi-pressure rail system; energy savings; throttle loss reduction

Citation: Guo, X.; Lengacher, J.; Vacca, A. A Variable Pressure Multi-Pressure Rail System Design for Agricultural Applications. *Energies* **2022**, *15*, 6173. https://doi.org/10.3390/en15176173

Academic Editors: Paolo Casoli and Massimo Rundo

Received: 20 July 2022
Accepted: 23 August 2022
Published: 25 August 2022

1. Introduction

In recent years, global warming and climate change have gained enormous attention and motivated many technological advancements. One key parameter is fossil fuel usage leading to greenhouse gas emission. In that regard, off-road mobile machineries, which include agriculture equipment, construction equipment, etc., play an important role [1]. However, the state-of-the-art hydraulic architectures for agricultural applications result in low system efficiency (20% on average [2]) and are unsuited for the efficiency requirements of the coming years. As a result, this study will focus on agricultural machine systems.

The typical usage of an agricultural tractor includes performing a certain function on a field, such as a planter, seeder, bailer, etc. Such implements have their own actuators which are typically powered by the high-pressure hydraulic system of the tractor through so-called hydraulic remote connections. The most common configuration of the overall hydraulic circuit can be represented with the simplified schematic of Figure 1 below. The tractor is equipped with a load-sensing (LS) hydraulic circuit, comprised of a variable displacement pump with a differential flow limiter (or LS regulator) and a load-sensing valve (referred to as an electro-hydraulic remote valve, EHR). Such a circuit is among the most energy efficient solutions for controlling multiple actuators [3]; the pump delivery pressure settles to the maximum actuator pressure (plus a pump margin p_{mar}), thus minimizing excessive system pressurization.

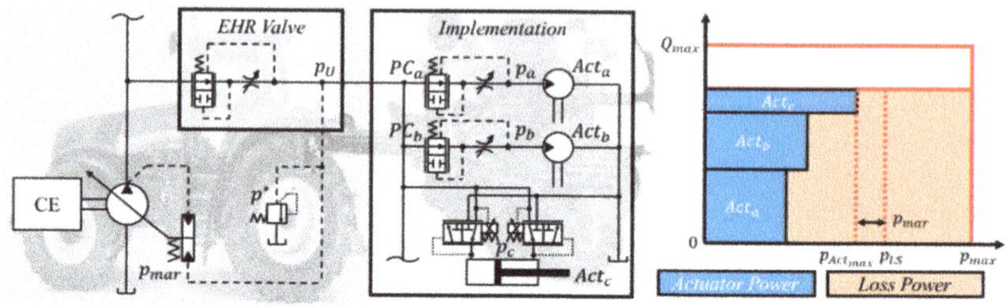

Figure 1. Hydraulic circuit for tractor and planter system with redundant flow control valve (**left**) and power loss (**Right**).

The implement hydraulic control system includes the actuators performing the agricultural functions and a local flow control valve at each actuator to regulate its velocity. Such an arrangement allows the placement of multiple actuators on a single remote line in the tractor, as shown in Figure 1 (left) and makes the control performance of the implement independent of the tractor brand or technology level.

Unfortunately, this arrangement voids the principle of operation of the LS system because the commanded flow rate at the remote valve is not satisfied by the LS pump, which instead must meet the demand of the implement valves. This is achieved by commanding the maximum flow at the EHR and letting the pressure compensators (PCs) of the implement reach the maximum pressure of the pump p^*. Therefore, the pump is in *pressure saturation* conditions no matter the actual load at the actuators. This typically leads to a very low energy efficiency of the hydraulic control system, as represented in the power plot of Figure 1 right, where the orange area represents energy loss due to throttling. If the system functioned as an LS system, the pump pressure would be at p_{LS}, the maximum actuator load $p_{Act_{max}}$, plus the pump margin p_{mar} to ensure correct functionality of the control valve, instead of p_{max}.

Due to the popularity of LS systems, the literature reports several methods for improve their energy efficiency. Bedotti et al. [4] attempted to reduce throttling losses by reducing the LS margin of the system. Reducing the margin risks compromising system performance, however, so ref. [4] settles on a safe margin reduction of 23%, yielding power savings of approximately 4%. Tian et al. [5] expanded on this and proposed a method of dynamically varying the LS margin of a pump, reducing throttling losses without risking system performance, but without addressing losses due to widely varied loads. Siebert et al. [6] put forward a system of reducing throttling losses by elevating the outlet pressure of actuators, thus lessening the losses while maintaining a constant pressure drop across the actuator. This solution saw a 44% reduction in throttling losses in a simulation. Another relevant technology is that of independent metering, wherein the inlet and outlet control orifice are decoupled and independently controlled [7], allowing for more flexible control and power optimization with precise dynamics [8].

None of the above solutions, however, solve the issue of the previously outlined control conflict among flow control valves, which is at the core of the inefficient tractor–implement operation due to pressure saturation of the supply pump. This pressure saturation could be avoided by sensing the load pressure at the actuator (p_a, p_b, or p_c, Figure 1 right), rather than at the EHR outlet (p_U). However, such a solution is normally not used because it does not achieve proper dynamic behavior (pump flow variation is not fast enough to cope with the actuator load variations), and it would make the tractor and planter technologies dependent on one another.

Therefore, to achieve higher energy efficiency in hydraulic control systems, alternative layout architectures should be considered. The literature usually reports on the displacement control actuation as one of the most efficient methods to control a function, as it eliminates throttling losses entirely [3,9]. However, this technology requires a dedicated flow supply (i.e., hydraulic pump) for each function, and therefore it is too costly and impractical for a tractor–implement system which must consider a multitude of functions. Successful application of displacement control technology was, however, possible in construction equipment [10–14]. A method for reducing the number of pumps was also proposed by Busquets et al. [15], where a control sharing protocol for the pump flow was designed based on typical excavator cycles, making it possible to control all excavator functions with four pumps. However, such a strategy is much more difficult to implement in agricultural systems as they can connect to implements with unknown drive cycles.

An alternative hydraulic actuation technology, not yet common in the field of off-road vehicles, are pressure-controlled flow supplies [3]. In such systems, the supply pump is not set to match a commanded flow, as in the system of Figure 1 (left), but rather to establish a commanded pressure. The simplest implementation of this design is the constant pressure (CP) system, in which the supply pump sets the pressure in a common pressure rail that connects the pump with the actuator, with control valves that control the flow from this rail [16]. The CP system decouples the actuators from the prime mover, allowing for power management strategies to be implemented at the prime mover. The CP system also achieves higher control bandwidth, as the actuator response usually functions on the control valve that is receiving flow from the rail and is independent of the supply pump. However, a basic CP system with one pressure rail suffers from excessive throttling losses, as the rail pressure must exceed the maximum load pressure, and flow from the rail to all other actuators at a lower pressure is throttled down. As described in [17], advanced CP systems address this problem in two possible ways: by controlling the displacement of the actuator directly, and thus the speed and torque, or by using hydraulic transformers to reduce the pressure delivered by the rail without throttling. The first strategy has been implemented successfully by Volvo [18] on a 30-ton excavator. This design utilized a novel variable-area cylinder and showed 34–50% gains in fuel efficiency. Several hydraulic transformer designs have been put forward, as seen in [19,20] but these systems have yet to see widespread adoption, due to the complexity, cost, and unproven reliability of the component.

The multi-pressure rail (MPR) system considered in this paper does not fall into either of the previously outlined categories for advanced CP systems, and instead proposes the use of more than one pressure rail. The idea was initially somewhat outlined by Lumkes et al. [21] who designed a general form of the MPR system and tested its feasibility on a backhoe test rig. The results proved functionality, but the work did not address specific controller design or power-saving potential. Dengler et al. in [22,23] proposed the addition of a fixed pressure medium pressure rail, calculating a theoretical increase of 20% efficiency over a state-of-the-art LS system in a simulation.

The most relevant prior work to the MPR system considered in this paper is the STEAM excavator developed by the IFAS team at RWTH Aachen, [24,25] and the hybrid hydraulic architecture (HHEA) by Li et al. [26]. Both systems used variations of MPR architectures for construction equipment. The STEAM excavator system makes extensive use of engine management strategies and energy recovered from overrunning loads where possible. Published measurements show a 30% reduction in fuel consumption. Although not demonstrated in an actual vehicle, the HHEA system conceptualizes the use of additional hydraulic and electric machines (one per each actuator) to essentially replicate the function of hydraulic transformers and reduce the throttle loss. This system should allow energy consumption to be reduced by 40% according to the simulation. Neither of these architectures vary the level of the instantaneous pressure in the rails, mostly due to the high dynamic features of the drive cycles typical of construction equipment. Opgenoorth et al. in [27] proposed an MPR architecture for an electrified excavator which included variable rail pressures, showing a 29% efficiency gain over a standard LS system in a simulation. However, no controller design or experimental validation were presented. Bertolin and Vacca in [28] provided a parametric study that evaluates how different cylinder areas affect the efficiency of the proposed system. It also considers different number of rails in the system and how to optimize their pressure levels to maximize system efficiency.

Finally, the authors of this paper previously put forward several options for the architecture and control of a variable pressure rail MPR system in [29]. This work used a physical test rig to prove the basic controllability of the system, and also to compare the relative effectiveness of one pump, two accumulators, and two pumps, no accumulator systems for such a system. The work was successful in identifying that while the variable pressure system was controllable with both supply systems, the two-pump system without accumulators had far better dynamic behaviors due to the absence of the additional capacitance from the accumulators.

The purpose of this paper is to formulate, design, and experimentally prove a variable pressure MPR system specific to an agricultural-implement system, which are characterized by a high number of actuators with low dynamic requirements and limited chances of energy recovery. The goal of the paper is to address the hydraulic system design as well as quantifying the energy savings that can be achieved considering an actual application.

This paper first presents the MPR system design considering the unique system characteristics in Section 2. In Section 3, the multi-level controller is introduced in detail. Following this, a test rig is designed, modeled, built, and tested to validate the stability and performance of the proposed MPR controller in Section 4. Moreover, Section 5 shows the two reference vehicles (a tractor and an implement) that are very popular in the field. Finally, a lumped parameter complete system model, including the reference tractor, planter, and the proposed MPR components, is built, validated, and used to predict the power saving and efficiency gain at the machine's representative working condition in Section 6. The results show up to 59% total power reduction at the pump shaft, corresponding to 89.8% system efficiency gain.

2. MPR System Description

Figure 2 shows the design of the MPR system with a generic number of pressure rails as considered in this work. It includes a supply system, pressure rails, pressure select and control valve sets (PSCV), and actuators. The supply system controls the pressure level in each rail. The rails connect the supply system to each PSCV. Then, the PSCVs connect the most appropriate rails to the inlet and outlet of each actuator to minimize throttling loss. The PSCVs also control the flow rate to match the commanded velocity.

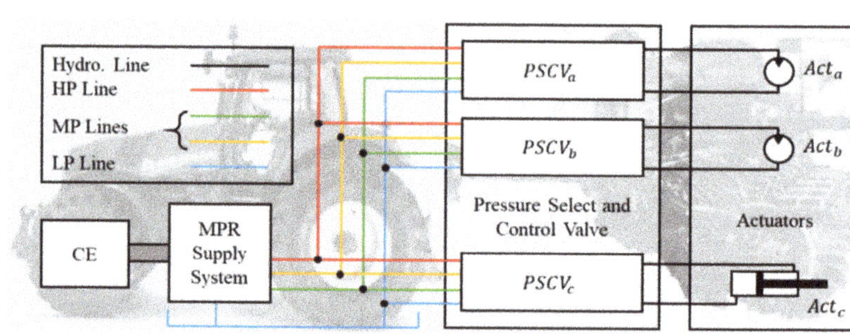

Figure 2. Generic design of the multi-pressure rail system with four pressure rails.

Figures 3 and 4 show an example of how the three-rail MPR system works for both hydraulic motors and cylinders with PSCV. First, the PSCV pressure selection stage (V1–V6) will be opened or closed to connect the line $A1$ and $B1$ to the pressure rails according to the command from the supervisory controller. Then, the metering stage control valve (VA and VB) will control the actuator speed tracking with the actuator controller. Using different combinations of inlet/outlet pressure levels, the MPR makes available to each actuator multiple output torque and force ranges. For example, considering an MPR system with three rails, a hydraulic motor can operate under six possible modes: high pressure at the inlet with low pressure at the outlet (HP-LP), HP-MP, MP-LP, LP-MP, LP-HP, and MP-HP. There are also three idle modes: LP-LP, MP-MP, and HP-HP. Due to the pressure losses in the lines and selection and metering valves, the pressure margin p_{mar} needs to be considered to leave enough pressure drop for proper functioning of the actuator. As a result, the torque range for each working mode is shown in Table 1.

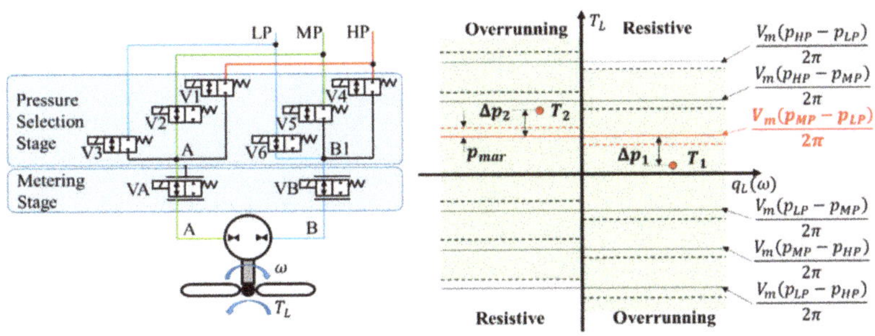

Figure 3. Working principle for MPR system with motor actuator.

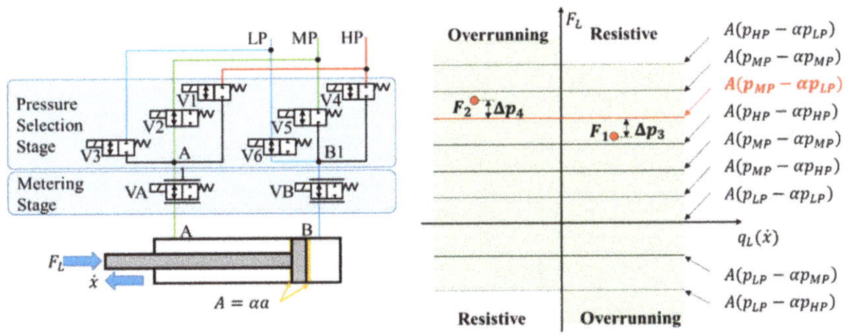

Figure 4. Working principle for MPR system with cylinder actuator.

Table 1. Working Mode Torque Range for Three-Rail MPR System.

Load Condition	Mode	Lower Limit (Nm)	Upper Limit (Nm)
Resistive	**MP-LP**	0	$\frac{V_m(p_{MP}-p_{LP}-p_{mar})}{2\pi}$
	HP-MP	$\frac{V_m(p_{MP}-p_{LP}-p_{mar})}{2\pi}$	$\frac{V_m(p_{HP}-p_{MP}-p_{mar})}{2\pi}$
	HP-LP	$\frac{V_m(p_{HP}-p_{MP}-p_{mar})}{2\pi}$	$\frac{V_m(p_{HP}-p_{LP}-p_{mar})}{2\pi}$
Overrunning	**LP-LP/MP-MP/HP-HP**	$\frac{V_m(p_{LP}-p_{MP}+p_{mar})}{2\pi}$	0
	LP-MP	$\frac{V_m(p_{MP}-p_{HP}+p_{mar})}{2\pi}$	$\frac{V_m(p_{LP}-p_{MP}+p_{mar})}{2\pi}$
	MP-HP	$\frac{V_m(p_{LP}-p_{HP}+p_{mar})}{2\pi}$	$\frac{V_m(p_{MP}-p_{HP}+p_{mar})}{2\pi}$
	LP-HP	Maximum System Torque	$\frac{V_m(p_{LP}-p_{HP}+p_{mar})}{2\pi}$

If the motor load is at the value T1 as shown in Figure 3, the system should work under mode MP-LP (the red line in the figure) to minimize the throttling loss to Δp_1. Clearly, this is a great advantage in comparison with the baseline system of Figure 1 (left) which puts all the actuators under the maximum pressure differential. In case of an overrunning load, such as T_2, the MPR has the potential to recover energy by connecting the motor inlet to LP and the outlet to MP. This allows the motor to work in pumping mode, thus inputting energy back to the system, under a minimal throttling loss Δp_2.

The hydraulic cylinder, illustrated in Figure 4 (pressure margin lines are omitted), works under a similar principle. The only difference is for cylinders with differential areas, the idle modes for hydraulic motors can also output force, such as force levels $A(p_{HP} - \alpha p_{HP})$, $A(p_{MP} - \alpha p_{MP})$, and $A(p_{LP} - \alpha p_{LP})$. As a result, there are nine force lines in total compared to the six torque lines of a symmetric actuator. When the cylinder has a resistive load, such as F_1 in Figure 4 (right), the best working mode is MP-LP (the red line in the figure) with the throttling loss Δp_3. With overrunning load F_2, the system prefers working in the same mode to recover energy to the MP rail with throttling loss Δp_4.

The case of unidirectional motors with no overrunning loads is further considered in this section to present general design considerations for the MPR system. This case is particularly relevant for agricultural tractor–implement systems, and in particular for the reference tractor–planter as it will be described in Section 4.

2.1. Number of Rails

The number of rails is a critical parameter in an MPR system, as it directly determines the number of modes available for the actuator connections. A two-rail system has only one mode, as each actuator can be connected only with the HP line and the LP line. In a three-pressure MPR, there are three modes: supply from medium pressure (MP) to LP rail (MP-LP), from HP to MP rail (HP-MP), and from HP to LP rail (HP-LP). The number of modes (n_{md}) is determined by the number of rails (n_R) as seen in Equation (1) with a unidirectional motor as the actuator.

$$n_{md} = \frac{n_R(n_R - 1)}{2} \quad (1)$$

Intuitively, the more rails there are in the system, the less throttling losses will occur. However, a high number of rails can be impractical, as it implies a high number of components (detrimental for costs and spatial requirements) and added control complexity. Therefore, the optimal number of rails comes from a compromise between energy loss and technical-economic feasibility. Considering all the benefits and costs of adding rails, the three-rail MPR system is the most suited choice for the reference system which will be introduced in Section 4. As a result, the MPR system design in this study will continue with three pressure rails. The simulation results that drove this decision can be seen in Section 6.

2.2. Supply System

Based on the analysis and experimental work in [29], to achieve the target of unidirectional, resistive load conditions present in most agricultural implements, the supply system shown in Figure 5 is most suitable due to its fast response in rail pressure control. This system uses one pump per pressure rail, allowing for a fast control response for both pressures independently. Most of the time this is the operating mode in which the system will run. Should a user require a higher flow than one pump can provide, however, a merging valve is included which connects both pumps to one rail. This setup allows for the pumps to be downsized somewhat, in turn causing them to typically run at higher displacements and efficiencies. Finally, the MP pump must be able to go over center in order to handle any cases in which flow coming from the HP rail to the MP rail exceeds the flow from the MP to LP rails.

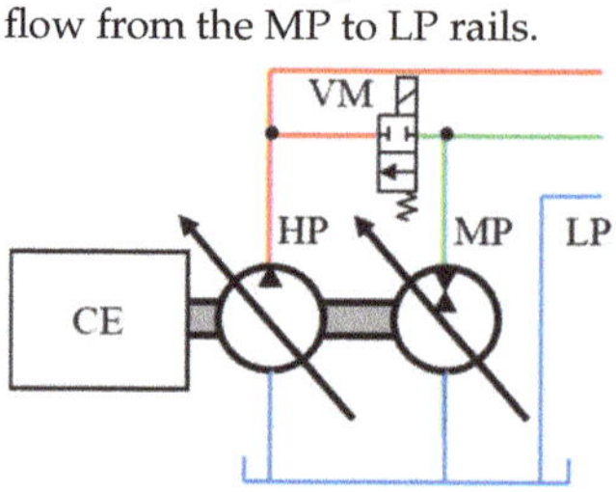

Figure 5. MPR supply system design.

To size the pumps, there are two options. For the first option, both pumps should be sized large enough to supply the full flow required by the users to eliminate any possibility of pump flow saturation for both rails. This sizing option will ensure the system always remains in normal MPR mode and no merging of flow is needed. However, the pumps will suffer from lower efficiency due to working at a lower displacement. For the second option, owing to the merging valve, the total flow from both pumps needs to satisfy the maximum required flow rate as shown in Equation (2). This provides design freedom on how to size HP and MP pumps and an optimization method is proposed in this work to determine the best size for both pumps.

$$V_{p_{HP}} + V_{p_{MP}} \geq \frac{q_{max}}{\omega_p} \tag{2}$$

To find the best size for both pumps, a cost function J is defined in Equation (3). This cost function evaluates only the average power loss during the reference machine's normal working condition, which represents the bulk of the machine's operating time throughout its lifetime. The first half of the cost function represents the summation of volumetric $q_{s_{p_k}}$ and hydromechanical losses $T_{s_{p_k}}$ for the pump k at corresponding working pressure $p_{pk} - p_{LP}$, and rotational speed ω_{p_k}. The second half calculates the throttling loss in the control valves for all n_{act} actuators at their flow rate q_j, and pressure drop across their control valve $p_{in} - \Delta p_i - p_{out}$. p_{in} and p_{out} are the inlet and outlet rail pressures that are selected by the pressure selection stage valve in Figure 4, and they are either HP, MP, or LP. The actuator load pressure Δp_i is the pressure drop across the actuator suction side and deliver side. With the cost Function (3) and Constrain (2), the best pump size that could satisfy the flow requirements and minimize the system loss could be found for the selected target machine.

$$J = \sum_{k=1}^{2} \left(q_{s_{pk}} \left(p_{pk} - p_{LP} \right) + T_{spk}\omega_{p_k} \right) + \sum_{j=1}^{n_{act}} q_j \left(p_{in} - \Delta p_i - p_{out} \right) \tag{3}$$

Due to limited resources, this study will use the same size of units as the reference machine (45 cc), which follows the first sizing option, and all the controller design and simulation results are with respect to that size.

2.3. Pressure Select and Control Valve Set

There are many ways to achieve these two functions. The most generic design is shown in Figure 6 (left). It includes six on/off valves (V1–V6) as the pressure select valves and two proportional valves (VA and VB) as the actuator control valves. Two pressure relief valves and check valve sets are used to protect the circuit from overpressure or cavitation. In this setup, the actuator can operate in both directions (A to B, or B to A) in both resistive and overrunning loading conditions, as both the inlet and return of the actuator could be connected to any rail. Moreover, both metering-in and metering-out control strategies are possible. This design is very flexible but also expensive, as it uses many solenoid-operated valves.

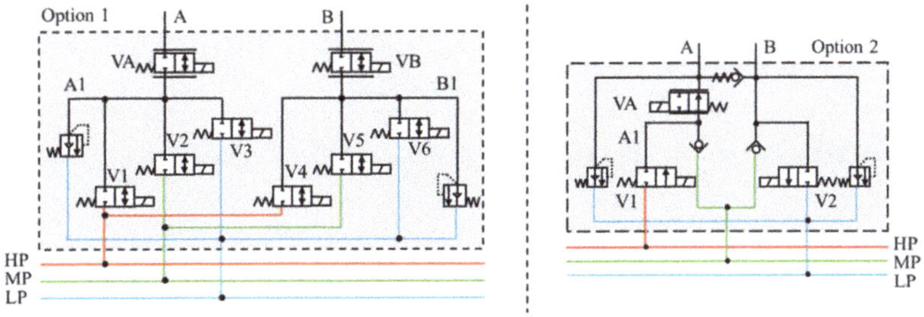

Figure 6. MPR pressure select and control valve set designs.

To better fit the valve set design to the actuator, the load characteristic (mentioned at the end of the state-of-the-art section) should be considered. Based on these characteristics, the architecture of Figure 6 (right) would be a more suitable design for this study. It only has two on/off valves (V1 and V2) with two check valves at the pressure select stage. The inlet side can connect to HP and MP pressure rails and the outlet can connect to MP and LP pressure rails. On top of that, it has only one metering valve (VA) on the A side for meter-in control. An anti-cavitation check valve is placed between the actuator work ports and two pressure relief and check valves are used to protect the circuit. This design can no longer achieve all working conditions as option 1 can, but is adequate for the previously discussed unidirectional actuators, without overrunning loads, and greatly reduces the number of solenoids, and thus cost and control complexity.

3. MPR Controller Design

The control structure for the proposed MPR system has two levels: the actuator level controllers and the supervisory level controller. The actuator level controllers include the actuator speed controller, the rail pressure controller, and the pump displacement controller. The supervisory controller is the online optimization controller that sets the HP and MP rail pressure to minimize total power loss due to fluid throttling.

The basic control structure is shown in Figure 7. The mechanical and hydraulic components of the tractor and its implement are in red and orange boxes, and the controllers are in green boxes. When the actuator command comes in, the supervisory controller records it and passes it on to the actuator speed controller. Following this, the actuator speed controller controls the metering valve of the actuator to track the actuator command. In the meantime, the actuator load pressure is measured and sent back to the supervisory controller, along with the actuator command, for use in the rail pressure online optimization. Next, the supervisory controller sends the optimized HP and MP rail pressure commands and the optimal working mode for each actuator to the rail pressure controller and the actuator selection valves. In addition, a feedforward displacement command is generated from the actuator speed and rail pressure commands and is then sent to the pump displacement controller. The rail pressure controller causes the rail pressure to follow the optimized command by sending the pressure feedback displacement command to the pump displacement controller. Finally, the pump displacement controller maneuvers the displacement of the HP and MP pumps.

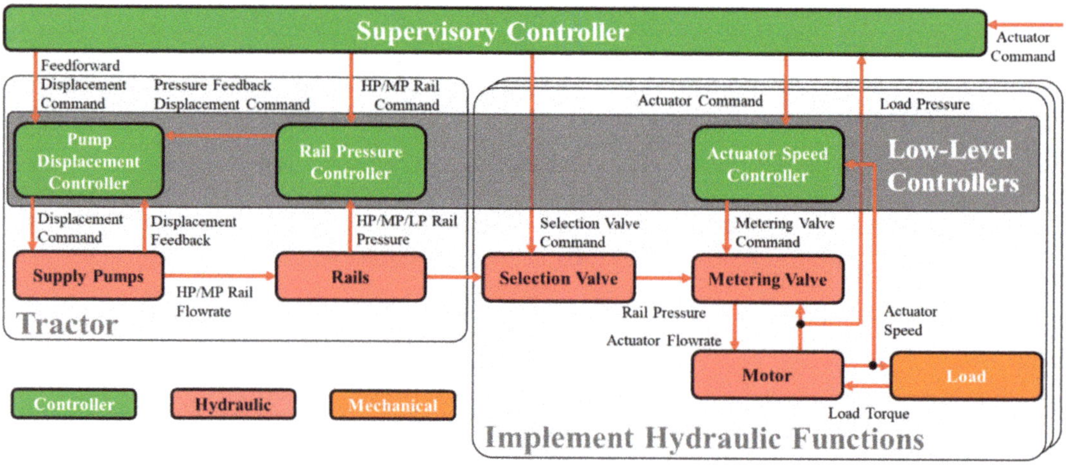

Figure 7. MPR control structure overview.

The detailed controller design is shown in the following sections.

3.1. Supervisory Controller

The supervisory controller will give the best HP and MP rail pressures and the corresponding mode for each actuator to minimize the throttling loss in the MPR system.

$$p_{HP} = \max(\Delta p_1, \Delta p_2, \ldots, \Delta p_n) + p_{mar} \tag{4}$$

To determine the HP rail pressure Equation (4) is used. The HP rail pressure is equal to the highest load pressure in the system plus a pressure margin of p_{mar} to overcome any losses in the line and the losses associated with the pressure drop necessary for the control valve to properly function.

As the HP rail pressure is easily determined, the only independent variable is the MP rail pressure (p_{MP}). To this end, an optimization method is used based on the cost function defined in Equation (5). This variable is constrained by Equations (7) and (8) which will be introduced later to maintain the pressure margin for proper actuator functionality. The three summations in the cost function represent the throttling loss for each actuator at all three modes. For example, when actuator i is working in mode MP-LP, its throttling loss is the product between its flow rate q_i and the pressure drops across the control valve, $p_{MP} - \Delta p_i - p_{LP}$. The last part of the cost function accounts for the power loss of the HP and MP supply pumps.

$$\begin{aligned} P(p_{MP}) = & \sum_{i=1}^{MP-LP} q_i \,(p_{MP} - \Delta p_i - p_{LP}) + \sum_{j=1}^{HP-MP} q_j \,(p_{HP} - \Delta p_j - p_{MP}) \\ & + \sum_{k=1}^{HP-LP} q_k \,(p_{HP} - \Delta p_k - p_{LP}) + q_{p_{HP}}(p_{HP} - p_{LP})(1 - \eta_{p_{HP}}) \\ & + q_{p_{MP}}(p_{MP} - p_{LP})(1 - \eta_{p_{MP}}) \end{aligned} \tag{5}$$

This cost function is discontinuous. When actuators work in a different combination of modes, Equation (5) has to be rearranged with different constraints for p_{MP}, which will lead to different local optimal choices of p_{MP} for difference combinations. In order to find the global minimum of this cost function, all the possible combinations must be evaluated. With a system that has n_R working rails and n_{act} actuators, the total number of possible working mode combinations n_{comb} is calculated in Equation (6).

$$n_{comb} = \left(\frac{n_R(n_R - 1)}{2} \right)^{n_{act}} \tag{6}$$

For a five-actuator system working with a three-pressure rail system, the n_{comb} is 243, which means the cost function needs to be evaluated 243 times at every instant to insure a global minimum for the cost function. This makes the optimization quite impractical for an online implementation. An incomplete list of possible combinations is shown in Table 2.

Table 2. Incomplete list of possible combinations for three-rail five-actuator MPR system.

Combi Nation	Actuators					Combi Nation	Actuators				
	1	2	3	4	5		1	2	3	4	5
1	MP-LP	MP-LP	MP-LP	MP-LP	MP-LP	82	HP-MP	MP-LP	MP-LP	MP-LP	MP-LP
2	MP-LP	MP-LP	MP-LP	MP-LP	HP-MP	83	HP-MP	MP-LP	MP-LP	MP-LP	HP-MP
3	MP-LP	MP-LP	MP-LP	MP-LP	HP-LP	84	HP-MP	MP-LP	MP-LP	MP-LP	HP-LP
4	MP-LP	MP-LP	MP-LP	HP-MP	MP-LP	85	HP-MP	MP-LP	MP-LP	HP-MP	MP-LP
...						...					

Fortunately, there are ways to reduce the list of combinations to be considered. First, organize actuators 1–5 with respect to their load pressure in decreasing order into actuators a–e. Following this, apply the rules below.

Rule 1: an actuator a should always work in mode HP-LP. After organizing the actuators, it is important to note that a-e is not necessarily in the same order as 1–5. From the organized list of load pressures, it is clear that actuator a, as the highest load pressure actuator, should always be working in mode HP-LP because the HP rail pressure is set based on this pressure. As a result, 162 combinations can be eliminated.

Rule 2: when an actuator moves to a lower mode, the actuators with a lower load pressure should never move to a higher mode. This is because there will always be a better combination to take its place. Figure 8 graphically describes this concept. In this setup, assuming the minimum pressure drop is guaranteed, the best combination is 229 (HP-LP, HP-LP, HP-MP, HP-MP, and MP-LP) with the optimized HP and MP pressure. The throttling loss for actuators d and e are $\Delta p_{d_{HP-MP}}$ and $\Delta p_{e_{MP-LP}}$, respectively. If actuators d or e were working in mode HP-LP instead of their current mode, this would result in combinations 231 (HP-LP, HP-LP, HP-MP, HP-MP, and HP-LP), 232 (HP-LP, HP-LP, HP-MP, HP-LP, and MP-LP), or 234 (HP-LP, HP-LP, HP-MP, HP-LP, and HP-LP) It is clear that from the plot, none of those combinations will result in lower throttling loss than combination 231, as the throttling losses of $\Delta p_{d_{HP-LP}}$ and $\Delta p_{e_{HP-LP}}$ are greater than $\Delta p_{d_{HP-MP}}$ and $\Delta p_{e_{MP-LP}}$. As a result, an additional 60 combinations are eliminated.

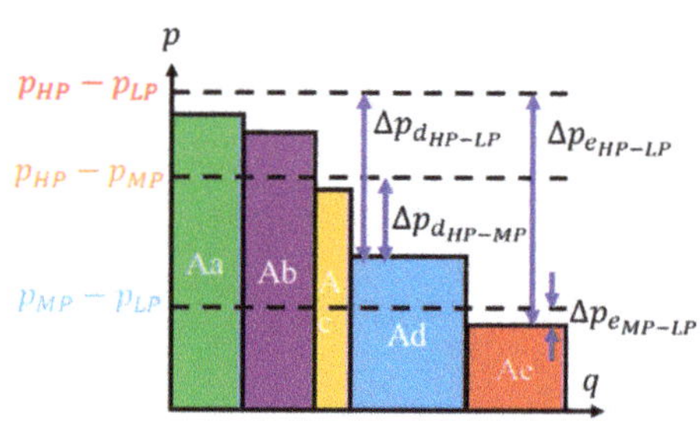

Figure 8. Pressure and flow rate diagram for MPR actuators after sorting.

Rule 3: at least one actuator should work in mode MP-LP. This rule is based on the recirculated excess flow in the MP rail. If no actuator is working in mode MP-LP, such as combination 203 (HP-LP, HP-MP, HP-MP, HP-MP, and HP-MP), there is no actuator taking flow from the MP rail, and all the flow coming from actuators working in mode HP-MP will need to be recirculated through the MP pump back to the shaft; this process generates losses. Therefore, it is certain that if the actuators working in mode HP-MP switch to mode MP-LP with correct MP pressure, such as combination 163 (HP-LP, MP-LP, MP-LP, MP-LP, and MP-LP), it will result in less losses for the system, as the pump loss due to recirculation is eliminated. As a result, five more combinations are eliminated.

To summarize, after applying the above rules, the final list of 16 available combinations is listed in Table 3.

Table 3. List of final combinations.

Combi Nation	Actuators					Combi Nation	Actuators				
	a	b	c	d	e		a	b	c	d	e
163	HP-LP	MP-LP	MP-LP	MP-LP	MP-LP	218	HP-LP	HP-LP	MP-LP	MP-LP	HP-MP
164	HP-LP	MP-LP	MP-LP	MP-LP	HP-MP	221	HP-LP	HP-LP	MP-LP	HP-MP	HP-MP
167	HP-LP	MP-LP	MP-LP	HP-MP	HP-MP	226	HP-LP	HP-LP	HP-MP	MP-LP	MP-LP
176	HP-LP	MP-LP	HP-MP	HP-MP	HP-MP	229	HP-LP	HP-LP	HP-MP	HP-MP	MP-LP
190	HP-LP	HP-MP	MP-LP	MP-LP	MP-LP	235	HP-LP	HP-LP	HP-LP	MP-LP	MP-LP
199	HP-LP	HP-MP	HP-MP	MP-LP	MP-LP	236	HP-LP	HP-LP	HP-LP	MP-LP	HP-MP
202	HP-LP	HP-MP	HP-MP	HP-MP	MP-LP	238	HP-LP	HP-LP	HP-LP	HP-MP	MP-LP
217	HP-LP	HP-LP	MP-LP	MP-LP	MP-LP	241	HP-LP	HP-LP	HP-LP	HP-LP	MP-LP

For each combination, the MP rail pressure must fall within the upper and lower constraints to ensure the combination works properly. The lower constraint is determined by Equation (7). The total pressure difference from the MP rail to the LP rail needs to be higher than the highest load working in mode MP-LP plus the pressure margin to ensure proper functioning of the control valve. The upper constraint is determined by Equation (8). The total pressure difference from the HP rail to the MP rail needs to be higher than the highest load pressure that is working in mode HP-MP plus the pressure margin. If the constraints for one combination are incompatible, then the combination is eliminated for that calculation. After rearranging the constraints, as shown in Equations (7) and (8), Table 4 shows the complete list of all lower and upper constraints for all combinations.

$$p_{MP} - p_{LP} \geq \max\left(\Delta p_{i_{MP-LP}}\right) + p_{mar} => p_{MP} \geq \max\left(\Delta p_{i_{MP-LP}}\right) + p_{mar} + p_{LP} \quad (7)$$

$$p_{HP} - p_{MP} \geq \max\left(\Delta p_{i_{HP-MP}}\right) + p_{mar} => p_{MP} \leq -\max\left(\Delta p_{i_{HP-MP}}\right) + p_{HP} - p_{mar} \quad (8)$$

Table 4. List of constrains for the final 16 combinations.

Combi Nation	Constraint *		Combi Nation	Constraint *		Combi Nation	Constraint *		Combi Nation	Constraint *	
	Lower	Upper		Lower	Upper		Lower	Upper		Lower	Upper
163	Δp_b	inf	190	Δp_c	$-\Delta p_b$	218	Δp_c	$-\Delta p_e$	235	Δp_d	inf
164	Δp_b	$-\Delta p_e$	199	Δp_d	$-\Delta p_b$	221	Δp_c	$-\Delta p_d$	236	Δp_d	$-\Delta p_e$
167	Δp_b	$-\Delta p_d$	202	Δp_e	$-\Delta p_b$	226	Δp_d	$-\Delta p_c$	238	Δp_e	$-\Delta p_d$
176	Δp_b	$-\Delta p_c$	217	Δp_c	inf	229	Δp_e	$-\Delta p_c$	241	Δp_e	inf

* All lower constraints need to add $p_{mar} + p_{LP}$, and all upper constraints need to add $p_{HP} - p_{mar}$.

Finally, with a closer look at the cost function (5), the value of the cost function is a linear function with the independent variable p_{MP}. This means if the derivative of the cost function with respect to p_{MP} is positive, the minimum p_{MP} should be used to achieve the lowest value of the cost function, which is at the lower constraint of p_{MP}, and vice versa. This further simplifies the optimization to just one evaluation of the sign of the derivative with respect to p_{MP} and one evaluation of the cost function at the corresponding constraint for each combination.

3.2. Real-Time Implementation of Supervisory Controller

The supervisory control logic shown above is operated at a constant frequency on a central ECU, checking continuously to ensure that the system stays in the optimal operating condition. The speed at which the supervisory controller runs can be far lower than that at which the actuator level controls run, 10 Hz and 100 Hz, respectively, for example. This real-time implementation does, however, introduce two control challenges that need to be addressed to ensure the system can operate stably.

First, the control logic proposed above assumed that the system reached a steady state immediately after switching modes and stayed there until the next switch. However, during real drive cycles, small fluctuations of actuator pressures may lead to switches between optimal modes and cause excessive switching between combinations if the value of the cost function for multiple combinations get very close. This could potentially make the system unstable. To overcome this issue, a power loss margin P_{mar} is added to the cost function comparison. The supervisory controller will allow a switch only if the new combination is outperforming the current combination in cost by P_{mar}. There is a possibility that the current combination will not work in the new time step. For example, the current combination at the new time step may have contradictory p_{MP} upper and lower constraints, which lead to no possible p_{MP} existing for the current combination. In this case, to ensure that the system functions correctly, the supervisory controller will force a switch to the new combination, regardless of P_{mar}. This way, switches will happen much less frequently at the cost to savings of P_{mar}, which is small compared to the total savings, and the benefits of system stability.

Second, during transient condition, such as multi-actuator switching, the system pressure experiences a discontinuity that will impose actuator load pressure oscillation for a short time. An example is shown in Figure 9. A simple step change is demonstrated in a single actuator load (top), along with a sample of four of the cost functions (middle), and the actuator's connection mode (bottom). The simulation used accounts for other actuators and cost functions, but only the relevant ones are shown. When the load change occurs, there are transient oscillations in the system that cause corresponding oscillations in the cost functions. The magnitude of these oscillations is instantaneously greater than P_{mar}, and if left unchecked, could lead to uncontrolled switching as seen in the figure. To resolve this, a brief cooldown time, t_{cd}, was added to the controller, forcing a minimum time (two seconds for the reference controller) between switches. As with the P_{mar} logic, an exception was implemented allowing the system to switch modes immediately if its current operating condition is unable to satisfy the pressure margin constraints.

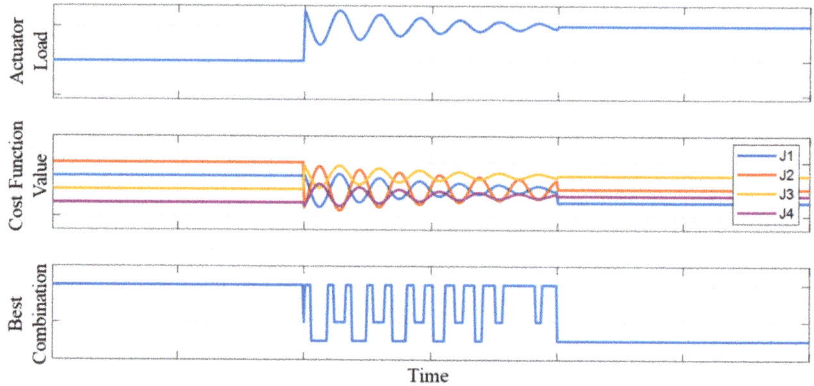

Figure 9. Conceptual example of actuator mode switching dynamic reaction.

3.3. Actuator Level Controllers

The actuator controller is a simple closed-loop PI (proportional integral) controller for actuator speed (Figure 10). The actuator command comes from the supervisory controller and the actuator speed is fed back from a speed sensor to close the control loop.

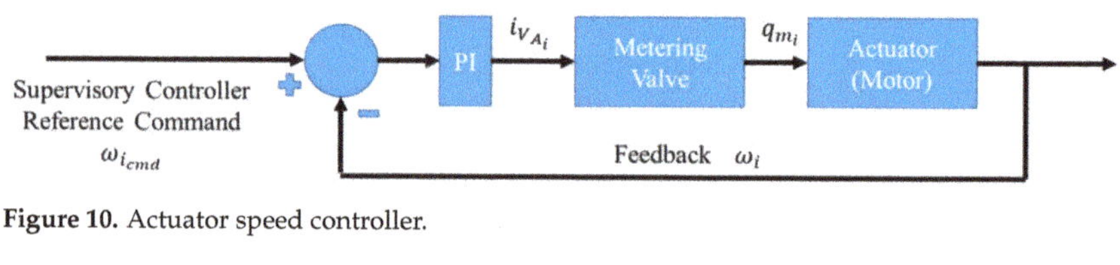

Figure 10. Actuator speed controller.

The rail pressure controller and pump displacement controller form a cascaded PI controller (Figure 11). The outer loop is the rail pressure controller that takes rail pressure commands from the supervisory controller and outputs the reference pump displacement command to the inner loop. The rail pressure measurements are used to close the control loop. The inner loop is a feedback–feedforward controller. In addition to the command from the outer loop, the feedforward pump displacement command is sent from the supervisory controller, which is a displacement estimation from the actuator modes and flow needed. The measured pump displacement is used to close the control loop.

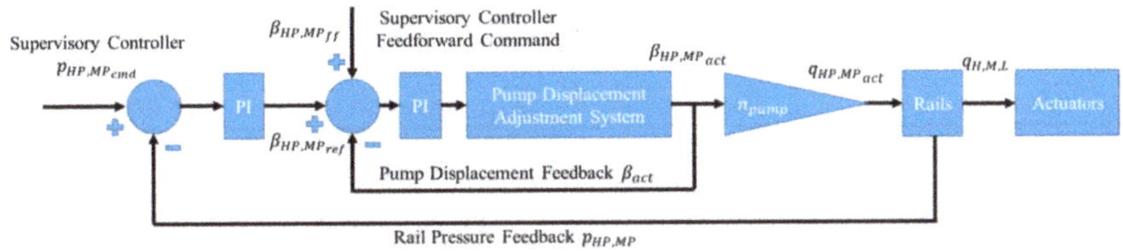

Figure 11. Rail pressure and pump displacement controller.

4. MPR Test Rig and Reference Machine

4.1. MPR Test Rig Design

A standalone test rig was designed to validate the effectiveness of the control scheme being used. The system was designed as shown in Figure 12. Key features in the design were variable load units, allowing the simulation of a wide variety of loading conditions, an unloading valve, enabling the switch between one and two pump supply systems, and a switching valve and accumulator system, enabling the function of the one pump system. The system also used the more general design for the valve set outlined above. This was done to ensure flexibility in future work with the system.

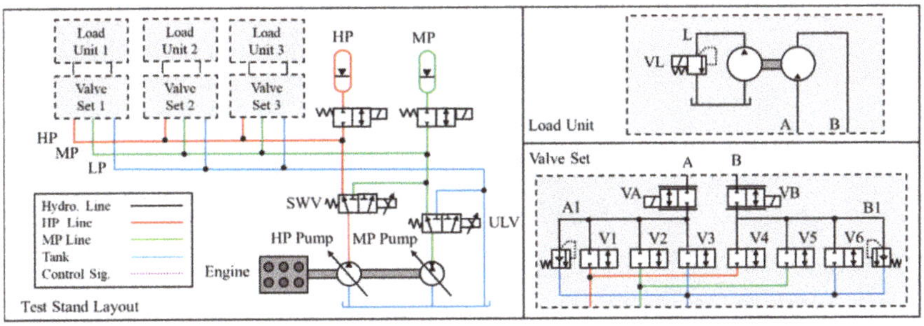

Figure 12. Hydraulic circuit for the MPR test rig.

The test rig design shown above was designed to be able to operate as both a one pump and two pump MPR system, by using the unloading valve (ULV) to isolate the MP pump and using the switching valve (SWV) to switch the HP pump between the rails, as it was initially designed for use on [29]. For the purposes of this paper, however, the system is used in the two-pump configuration.

The system is controlled using an NI CRio controller, which in turn uses Bosch Rexroth's BODAS pump controllers and pressure transducers to control the hydraulic components of the system. Figure 13 shows the completed test stand implementation.

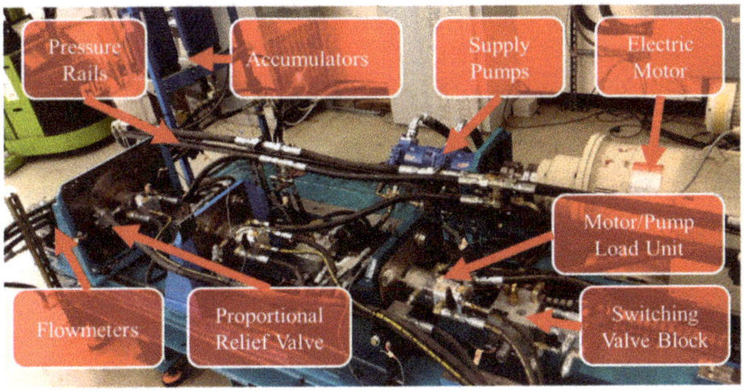

Figure 13. Completed test stand.

4.2. Reference Machine

To show the power savings introduced by the proposed MPR system and controller design, two reference machines (a tractor and an implement) are selected and modeled. The reference machines have been instrumented and tested in the field under normal working conditions to gather baseline data for building and validating the hydraulic model. Finally, the actual power consumption by the hydraulic system on the reference machines for the baseline machine and the proposed MPR machine are compared.

The reference tractor is a T8.390 New Holland tractor. The tractor and its simplified hydraulic circuit are shown in Figure 14. It has two sets of centralized load-sensing hydraulic systems. The first system has the first HP pump as a flow source and steering system, using the power beyond function and remote valves one to three as users. The flow first goes through the steering priority valve to satisfy the steering needs, then continues to supply the rest of the functions. The second system is powered by the second HP pump with the hitch function and EHR valves four to six. The implement is connected to the hydraulic system through the tractor EHR. An LP pump is used to power all low-pressure functions, such as the brake, clutch, differential lock, etc. The LP system, which is also a low power system, is not included in this study.

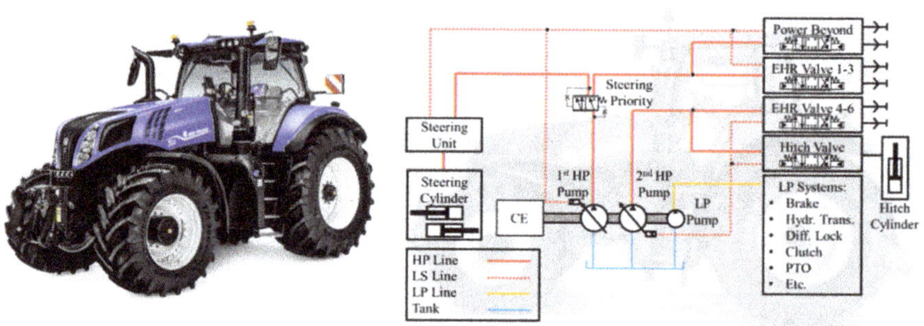

Figure 14. Reference tractor and its hydraulic circuit.

The reference implement is a Case Early Riser 2150 16-row planter. The planter and its simplified hydraulic circuit are shown in Figure 15. The planter has two hydraulic circuits. Circuit one contains a bulk fill fan system, fertilizer system, and compressor system. Circuit two contains an alternator system, vacuum system, weight management cylinder system (WMC), and hydraulic down pressure cylinder system (HDPC). These functions, except the compressor system, are continuously working during normal planter working conditions. In addition to the motor actuators, the two-cylinder systems also appear in the system. The cylinder systems follow the terrain and maintain a commanded downward force to ensure planting quality.

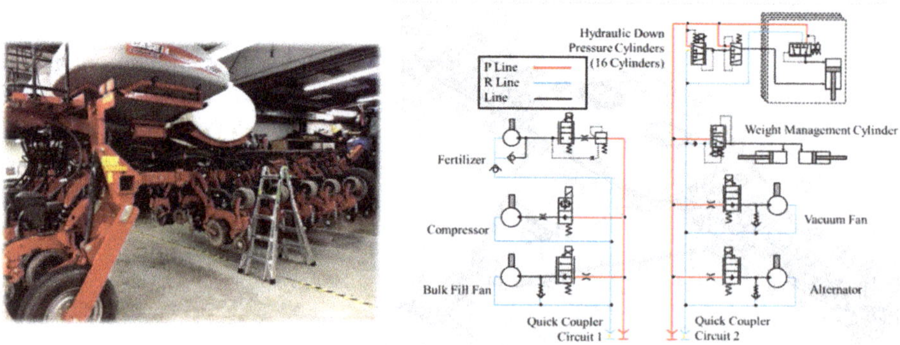

Figure 15. Reference planter and its hydraulic circuit.

5. System Modeling Methodology

This section presents the lumped parameter models used to estimate the effectiveness of the MPR solution in terms of controllability and energy savings. Specifically, the test rig model is used to validate the proposed MPR control performance. The tractor–planter model is instead used to predict the total power saving in realistic working conditions.

5.1. Test Rig Model

For easier understanding of the test rig modeling, the system is decomposed into individual components that could be modeled with a governing equation, such as hydraulic motor, pumps, valves, lines, etc. The exchange of information between the blocks is indicated in the block diagram in Figure 16. The inputs for the simulation are the actuator commanded speeds, commanded loads, and electric motor speed. They are directly fed into the MPR controller through the simulation. All other variables such as pressure, flow rate, torque, speed, and power are calculated internally in the simulation.

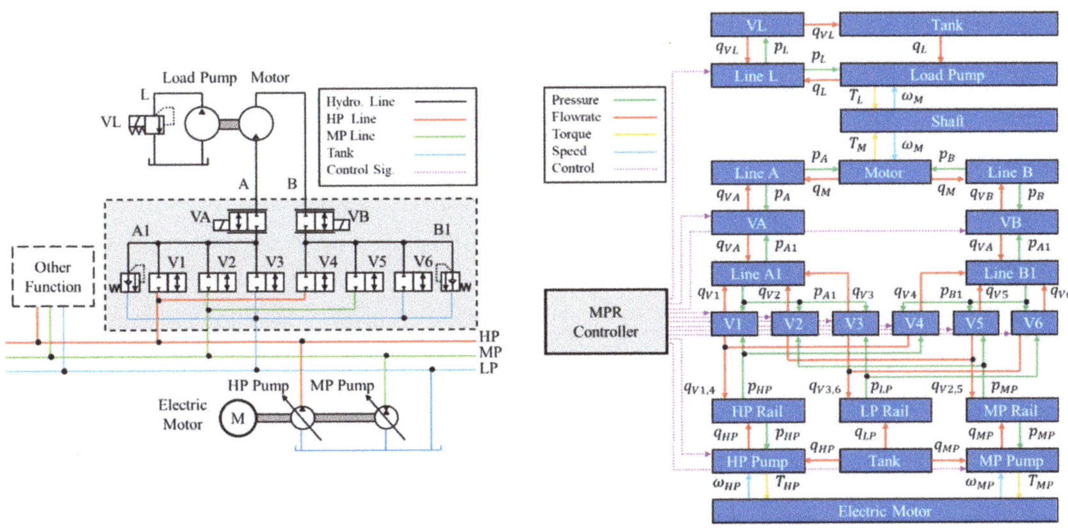

Figure 16. Block diagram for test rig modeling (left: test rig hydraulic circuit, right: modeling block diagram).

For simplicity, the electric motor in this model is assumed to be an ideal rotational speed source. The motor shaft speed ω_m is simulated using Equation (9). I_{shaft} is motor shaft inertia. T_L is the load torque on the motor shaft. T_m are the torque outputs by the motor.

$$I_{shaft}\dot{\omega_m} = T_m - T_L \tag{9}$$

Following this, the hydraulic unit (HP and MP pumps, motor, and load pump), effective flowrate q_{eff}, and torque T_{eff} are modeled using Equations (10) and (11). V is the unit maximum displacement. The volumetric loss q_s and hydromechanical loss T_s are determined using empirical lookup tables. The losses are related to unit fractional displacement β, pressure drop across the unit Δp, and unit speed ω. A more detailed description on the setup of the hydraulic unit model can be found in [30].

$$q_{eff} = \pm\beta V\omega - q_s(\beta, \Delta p, \omega) \tag{10}$$

$$T_{eff} = \pm\frac{\beta V \Delta p}{2\pi} - T_s(\beta, \Delta p, \omega) \tag{11}$$

Next, the pressure dynamics in each control volume (HP and MP rail, line A1, B1, A, B, and L) are modeled by the pressure build up Equation (12), where the K is the bulk modulus, Vol is the volume, $q_{in/out}$ is the flowrate that flows into or out of the control volume, and the $\frac{dV}{dt}$ is the control volume rate of change.

$$\dot{p} = \frac{K}{Vol}\left(q_{in} - q_{out} - \frac{dV}{dt}\right) \tag{12}$$

Finally, the flowrate q_V through valves (V1–6, A, B, and L) are modeled using the orifice Equation (13). The maximum orifice area A for each valve is different and it is calculated from the valve datasheet. The normalized opening of the valve y_V depends on the type of valve. For a proportional valve, the control input, which is normalized between 0 to 1, serves as the valve opening. For check valves and electrical on/off valves, the opening is either 0 or 1 depending on the working logic. For valves with hydraulic pilot pressure, a force balance between solenoid force, hydraulic force, and spring force is applied to calculate the valve opening.

$$q_V = \alpha_D A y_V \sqrt{\frac{2\Delta p}{\rho}} \tag{13}$$

A summary of the components and parameters used in this simulation model is shown in Table 5 below.

Table 5. Components and parameters used for test rig modeling.

Oil Density	850 kg/m^3
Discharge Coef.	0.7
Bulk Modulus	1.7e9 Pa
El. Motor Speed	1500 rpm
Supply Pumps	Bosch Rexroth A10VO $(EOC - P)$, max. displacement 45 cc/rev
Load Pump	Bosch Rexroth AZPV, 8 cc/rev, custom made
Load Motor	Bosch Rexroth AZMV, 8 cc/rev, custom made
Load Inertia	$0.001 \cdot kg \cdot m^2$
On/Off Valve	Bosch Rexroth $VEI - 16 - 10A - NC$ (5 bar Δp for 30 lpm q)
Proportional Valve	Bosch Rexroth KSVSR2 (14 bar Δp for 75 lpm q)
Load Valve	Bosch Rexroth VEP-5B-2S-P (max. cracking pressure 220 bar, flow rate pressure gradient 7.5 L/min/bar

5.2. Complete System Model

The complete tractor–planter model was built in the Simcenter Amesim environment. The model includes the tractor model and the planter model, in both the commercial solution and the proposed MPR component configuration. The simplified hydraulic circuit is shown in Figure 17. It is a 3-rail MPR system with 5 actuators. The 5 actuators are alternator motor, vacuum fan motor, bulk fill fan motor, fertilizer motor, and the cylinder systems.

The models for the tractor and planter were built following the same lumped parameter approach as the test rig model. Previous studies conducted by this team have presented the model development and its validation for the commercial tractor system [31,32] and tractor–implement system [33]. The baseline simulation results that will be presented in the next section are the straight output of these previously presented models. The relevant portions of the tractor model (engine, pumps, and control valves) and of the validated planter model (motors, cylinders, control valves, and corresponding load) will serve as the baseline for comparison with the MPR supply system and the MPR working actuator, respectively.

The test rig model (HP and MP rails, PSCV, and MPR controllers) will serve as the MPR component models which transfer power in between the tractor and planter models.

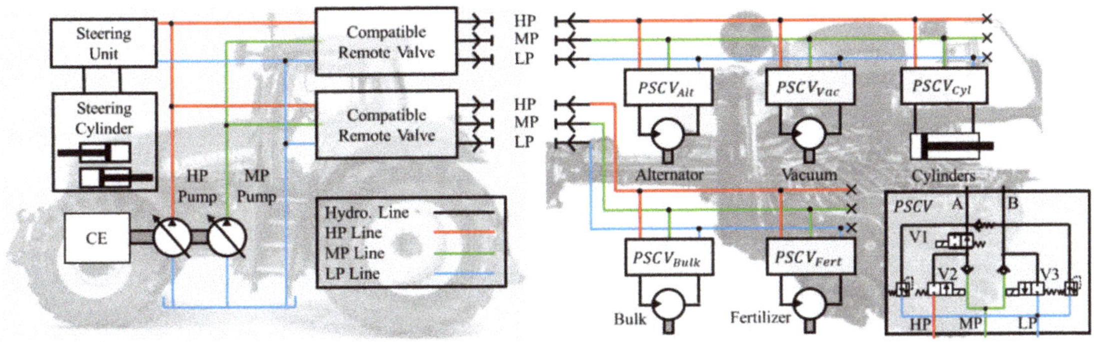

Figure 17. Simplified hydraulic circuit for the MPR system for the reference machines.

One important note regarding the model used is to consider the downforce cylinder system of the planter. An accurate modeling of the cylinder system presents complexity in determining the correct force inputs from the ground, as well as including a proper dynamic model for the bodies connected to the cylinder. To simplify this portion of the model, and ensure a realistic simulation, a simplified load module was built to recreate the cylinder flow/pressure dynamic using experimental data from real measurements as shown in Figure 18. This module will enforce the flow rate needed from the cylinder systems (which was measured from experiments) and use the same pilot-operated, electric-controlled, 3/2 proportional pressure-reducing valves to track the measured working pressure (also measured from experiments) in the cylinders (CH). Therefore, this load module consumes the flow rate and outputs the controlled cylinder pressure as measured from the experiments.

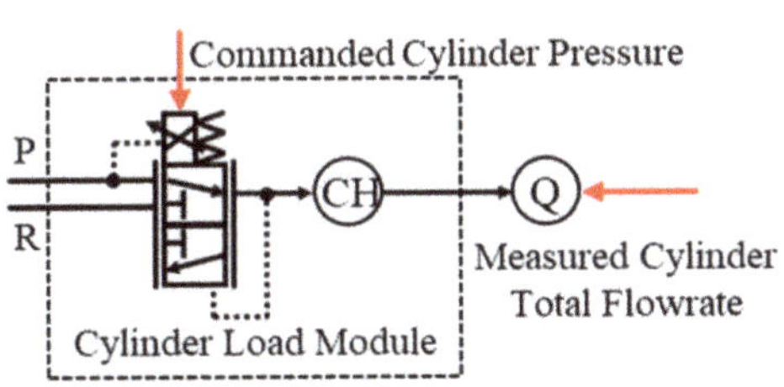

Figure 18. Cylinder systems model diagram.

6. Results and Discussion

This section firstly presents the simulation used to select a three-rail system, then discusses the validation of the test rig model, the planter model, and the power consumption prediction of the complete MPR tractor–implement model compared to the baseline solution.

6.1. Rail Number Analysis

A representative drive cycle was acquired for the selected reference machine, seen in Table 6, which displays typical usage. It was run through a streamlined version of the model discussed above, with most transient dynamics removed to focus on steady state performance.

Table 6. Representative drive cycle for the rail number analysis.

Normalized Cmd	Time	Alternator	Bulk Fill Fan	Fertilizer	Vacuum Fan	Cylinders
Normal	5–38%	60%	60%	~40%	~63%	Dynamic

Averaged over time, the simulation results are shown in Figure 19, with each additional rail improving loss reduction; by far the most remarkable improvement occurred following the addition of two to three pressure rails. These results motivated the selection of a three-rail system for this application, as the relatively minor reduction in losses for rails above three was judged not to be worth the additional control complexity.

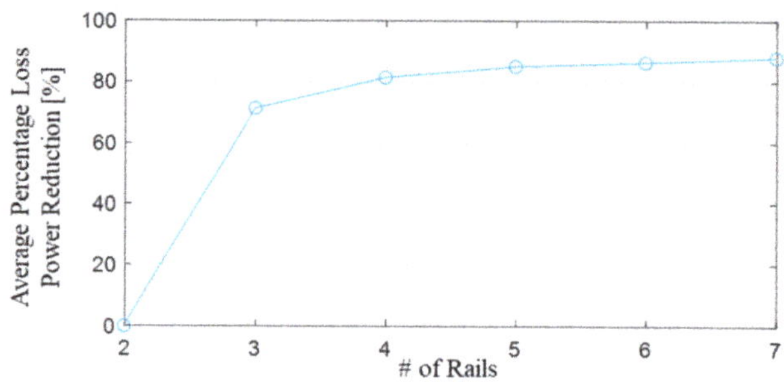

Figure 19. Loss reduction for number of rails.

6.2. Test Rig Model Validation Result

A drive cycle was generated for the test rig using the output of the supervisory controller to select loading conditions and command speeds for the three actuators to force the system into a sequence of mode switches. These switches were selected specifically to test the controllability and stability of the system. This drive cycle is outlined in Table 7.

Table 7. Validation drive cycle for MPR test rig.

Time	(A) 0–10 s		(B) 10–30 s		(C) 30–50 s		(D) 50–70 s	
	Command	Mode	Command	Mode	Command	Mode	Command	Mode
$\omega 1_{cmd}$ [rpm]	1200	3	1200	3	1200	3	600	2
$p1_L$ [bar]	160		160		160		20	
$\omega 2_{cmd}$ [rpm]	800	1	800	1	800	3	800	1
$p2_L$ [bar]	20		100		140		100	
$\omega 3_{cmd}$ [rpm]	600	1	600	2	600	1	1200	3
$p3_L$ [bar]	20		20		20		160	

The results from the test rig model are shown in Figure 20. The first row shows the switch from working condition B to C (started at 30 s) in the drive cycle of Table 7. The second row shows the last switch from working condition C to D (started at 50 s), which is the hardest switch. The plot on the left shows the actuators speed tracking, the middle plot shows the actuators load pressure and corresponding optimized rail pressures, and the right plot shows the actuators operating mode as determined by the supervisory controller.

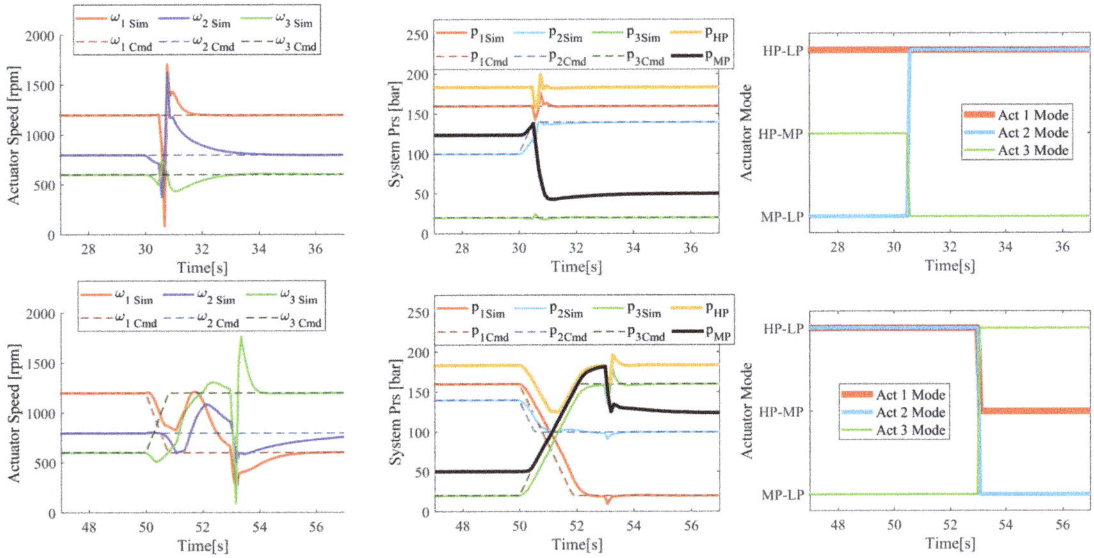

Figure 20. Test rig simulation results. Top row: mild switch; bottom row: hard switch.

Overall, the system behaves as expected and both the actuator level controller and supervisory controller perform correctly. The steady state command tracking is excellent, and the system is stable with a minimum number of mode switches. From the results, a speed peak for all actuators at the moment of the mode switch can be noticed. This occurs because when the switch happens, the local actuator speed controller, not being informed of the change in supply pressure, must react from its previous state. As a result, the sudden pressure change will create a flow spike and change the pressure drop across the motor. This aspect is accentuated in the test rig actuators because they have an extremely low mechanical inertia, being simply a pump directly connected to a hydraulic motor. Therefore, a short torque spike is reflected in a corresponding speed peak. The second reason is because the on/off valves have different opening and closing times. The on/off valves on the test rig have a closing time approximately three times longer than the opening time. When a switch happens, there will be a very short time during which the two rails are directly connected, causing the higher pressure rail flow to rush to the lower pressure rail, resulting in flow loss to the actuators and a subsequent drop followed by an overshoot in speed. A more sophisticated actuator-level control method could help alleviate some of these issues. The corresponding systems on the reference machine, however, are largely fan drives with very large inertias which will be much more robust against fluctuations in supply flow and pressure; as a result, the current level of control complexity is suitable for progressing to on-machine implementation.

The simulation drive cycle was then run on the physical test stand. The results were formatted using the simulation results and can be seen in Figure 21. The test stand results show good agreement with the simulated results, with the same speed peaks occurring due to low inertia and opening timing issues. One other clear discrepancy occurs during the difficult switch (bottom right panel from 50–54 s), due to factors not accounted for in the simulation. While the simulation shows the switch happening cleanly, in reality, additional dynamics result in the switch being divided into two separate switches for the test stand, with the first switch occurring immediately, and the second bringing the system to its optimal condition after the dynamic has settled out. This effect demonstrates the effectiveness of the anti-chatter logic, without which the system would have had numerous undesirable switches during this period.

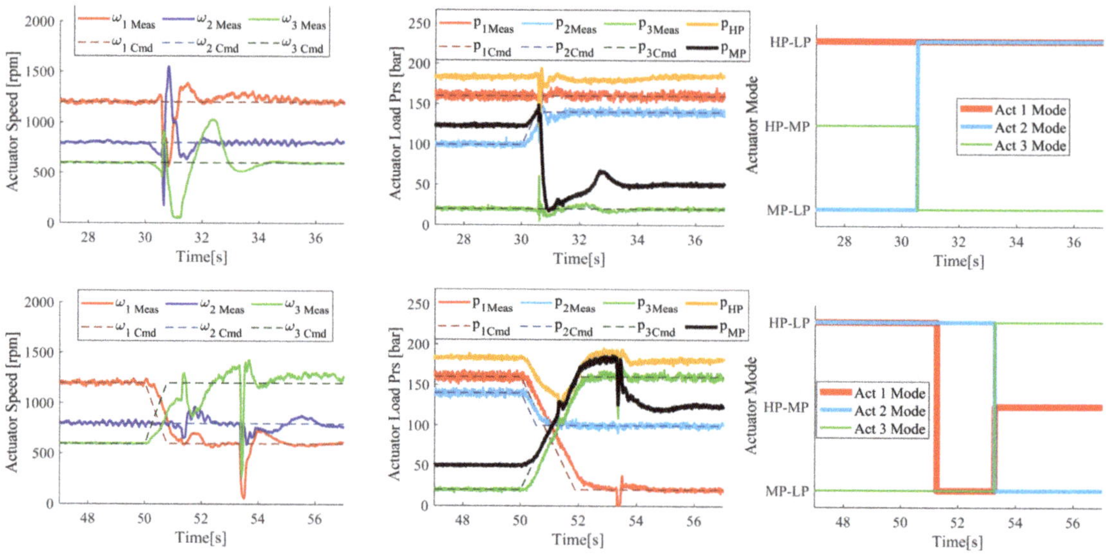

Figure 21. Test rig measurement results.

6.3. *Complete System Performance and Power Consumption Prediction Result*

This section presents the simulation result for the proposed MPR system applied to the tractor and the planter. All controllers are running at a fixed speed of 10 Hz. First, the drive cycle will be introduced. Then, system tracking and mode switching results will be presented to demonstrate system functionality. Finally, the pump shaft power consumption comparison will be introduced to show the power saving and efficiency gains.

The drive cycle includes two working conditions: normal and high-speed conditions. The normal and high-speed conditions are representative of two commonly used operating modes for this implement, according to the manufacturer, and are run back-to-back for the simulation. In reality, they are not run together, but this is a simplification only to show the operating conditions of both systems, and to provide a command switch to display transient behavior. According to the machine manufacturer, most users operate the planter in normal working conditions when planting. Therefore, power saving is most important in normal conditions. The drive cycle is shown in Table 8. For confidentiality, all parameters are normalized. This drive cycle was executed in the field during the baseline tests. The command for the alternator and the bulk fill fan is a constant motor speed. Fertilizer and vacuum fan systems do not have a direct motor speed command, so the measured flow rate of these systems is used to back-calculate the actual motor speed as the modeled system speed command. The cylinder system uses measured pressure values and flow rates from actual system operation for the load module to recreate the load condition, as discussed in Section 5.

Table 8. Representative drive cycle for the complete system model.

Normalized Cmd	Time	Alternator	Bulk Fill Fan	Fertilizer	Vacuum Fan	Cylinders
Normal	5–38%	60%	60%	~40%	~63%	Dynamic
High-Speed	60–90%	80%	80%	~60%	~70%	Dynamic

Figure 22 shows the results from the proposed three-rail MPR system for the reference machines. The vacuum and cylinder system has been chosen to show the system performance and controller activity. The left plot shows the results for the vacuum fan and the right plot shows the results for the cylinder systems from the same simulation. The top plot shows the actuator command tracking (left y-axis) and system mode (right y-axis), while the bottom plot shows the actuator load pressure with optimized rail pressures.

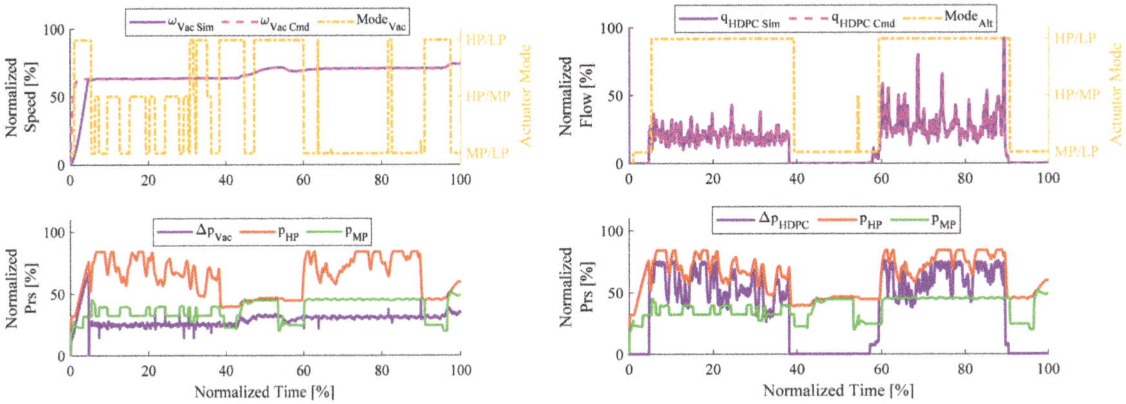

Figure 22. Complete system model result on MPR system performance.

From the results, all sub-systems track their command closely, as demonstrated by the vacuum. At mode switching, very little speed disturbance could be observed in the motor actuators. The tracking results prove that the proposed MPR system achieves at least the same level of performance as the baseline systems. In addition, from the bottom plot, the HP rail pressure closely follows the highest actuator pressure (which is the cylinder system pressure), the MP rail pressure is continuously adjusting with respect to the actuators load pressures, and the corresponding mode for each actuator is switching to ensure the power loss due to throttling is minimized. This shows that the supervisory controller works as expected to minimize system power loss.

In this drive cycle, when active, the cylinder system is always the highest load actuator, dominating the HP rail pressure. The high dynamic pressure variation leads to rapid mode switching in other actuators. This is more severe in normal working conditions. At high-speed working conditions, the cylinder system load is less dynamic, and the system becomes more stable as fewer switches occur. The designed controller is robust enough to ensure all systems stay stable and maintain good tracking, despite this fast dynamic.

Finally, Figure 23 shows the pump shaft power consumption comparison between the baseline and the MPR system. There is no doubt that the MPR system generates great power savings compared to the baseline simulation. The MPR system demonstrates that some of the mode switches reduce the system power. This is due to the pressure transient in chambers A_1 and B (Figure 6 right) at the moment of actuator mode switch. When it happens, for example when MP-LP switches to HP-LP, the pressure difference between the HP rail and the inlet chamber (which is currently at MP) creates a very short but large flow surge to charge the chamber pressure to the HP pressure. This flow peak, although it is short, shows its effect on the power consumption.

Table 9 summarizes the normalized total power, the efficiency for each simulation, and the corresponding power reduction and system efficiency gain as a percentage. As stated, the MPR system results in great power saving and efficiency gain.

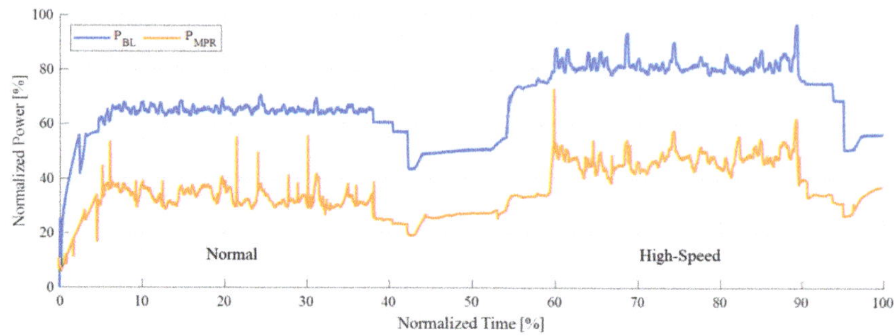

Figure 23. Complete system model result on power comparison between baseline, option one and option two.

Table 9. Complete system model result on average power saving and efficiency gain for baseline and MPR system.

	Baseline		MPR System	
	Normal	**High-Speed**	**Normal**	**High-Speed**
Total Power [%]	65.41	81.90	34.46	47.27
Reduction [%]	-	-	47.32	42.28
Efficiency [%]	20.19	27.89	38.32	48.32
Efficiency Gain [%]	-	-	89.80	73.25

7. Conclusions

This study applied the multi-pressure rail (MPR) hydraulic system to agricultural applications, such as tractors and towed implements, which normally run at long steady state conditions with unidirectional hydraulic motors as the main actuator. The detailed system design and controller designs were presented considering the application characteristic. The proposed MPR system could minimize the throttling loss at the actuator control valves by strategically choosing the pressures in HP and MP rails. This system represents a great alternative to the traditional centralized load sensing system for off-road agricultural machines.

The paper presented a complete MPR system design, including the number of rails, supply system, and pressure select and control valve set (PSCV). The optimal number of rails was first determined to be three, based on the tradeoff between power saving, system cost, and complexity for the chosen machine. The supply system was then selected to have two pumps to supply each rail independently. This design allowed for the elimination of the accumulators on the rails and enabled fast and accurate rail pressure control to help lower the throttling loss. Following this, two different PSCV designs were presented. A multi-level controller was designed to control the proposed MPR system. The supervisory controller controls the rail pressure for both rails and the actuator operation mode to minimize throttling loss, while the actuator controllers ensure the stability of the subsytems and command tracking. A detailed design was presented. One important note is that the proposed supervisory controller design follows the first sizing option for the supply system, meaning neither pump flow saturates. This condition holds true with the pump sizing and the actuator flow requirements of the reference vehicles. However, if the rails could suffer from flow saturation, i.e., insufficient pump flow, then the supervisory controller case elimination process and case selection logic would need to be modified, which is a great potential direction for future development.

To prove the stability and command tracking performance of the proposed MPR system, a test rig was designed to test the system design. After the test rig was designed, a lumped parameter model was built and a drive cycle that tested several critical mode switchings was constructed to predict the behavior of the test rig and aid in controller tuning. The test rig was then built and ran with the same drive cycle to validate the controller design. The results showed that the proposed MPR design and controller are stable, with exceptional performance during both steady state and transient conditions.

Finally, to predict the power saving and efficiency gain of the proposed MPR system compared to the baseline system, a high-fidelity lumped parameter model was built for the complete system, including the reference tractor, planter, and the proposed MPR system components. The model was fully validated, and a measured baseline drive cycle was used as a command input for each actuator in the complete system model. The results show that the proposed MPR system is not only able to provide the same level of performance, but also provides huge power savings and efficiency gains over the baseline system. For the most frequent normal working conditions, the MPR system achieved up to 59.4% total power reduction at the pump shaft, and 89.8% system efficiency gains.

Author Contributions: Conceptualization, X.G. and A.V.; methodology, X.G. and A.V.; software, X.G. and J.L.; validation, X.G. and J.L.; formal analysis, X.G.; experimental investigation, X.G. and J.L.; resources, A.V.; data curation, X.G. and J.L.; writing—original draft preparation, X.G. and J.L.; writing—review and editing, A.V.; visualization, X.G. and J.L.; supervision, A.V.; project administration, A.V.; funding acquisition, A.V. All authors have read and agreed to the published version of the manuscript.

Funding: This research was funded by the U.S. Department of Energy (DOE project DE-EE0009201, 'A New Approach for Increasing Efficiency of Agricultural Tractors and Implements').

Institutional Review Board Statement: Not applicable.

Informed Consent Statement: Not applicable.

Data Availability Statement: Not applicable.

Acknowledgments: Special thanks to Gary Kassen, Stefano Fiorati and Kena Shah at Case New Holland Industrial for the use and technical guidance on the reference machines. The authors would also like to acknowledge Enrique Busquets at Bosch Rexroth for making available hydraulic components and their operating features which helped both the experimental and simulation activities.

Conflicts of Interest: The authors declare no conflict of interest.

Nomenclature

Act	Actuator
EHR	Electro-Hydraulic Remote
HDPC	Hydraulic Down Pressure Cylinder
HP	High Pressure (Rail)
K	Actuator Load Pressure Order
LP	Low Pressure (Rail)
MP	Medium Pressure (Rail)
MPR	Multi-Pressure Rail
PSCV	Pressure Select and Control Valve Set
Vol	Volume
V	Valve
WMC	Weight Management Cylinder
A	Area
F	Force
J	Cost Function Value
K	Bulk Modulus
n	Number
P	Power
p	Pressure
Δp	Difference Pressure

q	Flowrate
T	Torque
V	Maximum Hydraulic Unit Displacement
x	Hydraulic Cylinder Position
y	Normalized Valve Opening
α	Hydraulic Cylinder Area Ratio
α_D	Discharge Coefficient
β	Hydraulic Unit Partial Displacement
η	Efficiency
ρ	Fluid Density
ω	Rotational Speed
Subscripts	
1, 2, 3, 4, 5	Actuator Label before Sorting
a, b, c, d, e	Actuator Label after Sorting
$comb$	Combination
cmd	Command
ff	Feedforward control
fb	Feedback control
i, j, k	Counter
in	PSCV inlet
L	Load
LS	Load Sensing
mar	Margin
md	Mode
p/m	Pump/Motor
out	PSCV outlet
R	Rail
Ref	Reference Command
s	Loss
U	Current System Load Sensing Pickup Location

References

1. Kean, A.J.; Sawyer, R.F.; Harley, R.A. A Fuel-Based Assessment of Off-Road Diesel Engine Emissions. *J. Air Waste Manag. Assoc.* **2000**, *50*, 1929–1939. [CrossRef] [PubMed]
2. Love, L.J.; Lanke, E.; Alles, P. *Estimating the Impact (Energy, Emissions and Economics) of the US Fluid Power Industry*; Oak Ridge National Laboratory: Oak Ridge, TN, USA, 2012.
3. Vacca, A.; Franzoni, G. *Hydraulic Fluid Power: Fundamentals, Applications, and Circuit Design*; Wiley: Hoboken, NJ, USA, 2021.
4. Bedotti, A.; Campanini, F.; Pastori, M.; Riccò, L.; Casoli, P. Energy saving solutions for a hydraulic excavator. *Energy Procedia* **2017**, *126*, 1099–1106. [CrossRef]
5. Tian, X.; Stump, P.; Vacca, A.; Fiorati, S.; Pintore, F. Power-Saving Solutions for Pre-Conpensated Load-Sensing Systems on Mobile Machines. *Trans. ASABE* **2021**, *64*, 1435–1448. [CrossRef]
6. Siebert, J.; Wydra, M.; Geimer, M. Efficiency Improved Load Sensing System—Reduction of System Inherent Pressure Losses. *Energies* **2017**, *10*, 941. [CrossRef]
7. Eriksson, B.; Palmberg, J.-O. Individual metering fluid power systems: Challenges and opportunities. *Proc. Inst. Mech. Eng. Part I J. Syst. Control. Eng.* **2011**, *225*, 196–211. [CrossRef]
8. Lyu, Z.; Chen, Z.; Yao, B. Energy Saving Motion Control of Independent Metering Valves and Pump Combined Hydraulic System. *IEEE ASME Trans. Mechatron.* **2019**, *24*, 1909–1920. [CrossRef]
9. Murrenhoff, H.; Sgro, S.; Vukovic, M. An overview of energy saving architectures for mobile applications. In Proceedings of the 9th International Fluid Power Conference, Aachen, Germany, 24–26 March 2014.
10. Williamson, C.; Zimmerman, J.; Ivantysynova, M. Efficiency Study of an Excavator Hydraulic System Based on Displacement-Controlled Actuators. In Proceedings of the ASME/BATH Symposium on Fluid Power and Motion Control, Bath, UK, 14–16 September 2008.
11. Zimmerman, J. Toward Optimal Multi-Actuator Displacement Controlled Mobile Hydraulic Systems, West Lafayette. Ph.D. Thesis, Purdue University, West Lafayette, IN, USA, 2012.
12. Daher, N.; Ivantysynova, M. Yaw stability control of articulated frame off-highway vehicles via displacement controlled steer-by-wire. *Control Eng. Pract.* **2015**, *45*, 46–53. [CrossRef]

13. Zhang, S.; Li, S.; Minav, T. Control and Performance Analysis of Variable Speed Pump-Controlled Asymmetric Cylinder System uder Four-Quadrant Operation. *Actuators* **2020**, *9*, 123. [CrossRef]
14. Fassbender, D.; Brach, C.; Minav, T. Using Displacement Control for Single Cylinders on an Electric Mobile Machine—Improved Efficiency Versus Increased Component Costs. In Proceedings of the 13th International Fluid Power Conference, 13. IFK, Aachen, Germany, 13–15 June 2022.
15. Busquets, E. Advanced Control Algorithms for Compact and Highly Efficient Displacement-Controlled Multi-Actuator and Hydraulic Hybrid Systems, West Lafayette. Ph.D. Thesis, Purdue University, West Lafayette, IN, USA, 2016.
16. Murrenhoff, H. *Regelung von Verstellbaren Verdrängereinheiten am Konstant-Drucknetz*; RWTH: Aschen, Germany, 1983.
17. Dreher, T. The Capability of Hydraulic Constant Pressure System with a Focus on Mobile Machines. In Proceeding of the 6th FPNI-PhD Symp., West Lafayette, IN, USA, 15–19 June 2010.
18. Heybroek, K.; Sahlman, M. A hydraulic hybrid excavator based on multi-chamber cylinders and secondary control—Design and experimental validation. *Int. J. Fluid Power* **2018**, *19*, 91–105. [CrossRef]
19. Vael, G.E.M.; Achten, P.A.J.; Fu, Z. *The Innas Hydraulic Transformer the Key to the Hydrostatic Common Pressure Rail*; International: Warrendale, PA, USA, 2000. [CrossRef]
20. Shen, W.; Huang, H.; Pang, Y.; Su, X. Review of the Energy Saving Hydraulic System Based on Common Pressure Rail. *IEEE Access* **2017**, *5*, 655–669. [CrossRef]
21. Lumkes, J.; Andruch, J. Hydraulic Circuit for Reconfigurable and Efficient Fluid Power Systems. In Proceedings of the 12th Scandinavian International Conference on Fluid Power, Tampere, Finland, 18–20 May 2011.
22. Dengler, P.; Groh, J.; Geimer, M. Valve control concepts in a constant pressure system with an intermediate pressure line. In Proceedings of the 21st International Conference on Hydraulics and Pneumatics, Ostrava, Czech Republic, 1–3 June 2011.
23. Dengler, P.; Geimer, M.; Baum, H.; Schuster, G.; Wessing, C. Efficiency improvement of a constant pressure system using an intermediate pressure line. In Proceedings of the 8th International Fluid Power Conference, Dresden, Germany, 26–28 March 2012.
24. Vukovic, M.; Leifeld, R.; Murrenhoff, H. STEAM—A Hydraulic Hybrid Architecture for Excavators. In Proceedings of the 10th International Fluid Power Conference, Dresden, Germany, 8–10 March 2016.
25. Vukovic, M.; Leifeld, R.; Murrenhoff, H. Reducing Fuel Consumption in Hydraulic Excavators—A Comprehensive Analysis. *Energies* **2017**, *10*, 687. [CrossRef]
26. Siefert, J.; Li, P.Y. Optimal Control and Energy-Saving Analysis of Common Pressure Rail Architectures: HHEA and STEAM. In Proceedings of the BATH/ASME 2020 Symposium on Fluid Power and Motion Control, Virtual, 9–11 September 2020.
27. Opgenoorth, A.; Hass, C.; Frischkorn, K.; Schmitz, K. Hydraulische Mehrdrucksysteme für mobile Arbeitsmaschinen mit elektrischen Antrieben. In Proceedings of the Hybride und Energieeffiziente Antriebe für Mobile Arbeitsmaschinen: 8. Fachtagung, Karlsruhe, Germany, 23 February 2021.
28. Bertolin, M.; Vacca, A. A Parametric Study on Architectures Using Common-Pressure Rail Systems and Multi-Chamber Cylinders. In Proceedings of the IEEE Global Fluid Power Society PhD Symposium, Napoli, Italy, 12–14 October 2022.
29. Guo, X.; Madau, R.; Lengacher, J.J.; Vacca, A.; Cardoso, R. Multi-Pressure Rail System Design with Variable Pressure Control Strategy. In Proceedings of the 13th International Fluid Power Conference, 13. IFK, Aachen, Germany, 13–15 June 2022.
30. Guo, X.; Vacca, A. Advanced Design and Optimal Sizing of Hydrostatic Transmission Systems. *Actuators* **2021**, *10*, 243. [CrossRef]
31. Tian, X.; Gomez, J.C.; Vacca, A.; Fiorati, S.; Pintore, F. Analysis of Power Distribution in the Hydraulic Remote System of Agricultural Tractors Through Modelling and Simulations. In Proceedings of the ASME/BATH 2019 Symposium on Fluid Power and Motion Control, Longboat Key, FL, USA, 7–9 October 2019.
32. Tian, X.; Vacca, A.; Fiorati, S.; Pintore, F. An Analysis of the Energy Consumption in the High-Pressure System of an Agricultural Tractor through Modeling and Experiment. In Proceedings of the 77th International Conference on Agricultural Engineering, Hannover, Germany, 9 November 2019; pp. 9–18. [CrossRef]
33. Tian, X.; Guo, X.; Stump, P.; Vacca, A.; Fiorati, S.; Pintore, F. New Hydraulic Control Technologies for Improving the Energy Efficiency of the Hydraulic System of Agricultural Tractors and Their Implements. In Proceedings of the IEEE Global Fluid Power Society PhD Symposium, Napoli, Italy, 12–14 October 2022.

energies

MDPI

Article

A Concept of Risk Prioritization in FMEA of Fluid Power Components

Joanna Fabis-Domagala † and Mariusz Domagala *,†

Faculty of Mechanical Engineering, Cracow University of Technology, Al. Jana Pawla II 37, 31-864 Cracow, Poland
* Correspondence: domagala@mech.pk.edu.pl
† These authors contributed equally to this work.

Abstract: FMEA is a widely used tool for decades and is also used as an industrial standard. However, there are two main drawbacks of this analysis that have been specified from the beginning. The first one is risk prioritization, which is expressed by a risk priority number (RPN). The RPN is a product of three factors with equal weight: severity (S), occurrence (O), and detection (D), which may produce equal risk priority for different combinations of S, O, and D. The second is the uncertainties caused by converting linguistic terms into quantitative data. The essential data used in the FMEA strongly depend on subjective experts' opinions, knowledge, and experience. For decades, various attempts of overcoming these weaknesses have been made, not only by academics but also by industry. The Automotive Industry Action Group (AIAG) and Verband der Automobilindustrie (VDA) have created an FMEA handbook that defines action priority (AP) depending on the combination of severity, occurrence, and detection numbers. This study presents an alternative to risk prioritization in FMEA based on failures of the tasks which analyzed systems perform. The fundamental factors S, O, and D have been redefined in a way to minimize uncertainties. The proposed method has been implemented in the flow control valve and can be easily applied in mechanical engineering applications.

Keywords: FMEA; fluid power; flow control valve; risk prioritization

Citation: Fabis-Domagala, J.; Domagala, M. A Concept of Risk Prioritization in FMEA of Fluid Power Components. *Energies* **2022**, *15*, 6180. https://doi.org/10.3390/en15176180

Academic Editors: Paolo Casoli and Massimo Rundo

Received: 28 July 2022
Accepted: 20 August 2022
Published: 25 August 2022

1. Introduction

Fluid power systems are not the latest achievement in engineering but are still widely used in industrial drive systems. In some cases, they cannot be replaced by other systems due to their unique features that are unachievable by other systems. Wide application, mainly in the highly demanding applications in which safety has the highest priority, requires improving reliability. Failures of fluid power components have a complex nature due to the interaction between high-pressure fluid with solid and chemical agent contaminants and structural parts. The mentioned reasons and the relative elements' motion make the fluid power component's likelihood of failure high. The research on failures and reliability of fluid power systems or their components is made with the use of various tools and methods. Ref. [1] presents work on monitoring and modeling gradual failure on a typical fluid power system. Y. Lee et al. [2] have investigated a failure of the hydraulic system which led to a fire on the wind turbine. Refs. [3–5] deal with the failure of hydraulic pumps. Research on failures, their analysis, and their influence on system operation are also conducted for fluid power components [6–8]. Watton J. [9] has prepared a complete compendium of analysis of fluid power system failures, while [10] was focused on the reliability of mechanical parts, including the fluid power system's components, and led to the base failure rate.

The reliability of fluid power components can be increased by any means, including failure modes and effects analysis (FMEA) [11,12], fault tree analysis (FTA) [13], root cause analysis (RCA) [14], and their extensions or modifications. FMEA is one of the most commonly used tools for improving system quality; however, the risk assessment has been criticized [15] and has been found as a method weakness. The traditional FMEA uses a risk priority number (RPN), which is a product of severity (S), occurrence (O), and detection (D). The equal weight of those factors may give the same results for a different combination and can be very confused regarding safety. Although the traditional FMEA was formally defined in the industrial standards in the early 1960s [16], later [17] the industry also noticed the weakness in the risk evaluation. The 2008 FMEA handbook [18] says that the RPN should no longer be recommended practice to define the need for action. Additionally, it says that for failure modes with severity numbers 9 or 10, the risk must be adequately addressed. The latest release [19] introduced the action priority (AP) rating table in which the AP depends on a combination of S, O, and D, where the severity plays the dominant role. Table 1 shows AP classification table. The failure modes are classified into three categories of AP: high risk (H), medium risk (M), and low risk (L) for design or process FMEA.

Table 1. The AP classification [19].

S 9–10										
D\O	10	9	8	7	6	5	4	3	2	1
10	H	H	H	H	H	H	H	H	H	L
9	H	H	H	H	H	H	H	H	H	L
8	H	H	H	H	H	H	H	H	H	L
7	H	H	H	H	H	H	H	H	H	L
6	H	H	H	H	H	H	H	M	M	L
5	H	H	H	H	H	H	H	M	M	L
4	H	H	H	H	H	H	H	L	L	L
3	H	H	H	H	H	H	H	L	L	L
2	H	H	H	H	H	H	H	L	L	L
1	H	H	H	H	H	M	M	L	L	L

S 7–8										
D\O	10	9	8	7	6	5	4	3	2	1
10	H	H	H	H	H	H	H	M	M	L
9	H	H	H	H	H	H	H	M	M	L
8	H	H	H	H	H	H	H	M	M	L
7	H	H	H	H	H	H	H	M	M	L
6	H	H	H	H	H	M	M	M	M	L
5	H	H	H	H	H	M	M	M	M	L
4	H	H	H	H	H	M	M	L	L	L
3	H	H	H	H	H	M	M	L	L	L
2	H	H	H	H	H	M	M	L	L	L
1	H	H	H	M	M	M	M	L	L	L

S 4–6										
D\O	10	9	8	7	6	5	4	3	2	1
10	H	H	H	M	M	M	M	L	L	L
9	H	H	H	M	M	M	M	L	L	L
8	H	H	H	M	M	M	M	L	L	L
7	H	H	H	M	M	M	M	L	L	L
6	H	H	H	M	M	L	L	L	L	L
5	H	H	H	M	M	L	L	L	L	L
4	M	M	M	M	M	L	L	L	L	L
3	M	M	M	M	M	L	L	L	L	L
2	M	M	M	M	M	L	L	L	L	L
1	M	M	M	L	L	L	L	L	L	L

S 2–3										
D\O	10	9	8	7	6	5	4	3	2	1
10	M	M	M	L	L	L	L	L	L	L
9	M	M	M	L	L	L	L	L	L	L
8	M	M	M	L	L	L	L	L	L	L
7	M	M	M	L	L	L	L	L	L	L
6	M	M	M	L	L	L	L	L	L	L
5	M	M	M	L	L	L	L	L	L	L
4	L	L	L	L	L	L	L	L	L	L
3	L	L	L	L	L	L	L	L	L	L
2	L	L	L	L	L	L	L	L	L	L
1	L	L	L	L	L	L	L	L	L	L

S 1										
D\O	10	9	8	7	6	5	4	3	2	1
10	L	L	L	L	L	L	L	L	L	L
9	L	L	L	L	L	L	L	L	L	L
8	L	L	L	L	L	L	L	L	L	L
7	L	L	L	L	L	L	L	L	L	L
6	L	L	L	L	L	L	L	L	L	L
5	L	L	L	L	L	L	L	L	L	L
4	L	L	L	L	L	L	L	L	L	L
3	L	L	L	L	L	L	L	L	L	L
2	L	L	L	L	L	L	L	L	L	L
1	L	L	L	L	L	L	L	L	L	L

It seems to be a step forward from the previous release because it removes the subjectiveness of risk assessment. However, even for the highest values of severity (S 9–10), the risk is defined as medium (M) for detection (D 5–6) and occurrence (O 2–3) or low (L) for detection (D 1–3) and occurrence (O 1-3).

Other drawbacks are the conversion of linguistic terms into quantitative data and uncertainties related to subjective opinions, experience, and knowledge of experts evaluating systems.

Different methods and tools are used to overcome the weaknesses mentioned earlier. The traditional risk assessment can be extended with additional factors [20] or relevant weights [21]. Another approach is implementing other perspectives on the risk, such as the customer's [22] or maintenance [23] perspective.

There are numerous methods that are used to minimize the uncertainties of the FMEA process. The most common approaches are the grey theory set [24], linguistic theory [25], fuzzy sets [26], and reasoning theory [27].

Liu et al. [28], in [28], have prepared a complex and methodical literature review of methods and tools currently used in the FMEA in both risk evaluation and elimination of uncertainties. The recent studies on overcoming traditional FMEA's drawbacks focus on implementing new methods or combining those mentioned earlier. Ref. [29] presents an approach of FMEA in which failure modes are classified based on the combination of risk factors in pairs: S and O, S and D, and O and D. The risk factors pair results were analyzed by using grey relation analysis. Yu et al. [30] have implemented the cloud model theory to minimize the linguistic uncertainties and have used the VIKOR model to determine the risk priority. The authors in [31] have modified the FMEA framework for IT according to the recommendation from the relevant literature review. Ref. [32] has utilized the cloud model to improve the FMEA. A three-stage fuzzy risk assessment based on FMEA has been proposed by Yelda et al. [33]. In ref. [34] the authors have proposed an approach in which the FMEA method has been modified by integrating the fuzzy rule base (FRB) and grey relations theory (GRT) to overcome the traditional FMEA methods' drawbacks. The uncertainty dealt with a different knowledge background of experts has been presented in [35], where prospect theory has been implemented. Another approach to converting linguistic terms into quantitative data is in [36]. The fuzzy numbers and traditional measurement of alternatives and ranking according to compromise solution (MARCOS) methods have been used. Meanwhile, Shi et al. [37] have proposed integrating hesitant linguistic preference relations (HLPRs) and an extended dynamic consensus model in FMEA. Another approach to the modification of FMEA is taking into consideration influences of failure modes and the attenuation effect of such influences in the system [38].

The drawbacks of FMEA still draw the attention of scholars and remain unsolved.

Despite the rich assortment of implemented methods and tools, the risk assessment in FMEA is still problematic. Approaches available in the literature are excessively complicated or too computationally expensive to be able to be used in the industrial standard. The recent practical implementation of FMEA by AIAG and VDA is helpful in removing subjective decisions in risk assessment. However, the implemented risk ranking may lead to misvalued failure modes in the aspect of safety. The uncertainty linked to the subjectiveness of experts' knowledge and background is still high. Our main motivation was to modify FMEA in a way to be similar to traditional analysis without using complex tools and methods and concurrently minimize typical uncertainties. The main idea of our approach is the assumption that any system is created to perform the intended task(s) and failures that can occur may disorganize that process. For that reason, we categorize failures that may occur for individual components as those that stop the system from performing a system task and those that still allow the main task to be completed but with major or minor malfunctions. We propose assessing the system's risk based on the combination of failures and related components in the aspect of the ability to perform the intended task(s). Once the failures for related components are categorized, they are prioritized by the product of occurrence (O) and detection (D) factors. In the proposed approach, the occurrence (O) has been defined as a function of subfactors that determine the likelihood of failure occurrences such as the manufacturing process, duty time, and functions performed by individual components. The detection (D) remains the same as in traditional FMEA. The presented approach is an extension of our previous work [39], in which a similar approach has been proposed. The novelty of this approach is the risk categorization and the method which allows for evaluating the occurrence (O) factor. In contrast to our previous study, in which we used a base failure rate to evaluate the occurrence, in this approach, it was defined based on functions performed by components. To distinguish those failures which can be created for other reasons than during normal operation, we have proposed to use additional modification subfactors. The occurrence (O) and detection (D) factors are equally valued. The proposed method was implemented in the risk assessment for the FMEA of the flow control valve.

2. Methodology

The presented methodology by principle is very similar to the traditional FMEA framework. Almost all steps are analogical. As presented in our previous studies [40], the potential failures are typical for fluid power components used in the conventional FMEA.

2.1. Assumptions

The primary purpose of this study was a qualitative analysis of failures and their end effects on components of fluid power systems. Failures that may occur in them are complex, and primary failure may only trigger the final failure. Therefore, we assume only primary forms of failure, and the presented analysis was carried out only for primary mechanical failures. We also assume that only a few essential data are available for the investigated component. Detailed technical data and manufacturing/assembly processes are inaccessible.

Another assumption is that the failures can be detected without a sophisticated diagnostic system or disassembly. Only basic measurements such as the pressure and flow rate are available.

2.2. Method

One of the stages of the FMEA is the functional analysis of the system, which includes the identification of the basic elements of the system, their potential failures, and causes. However, formal rules for this step are not strictly defined and may differ depending on the considered case. The proposed method assumes that any system is created to fulfill one or more tasks. Therefore, if any failure occurs, the system may not be able to perform the intended task or only perform it partially with some degradation.

We assume that system S consists of a set of components $c_i \in C, i = \{1 \ldots n\}$, which can be connected to each other in any way (parallel, series, or series-parallel). Each component c_i performs a specific function or several functions $f_j \in F, j = \{1, 2\}$ in the system S. The set of component functions f_i can be determined as follows:

- f_1: main component functions;
- f_2: auxiliary functions.

Each components c_i may have a certain state determined by the failure $fa \in Fa, fa = \{fc, fm, fn\}$, which can be determined as follows:

- fc: failures, which stop s component from executing its function;
- fm: failures, which allow a component to execute its function with a major degradation;
- fn: failures, which allow a component to execute its function with a minor degradation.

We introduce the failure importance factor f_i as:

$$f_i = \begin{cases} 3, & \text{for } f_c \\ 2, & \text{for } f_m \\ 1, & \text{for } f_n \end{cases} \tag{1}$$

Among components c_i, when creating a system we can distinguish those:

- that are essential to perform the intended system's task: main components c_e;
- that are supplementing main components and ensure the proper system operation: major components c_m;
- whose failure has little effect on the main task of the system: minor components c_n;
- additional components whose failure does not affect the main task of the system c_a.

We introduce the component importance factor c_{im} as:

$$c_{im} = \begin{cases} 4, & \text{for } c_e \\ 3, & \text{for } c_m \\ 2, & \text{for } c_n \\ 1, & \text{for } c_a \end{cases} \tag{2}$$

Additionally, we assume that failures that may appear for some components may influence other components' failures. For that reason, we have implemented a component failure factor c_{in} which can be described as:

$$c_{in} = \frac{k}{n} \tag{3}$$

where:

k is the number of components that a given component may influence;

n is the total number of components.

The system severity S_e we can define as the ability to perform the intended task. We assume that risk assessment is evaluated primarily by severity Se, which can be expressed as:

$$S_e = (c_{im} + c_{in}) \cdot f_i \tag{4}$$

The severity S_e is valued with numbers 1–15.

The occurrence O we define as an estimator of failure for corresponding components. We assumed that the main contributor to the likelihood of component failure is the number of functions (f_f factor) the component realizes with modification factors λ_i:

$$O = \alpha\left(f_f + \sum \lambda_i\right) \tag{5}$$

where:

$$f_f = \frac{\Sigma f_1}{\Sigma f_1 + \Sigma f_2} \tag{6}$$

$$\lambda_i = \lambda_{mo} + \lambda_{ma} + \lambda_{dt} + \lambda_{ev} \tag{7}$$

The factors above arise from the analysis of fluid power components [10], which states that the reliability of fluid power components depends on:

- relative motion λ_{mo};
- tolerances and surface quality which requires more sophisticated manufacturing processes λ_{ma};
- duty time λ_{dt};
- environmental conditions λ_{ev}.

Each modification factor λ_i has equal weight and may reach a maximal value equal to 0.25. The function factor f_f maximal value is 4. The evaluation of all mentioned factors can be set arbitrarily and then renormalized to the required range.

Since the mentioned study defines failure rate, it was necessary to introduce a corrective factor α, which allows distinguishing failures which occur during regular operation or in other circumstances. The coefficient is valued in the following way:

- normal operation: 2;
- operation outside specified specification: 1.5;
- maintenance: 1;
- assembly: 0.5.

The detection D has been defined and evaluated in the following way:

- detection of failure is almost certain: 1;
- detection is very high: 2;
- detection is high: 3;
- detection is moderately high: 4;
- detection is moderate: 5;
- detection is low: 6;
- detection is very low: 7;
- detection is remote: 8;
- detection is very remote: 9;

- detection is nearly not possible: 10.

According to our primary hypothesis, the risk assessment is realized based on severity S_e, which defines whether the system is able to perform intended tasks. The prioritization of failures inside task failures is performed by product occurrence O and detection D.

3. Case Study

The investigated fluid power component is a two-way flow control valve presented in Figure 1. The valve's purpose is to maintain a constant flow rate regardless of the pressure difference between ports A and B. When the pressure line is connected to port A, the flow rate is controlled by the throttling gap between piston C21 and sleeve C3. The solenoid sets the gap by setting the position of the piston C21. The constant flow rate is provided by a pressure compensator (piston C22 and nozzle C4).

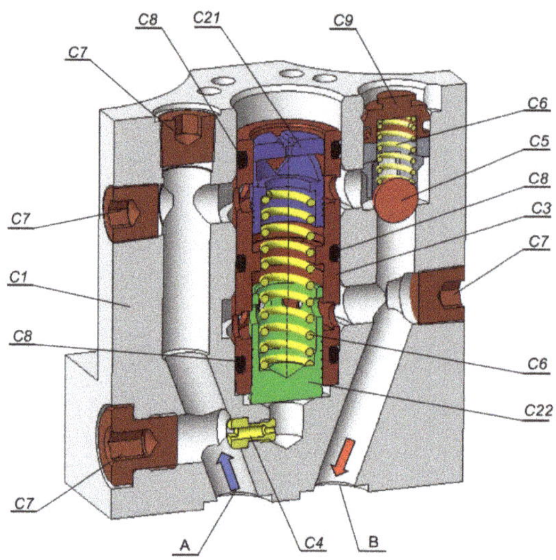

Figure 1. Flow control valve: 2-way flow control valve type UDRD6 by Ponar Wadowice: C1—body, C21—piston 1, C22—piston 2, C3—sleeve, C4—nozzle, C5—ball, C6—spring, C7—plug, C8—sealing ring, C9—check valve seat, A and B—valve port.

We can distinguish the following components whose failure will disturb the main task of the valve:

- body (c_1);
- piston 1 (c_{21});
- piston 2 (c_{22});
- sleeve (c_3);
- nozzle (c_4);
- spring (c_6);
- plug (c_7);
- sealing ring (c_8).

The following components, when failures occur, may still complete the primary task but with some limitations:

- ball (c_5);
- check valve seat (c_9);
- oil (c_{10}), it should be understood as a component that transfers pressure energy.

Failures that leave the valve unable to perform the task are:

- fracture (fc_1);

- yielding (fc_2);
- seizing (fc_3);
- aging (fc_4);
- hardening (fc_5);
- thread stripping (fc_6);
- clogging (fc_7);
- extreme contamination (fc_8).

Failures that still allow the valve to perform the task but with a major degradation are:

- erosion (fm_1);
- wear (fm_2);
- aeration (fm_3);
- contamination (fm_4);
- chemical action (fm_5).

Failures that still allow the valve to perform the task but with a minor degradation are the same as the aforementioned. The factor which defines whether the failure belongs to major or minor degradation is its intensity. For simplicity, we will not consider failures with minor degradation. Functions that are performed by individual components are presented in Table 2.

Table 2. Functions performed by valve components.

Component c_i	Primary Function f_1	Secondary Function f_2
body (c_1)	connecting with the system fixing piston/sleeve position seals flow management	fixing check valve mounting seat for spring seat for nozzle
piston 1 (c_{21})	flow rate setting sealing	- -
piston 2 (c_{22})	flow rate control sealing	- -
sleeve (c_3)	guiding pistons sealing	- -
nozzle (c_4)	pressure control	-
ball (c_5)	-	closing/opening flow
spring (c_6)	positioning	-
plug (c_7)	closing flow ducts	-
sealing ring (c_8)	sealing	-
check valve seat (c_9)	- -	guiding ball guiding spring
oil (c_{10})	energy transfer -	greasing cooling

Task failures for the corresponding components are presented in Table 3.

Table 3. Task failures.

Valve Function	Component	Failure
The primary task is not realized	body c_1	fracture f_{c1} yielding f_{c2} thread stripping f_{c6}
	piston 1 c_{21}	fracture f_{c1} yielding f_{c2} seizing f_{c3}
	piston 2 c_{22}	fracture f_{c1} yielding f_{c2} seizing f_{c3}
	sleeve c_3	fracture f_{c1} yielding f_{c2} seizing f_{c3}
	nozzle c_4	fracture f_{c1} clogging f_{c7}
	spring c_6	fracture f_{c1} yielding f_{c2}
	plug c_7	fracture f_{c1} thread stripping f_{c6}
	sealing ring c_8	fracture f_{c1} aging f_{c4} hardening f_{c5}
	oil c_{10}	extreme contamination f_{c8}
The major degradation of the primary task	body c_1 piston 1 c_{21} piston 2 c_{22} sleeve c_3 plug c_7	erosion f_{m1} wear f_{m2} chemical action f_{m5}
	ball c_5	fracture f_{c1} seizing f_{c3} erosion f_{m1} wear f_{m2} chemical action f_{m5}
	spring c_6	chemical action f_{m5}
	sealing ring c_8	wear f_{m2} chemical action f_{m5}
	check valve seat c_9	fracture f_{c1} seizing f_{c3} yielding f_{c2} thread stripping f_{c6} erosion f_{m1} wear f_{m2} chemical action f_{m5}
	oil c_{10}	aeration f_{m3} contamination f_{m4} chemical action f_{m5}

4. Results

Tables 4 and 5 present the risk assessment of failures and corresponding components. Table 4 presents failures and corresponding component ratings when the primary task of the valve is not realized. The rank is created based on the severity number S_e. For the failures with equal S_e, the rank is evaluated based on a product of occurrence (O) and detection (D). Table 5 presents failures and corresponding component ratings when the primary task of the valve is maintained with major deficiencies.

The results of failures with minor degradation are essentially the same as those presented in Table 4. The only difference is failure intensity.

Table 4. The rank of failures when the primary task is not completed.

Rank		**Severity**					**Occurrence**						**Detection**	
		c_{im}	c_{in}	f_i	S_e	f_f	λ_{mo}	λ_{ma}	λ_{dt}	λ_{ev}	α	O	D	$O \cdot D$
1	$c_{10} fc_8$	4	0.81818	3	14.4545	1.2	0	0	0	0.125	1.5	1.9875	4	7.95
2	$c_1\ fc_2$	4	0.54545	3	13.6363	4	0.0625	0.125	0.25	0.25	1.5	7.0313	6	42.1875
3	$c_3\ fc_3$	4	0.54545	3	13.6363	1.6	0.0625	0.25	0.25	0.0625	2	4.45	9	40.05
4	$c_{21}\ fc_1$	4	0.54545	3	13.6363	1.6	0.125	0.25	0.1875	0.0625	1.5	3.3375	9	30.0375
	$c_{21}\ fc_2$	4	0.54545	3	13.6363	1.6	0.125	0.25	0.1875	0.0625	1.5	3.3375	9	30.0375
	$c_3\ fc_1$	4	0.54545	3	13.6363	1.6	0.0625	0.25	0.25	0.0625	1.5	3.3375	9	30.0375
	$c_3\ fc_2$	4	0.54545	3	13.6363	1.6	0.0625	0.25	0.25	0.0625	1.5	3.3375	9	30.0375
5	$c_1\ fc_1$	4	0.54545	3	13.6363	4	0.0625	0.125	0.25	0.25	1.5	7.0313	3	21.0938
6	$c_{21}\ fc_3$	4	0.54545	3	13.6363	1.6	0.125	0.25	0.25	0.0625	2	4.575	6	27.45
7	$c_1\ fc_6$	4	0.54545	3	13.6363	4	0.0625	0.125	0.25	0.25	0.5	2.3438	2	4.6875
8	$c_{22}\ fc_1$	4	0.45454	3	13.3636	1.6	0.125	0.25	0.25	0.0625	1.5	3.4313	9	30.8813
9	$c_{22}\ fc_2$	4	0.45454	3	13.3636	1.6	0.125	0.25	0.25	0.0625	1.5	3.4313	9	30.8813
	$c_{22}\ fc_3$	4	0.45454	3	13.3636	1.6	0.125	0.25	0.25	0.0625	2	4.575	6	27.45
10	$c_6\ fc_2$	4	0.36363	3	13.0909	0.8	0.0625	0.0625	0.25	0.0625	1.5	1.8563	9	16.7063
11	$c_6\ fc_1$	4	0.36363	3	13.0909	0.8	0.0625	0.0625	0.25	0.0625	1.5	1.8563	7	12.9938
12	$c_4\ fc_7$	4	0.18181	3	12.5454	0.8	0.0625	0.125	0.1875	0.0625	2	2.475	9	22.275
13	$c_4\ fc_1$	4	0.18181	3	12.5454	0.8	0.0625	0.125	0.1875	0.0625	1.5	1.8563	6	11.138
14	$c_8\ fc_4$	4	0.09090	3	12.2727	0.8	0.0625	0.0625	0.25	0.0625	2	2.475	9	22.275
15	$c_8\ fc_5$	4	0.09090	3	12.2727	0.8	0.0625	0.0625	0.25	0.0625	2	2.475	9	22.275
	$c_8\ fc_1$	4	0.09090	3	12.2727	0.8	0.0625	0.0625	0.25	0.0625	2	2.475	6	14.85
16	$c_7\ fc_1$	4	0.09090	3	12.2727	0.8	0.0625	0.0625	0.0625	0.25	0.5	0.6183	1	0.6188
	$c_7\ fc_6$	4	0.09090	3	12.2727	0.8	0.0625	0.0625	0.0625	0.25	0.5	0.6183	1	0.6188

Table 5. The ranks of functional failures with major degradation.

Rank		Severity				Occurrence							Detection	
		c_{im}	c_{in}	f_i	S_e	f_f	λ_{mo}	λ_{ma}	λ_{dt}	λ_{ev}	α	O	D	$O \cdot D$
1	$c_8\, fm_2$	4	0.0909	3	12.2727	0.8	0.0625	0.0625	0.25	0.0625	2	2.4750	10	24.75
	$c_8\, fm_5$	4	0.0909	3	12.2727	0.8	0.0625	0.0625	0.25	0.0625	2	2.4750	10	24.75
2	$c_{10}\, fm_3$	4	0.8182	2	9.6364	1.2	0.0625	0.0625	0.25	0.25	2	3.65	4	14.6
	$c_{10}\, fm_4$	4	0.8182	2	9.6364	1.2	0.0625	0.0625	0.25	0.25	2	3.65	4	14.6
	$c_{10}\, fm_5$	4	0.8182	2	9.6364	1.2	0.0625	0.0625	0.25	0.25	2	3.65	4	14.6
3	$c_1\, fm_1$	4	0.5455	2	9.0909	4	0.0625	0.1250	0.25	0.25	2	9.3750	10	93.75
	$c_1\, fm_2$	4	0.5455	2	9.0909	4	0.0625	0.1250	0.25	0.25	2	9.3750	10	93.75
	$c_1\, fm_5$	4	0.5455	2	9.0909	4	0.0625	0.1250	0.25	0.25	2	9.3750	10	93.75
4	$c_{21}\, fm_5$	4	0.5455	2	9.0909	1.6	0.1250	0.25	0.25	0.0625	2	4.5750	10	45.75
5	$c_{21}\, fm_1$	4	0.5455	2	9.0909	1.6	0.1250	0.25	0.1875	0.0625	2	4.45	10	44.5
	$c_{21}\, fm_2$	4	0.5455	2	9.0909	1.6	0.1250	0.25	0.1875	0.0625	2	4.45	10	44.5
	$c_3\, fm_1$	4	0.5455	2	9.0909	1.6	0.0625	0.25	0.25	0.0625	2	4.45	10	44.5
	$c_3\, fm_2$	4	0.5455	2	9.0909	1.6	0.0625	0.25	0.25	0.0625	2	4.45	10	44.5
	$c_3\, fm_5$	4	0.5455	2	9.0909	1.6	0.0625	0.25	0.25	0.0625	2	4.45	10	44.5
6	$c_{22}\, fm_1$	4	0.4545	2	8.9091	1.6	0.1250	0.25	0.25	0.0625	2	4.575	10	45.75
	$c_{22}\, fm_2$	4	0.4545	2	8.9091	1.6	0.1250	0.25	0.25	0.0625	2	4.575	10	45.75
	$c_{22}\, fm_5$	4	0.4545	2	8.9091	1.6	0.1250	0.25	0.25	0.0625	2	4.575	10	45.75
7	$c_6\, fm_5$	4	0.3636	2	8.7273	0.8	0.0625	0.0625	0.125	0.0625	2	2.225	10	22.25
8	$c_4\, fm_1$	4	0.1818	2	8.3636	0.8	0.0625	0.125	0.1875	0.0625	2	2.475	10	24.75
	$c_4\, fm_2$	4	0.1818	2	8.3636	0.8	0.0625	0.125	0.1875	0.0625	2	2.475	10	24.75
	$c_4\, fm_5$	4	0.1818	2	8.3636	0.8	0.0625	0.125	0.1875	0.0625	2	2.475	10	24.75
9	$c_9\, fc_1$	2	0.2727	3	6.8182	0.4	0.0625	0.125	0.125	0.0625	1.5	1.1625	10	11.625
	$c_9\, fc_2$	2	0.2727	3	6.8182	0.4	0.0625	0.125	0.125	0.0625	1.5	1.1625	10	11.625
10	$c_9\, fc_6$	2	0.2727	3	6.8182	0.4	0.0625	0.125	0.125	0.0625	0.5	0.3875	10	3.875
11	$c_5\, fc_3$	2	0.1818	3	6.5455	0.8	0.1250	0.125	0.125	0.0625	2	2.475	10	24.75
12	$c_5\, fc_1$	2	0.1818	3	6.5455	0.8	0.1250	0.125	0.125	0.0625	1.5	1.8563	10	18.5625
13	$c_7\, fm_1$	3	0.0909	2	6.1818	0.8	0.0625	0.0625	0.0625	0.25	2	2.475	10	24.75
	$c_7\, fm_2$	3	0.0909	2	6.1818	0.8	0.0625	0.0625	0.0625	0.25	2	2.475	10	24.75
	$c_7\, fm_5$	3	0.0909	2	6.1818	0.8	0.0625	0.0625	0.0625	0.0625	2	2.1	10	21
14	$c_9\, fm_1$	2	0.2727	2	4.5455	0.4	0.0625	0.125	0.125	0.0625	2	1.55	10	15.5
	$c_9\, fm_2$	2	0.2727	2	4.5455	0.4	0.0625	0.125	0.125	0.0625	2	1.55	10	15.5
	$c_9\, fm_5$	2	0.2727	2	4.5455	0.4	0.0625	0.125	0.125	0.0625	2	1.55	10	15.5
15	$c_5\, fm_1$	2	0.1818	2	4.3636	0.8	0.1250	0.125	0.125	0.0625	2	2.475	10	24.75
	$c_5\, fm_2$	2	0.1818	2	4.3636	0.8	0.1250	0.125	0.125	0.0625	2	2.475	10	24.75
	$c_5\, fm_5$	2	0.1818	2	4.3636	0.8	0.1250	0.125	0.125	0.0625	2	2.475	10	24.75
	$c_5\, fm_5$	2	0.1818	2	4.3636	0.8	0.1250	0.125	0.125	0.0625	0.5	0.6188	10	6.1875

5. Discussion

The ranking method in this study allowed for the segregation of those failures that can potentially occur and leave the valve component unable to perform the intended task. However, some failures have the same value of severity (S_e) inside the assumed categories. The adopted product of occurrence (O) and detection (D) as an additional evaluating criterion did not allow for distinguishing their ranks. This indicates that another factor apart from those presented in this study would need to be employed. Another solution might be a modification of the detection factor (D), which remained unchanged, and problems with uncertainties related to linguistic conversion to quantitative data remain relatively high. The presented method uses factors based on failure history and research of similar but not strictly identical objects; therefore, it is suitable during the design process or initial stage of failure analysis. Additionally, it can be combined with other tools and methods such as root cause analysis (RCA) or fault tree analysis (FTA). As was mentioned earlier, a ranking for failures with minor degradation of the valve's intended function is the same as for major degradation. The only difference is the intensity of failure. As the majority of failures in Table 5 are caused during the normal operational condition, it is certain the degradation of the valve's main task will depend on the intensity of wear or erosion of valve components.

Another feature that can be observed is the source of the failures which stop the valve from performing its task. All of those failures are mainly caused during assembling, maintenance work, or operation outside specifications. Analogously, those failures which cause major or minor degradation of the valve are mainly produced during normal operation.

The presented method can be easily extended to any fluid power components, however, we are unable to perform FMEA for electrical/electronic components. It can be used only for purely mechanical or hydromechanical components. Proposed subfactors in occurrence (O) were defined only as typical for mechanical or hydromechanical failures.

The severity values S_e can be set arbitrary with any values; the proposed methods only prioritize failures for related components inside categorized failures on those which stop the valve from performing its task and those which cause a major or minor malfunction.

Comparison with the Traditional FMEA Risk Assessment

The Table 6 presents a risk analysis using the traditional approach for selected failures and related components by invited experts. The risk has been expressed as the RPN (product of severity (S), occurrence (O), and detection (D)) and by using action priority (AP).

Table 6. The RPN and AP.

Failure/Component	S	O	D	RPN	AP
$c_1\ fc_1$	10	3	3	90	L
$c_7\ fc_6$	10	1	1	10	L
$c_1\ fc_2$	10	3	6	180	H
$c_{21}\ f_{c1}$	10	3	9	270	H
$c_{21}\ fc_2$	10	3	9	270	H
$c_{21}\ fc_3$	10	5	6	300	H
$c_{22}\ fc_1$	10	3	9	270	H
$c_{22}\ fc_2$	10	3	9	270	H
$c_{22}\ fc_3$	10	5	6	300	H
$c_{21}\ fm_1$	8	7	10	560	H
$c_{21}\ fm_2$	8	7	10	560	H
$c_{22}\ fm_1$	8	7	10	560	H
$c_{22}\ fm_2$	8	7	10	560	H

According to the table, the highest RPN value was reached for failure wear fm_2 and erosion fm_1 for the components pistons 1 c_{21} and piston 2 c_{22}. For both, the action priority (AP) is high (H). This is overestimating those failures, which can cause a major or minor malfunction in the valve. In contrast, the failure fracture fc_1 for the body c_1 and thread stripping fc_6 for plug c_7 are underestimated. Their RPN values are lower than other failures, and action priority (AP) is low (L). Those failures, when they occur, stop the valve's operation.

6. Conclusions

Despite the continuous improvement of FMEA and its new standards, many problems remain unresolved. The proposal of risk assessment presented in this paper and the assessment of the likelihood of potential defects allows minimizing the uncertainty of the gathered data.

The proposed method categories failures and relevant components in the aspect of their influence on the ability to perform the intended task. The method presented in this paper is universal and can be used in any component of the fluid power system.

The proposed method also identifies failures that can cause critical failure to the system, which would be neglected by the traditional FMEA method or even the latest one implemented in the industry.

The proposed methods do not require computationally extensive tools or simulational methods. In principle, it is very similar to the traditional FMEA method. The proposed failures and component categorization make it so the proposed method can be adjusted to individual needs and valued with arbitrary weights.

The presented method has shown its superiority over the recent AIAG and VDA FMEA standard and the traditional FMEA framework. Some of the failures that can occur on main valve components would not be recognized as critical to the system.

An extension of the presented method might be a redefinition of the detection factor to minimize the still relatively high uncertainties and implement new risk factors such as the cost of manufacturing or repairs.

Author Contributions: Conceptualization, J.F.-D.; methodology, M.D.; software, M.D.; validation, J.F.-D.; formal analysis, J.F.-D.; investigation, M.D.; resources, M.D.; data curation, M.D.; writing—original draft preparation, M.D.; writing—review and editing, M.D.; visualization, J.F.-D.; supervision, J.F.-D.; project administration, J.F.-D.; funding acquisition, J.F.-D. All authors have read and agreed to the published version of the manuscript.

Funding: This research received no external funding.

Institutional Review Board Statement: Not applicable.

Informed Consent Statement: Not applicable.

Data Availability Statement: Not applicable.

Conflicts of Interest: The authors declare no conflict of interest.

Abbreviations

The following abbreviations are used in this manuscript:

MDPI	Multidisciplinary Digital Publishing Institute
DOAJ	Directory of Open Access Journals
FMEA	Failure Modes and Effect Analysis
RCA	Root Cause Analysis
FTA	Fault Tree Analysis
RPN	Risk Priority Number
AIAG	Automotive Industry Action Group
VDA	and der Automobilindustrie VDA
AP	Action Priority

References

1. Zhao, X.; Zhou, C.; Zhao, L.; Zhang, S.; Wen, P.; Jiang, J. Condition evaluation model of fluid power system in gradual failure based on data envelopment analysis. *Comput. Fluids* **2015**, *110*, 219–226. [CrossRef]
2. Lee, Y.; Lee, G.; Yang, J.; Park, J.; Baek, D. Failure analysis of a hydraulic power system in the wind turbine. *Eng. Fail. Anal.* **2020**, *107*, 104218. [CrossRef]
3. Ma, Z.; Wang, S.; Shi, J.; Li, T.; Wang, X. Fault diagnosis of an intelligent hydraulic pump based on a nonlinear unknown input observer. *Chin. J. Aeronaut.* **2018**, *31*, 385–394. [CrossRef]
4. Li, T.; Wang, S.; Zio, E.; Shi, J.; Ma, Z. A numerical approach for predicting the remaining useful life of an aviation hydraulic pump based on monitoring abrasive debris generation. *Mech. Syst. Signal Process.* **2018**, *136*, 106519. [CrossRef]
5. Hast, D.; Findeisen, R.; Streif, S. Detection and isolation of parametric faults in hydraulic pumps using a set-based approach and quantitative–qualitative fault specifications. *Control Eng. Pract.* **2015**, *40*, 61–70. [CrossRef]
6. Gianni, N.; Tito, M. Failure of a heavy-duty hydraulic cylinder and its fatigue re-design. *Eng. Fail. Anal.* **2011**, *18*, 1030–1036.

7. Roquet, P.; Gamez-Montero, P.J.; Castilla, R.; Raush, G.; Codina, E. A Simplified Methodology to Evaluate the Design Specifications of Hydraulic Components. *Appl. Sci.* **2018**, *8*, 1612. [CrossRef]
8. Guo, Y.; Xiong, G.; Zeng, L.; Li, Q. Modeling and Predictive Analysis of Small Internal Leakage of Hydraulic Cylinder Based on Neural Network. *Energies* **2021**, *14*, 2456. [CrossRef]
9. Watton, J. *Modelling, Monitoring and Diagnostic Techniques for Fluid Power Systems*; Springer: London, UK, 2007.
10. *Handbook of Reliability Prediction for Mechanical Equipment*; Naval Surface Warfare Center: West Bethesda, MD, USA, 2011.
11. Stirrup, T.S.; Chamberlin, H.C.; Curran, K.L.F. *Scaled Wind Farm Technology Hydraulic System Failure Modes and Effects Analysis*; Sandia National Laboratories: Albuquerque, NM, USA; Livermore, CA, USA, 2015.
12. Fabis-Domagala, J.; Domagala, M.; Momeni, H. A Matrix FMEA Analysis of Variable Delivery Vane Pumps. *Energies* **2021**, *14*, 1741. [CrossRef]
13. Zhang, F.; Cheng, L.; Gao, Y.; Xu, X.; Wang, Y. Fault tree analysis of a hydraulic system based on the interval model using latin hypercube sampling. *J. Intell. Fuzzy Syst.* **2019**, *37*, 8345–8355. [CrossRef]
14. Hemati, A.; Shooshtari, A.A. Gear Pump Root Cause Failure Analysis Using Vibrations Analysis and Signal Processing. *J. Fail. Anal. Preven.* **2020**, *20*, 1815–1818. [CrossRef]
15. Liu, H.C.; Liu, L.; Liu, N. Risk evaluation approaches in failure mode and effects analysis: A literature review. *Expert Syst. Appl.* **2013**, *40*, 828–838. [CrossRef]
16. *MIL-P 1629*; USA Military Standard, Procedure for Performing a Failure Mode, Effects and Criticality Analysis (MIL-P 1629, USA). Military Specifications and Standards: Washington, DC, USA, 1949.
17. *J1739 200901*; Potential Failure Mode and Effects Analysis in Design (Design FMEA), Potential Failure Mode and Effects Analysis in Manufacturing and Assembly Processes (Process FMEA). SAE International: Warrendale, PA, USA, 2009.
18. AIAG. *Potential Failure Mode and Effect Analysis (FMAE)*, 4th ed.; AIAG:Automotive Industry Action Group, 2008. Available online: https://www.aiag.org/store/publications/details?ProductCode=FMEA-4 (accessed on 20 July 2022).
19. AIAG & VDA FMEA Handbook. AIAG: Automotive Industry Action Group. 2019. Available online: https://www.aiag.org/store/publications/details?ProductCode=FMEAAV-1 (accessed on 20 July 2022).
20. Anes, V.; Henriques, E.; Freitas, M.; Reis, L. A new risk prioritization model for failure mode and effects analysis. *Qual. Reliab.Eng. Int.* **2018**, *34*, 516–528. [CrossRef]
21. Tang, Y.; Zhou, D.; Chan, F.T.S. AMWRPN: Ambiguity Measure Weighted Risk Priority Number Model for Failure Mode and Effects Analysis. *IEEE Access* **2018**, *6*, 27103–27110. [CrossRef]
22. Koomsap, P.; Charoenchokdilok, T. Improving risk assessment for customer-oriented FMEA. *Total Qual. Manag. Bus. Excell.* **2018**, *29*, 1563–1579. [CrossRef]
23. Pancholi, N., Bhatt M., FMECA-based maintenance planning through COPRAS-G and PSI. *J. Qual. Maint. Eng.* **2018**, *24*, 224–243. [CrossRef]
24. Li, Z.; Chen, L. A novel evidential FMEA method by integrating fuzzy belief structure and grey relational projection method. *Eng. Appl. Artif. Intell.* **2019**, *77*, 136–147. [CrossRef]
25. Shi, S.; Fei, H.; Xu, X. Application of a FMEA method combining interval 2-tuple linguistic variables and grey relational analysis in preoperative medical service process. *IFAC-PapersOnLine* **2019**, *52*, 1242–1247. [CrossRef]
26. Filo, G.; Fabis-Domagala, J.; Domagala, M.; Lisowski, E.; Momeni, H. The idea of fuzzy logic usage in a sheet-based FMEA analysis of mechanical systems. *MATEC Web Conf.* **2018**, *183*, 03009. [CrossRef]
27. Qin, J.; Xi, Y.; Pedrycz, W. Failure mode and effects analysis (FMEA) for risk assessment based on interval type-2 fuzzy evidential reasoning method. *Appl. Soft Comput.* **2020**, *89*, 10613. [CrossRef]
28. Liu, H.C.; Chen, X.; Dua, C.; Wang, Y. Failure mode and effect analysis using multi-criteria decision making methods: A systematic literature review. *Comput. Ind. Eng.* **2019**, *135*, 881–897. [CrossRef]
29. Ouyang, L.; Che, Y.; Yan, L.; Park, C. Multiple perspectives on analyzing risk factors in FMEA. *Comput. Ind.* **2022**, *141*, 103712. [CrossRef]
30. Yu, J.; Wu, S.; Chen, H.; Yu, Y.; Fan, H.; Liu, J. Risk assessment of submarine pipelines using modified FMEA approach based on cloud model and extended VIKOR method. *Process. Saf. Environ. Prot.* **2021**, *155*, 555–574.
31. Apol Pribadi S.; Nina Fadilah N. The consistency analysis of failure mode and effect analysis (FMEA) in information technology risk assessment. *Heliyon* **2020**, *6*, e03161. [CrossRef]
32. Lipeng, W.; Fang, Y.; Fang, W.; Zijun L. FMEA-CM based quantitative risk assessment for process industries—A case study of coal-to-methanol plant in China. *Process. Saf. Environ. Prot.* **2021**, *149*, 299–311. [CrossRef]
33. Yelda, Y.; Gülin Feryal, C. A FMEA based novel intuitionistic fuzzy approach proposal: Intuitionistic fuzzy advance MCDM and mathematical modeling integration. *Expert Syst. Appl.* **2021**, *183*, 115413. [CrossRef]
34. Shamsu, H.; Jin, W.; Christos, K.; Musa, B. Modified FMEA hazard identification for cross-country petroleum pipeline using Fuzzy Rule Base and approximate reasoning. *J. Loss Prev. Process. Ind.* **2022**, *74*, 104616. [CrossRef]
35. Yan, Z.; Chuanhao, F.; Hengjie, Z. From diversity to consensus: Impacts of opinion evolution and psychological behaviours in failure mode and effect analysis. *Appl. Soft Comput.* **2022**, *128*, 109399. [CrossRef]
36. Soumava, B.; Shankar, C. Failure analysis of CNC machines due to human errors: An integrated IT2F-MCDM-based FMEA approach. *Eng. Fail. Anal.* **2021**, *130*, 105768. [CrossRef]

37. Shi, H.; Liu, Z.; Liu, H.C. A new linguistic preference relation-based approach for failure mode and effect analysis with dynamic consensus reaching process. *Inf. Sci.* **2022**, 977–993. [CrossRef]
38. Qun, W.; Guozhu, J.; Yuning, J.; Wenyan, S. A new approach for risk assessment of failure modes considering risk interaction and propagation effects. *Reliab. Eng. Syst. Saf.* **2021**, *216*, 108044. [CrossRef]
39. Fabis-Domagala, J.; Domagala, M.; Momeni, H. A Concept of Risk Prioritization in FMEA Analysis for Fluid Power Systems. *Energies* **2021**, *14*, 6482. [CrossRef]
40. Fabis-Domagala, J.; Momeni, H.; Filo, G.; Bikass, S.; Lempa, P. Matrix FMEA Analysis of the flow conrol valve. *QPI* **2019**, *1*, 590–595.

energies

MDPI

Article

Data-Driven Condition Monitoring of a Hydraulic Press Using Supervised Learning and Neural Networks

Faried Makansi * and Katharina Schmitz

RWTH Aachen University, Institute for Fluid Power Drives and Systems (ifas), 5074 Aachen, Germany
* Correspondence: faried.makansi@ifas.rwth-aachen.de

Abstract: The automated evaluation of machine conditions is key for efficient maintenance planning. Data-driven methods have proven to enable the automated mapping of complex patterns in sensor data to the health state of a system. However, generalizable approaches for the development of such solutions in the framework of industrial applications are not established yet. In this contribution, a procedure is presented for the development of data-driven condition monitoring solutions for industrial hydraulics using supervised learning and neural networks. The proposed method involves feature extraction as well as feature selection and is applied on simulated data of a hydraulic press. Different steps of the development process are investigated regarding the design options and their efficacy in fault classification tasks. High classification accuracies could be achieved with the presented approach, whereas different faults are shown to require different configurations of the classification models.

Keywords: condition monitoring; fault detection and diagnosis; industrial hydraulics; hydraulic press; supervised learning; neural networks; feature extraction; feature selection

Citation: Makansi, F.; Schmitz, K. Data-Driven Condition Monitoring of a Hydraulic Press Using Supervised Learning and Neural Networks. *Energies* **2022**, *15*, 6217. https://doi.org/10.3390/en15176217

Academic Editors: Paolo Casoli and Massimo Rundo

Received: 29 July 2022
Accepted: 23 August 2022
Published: 26 August 2022

1. Introduction

Over the lifetime of an industrial machine, wear and tear of various components of the system is inevitable. Especially in applications that involve the transmission of high power under rough conditions, as typically present in hydraulics, the health state of machine elements will degrade over time. In order to maintain high machine availability and process quality, maintenance actions are required. However, the task of planning maintenance actions can be challenging for machine operators when a balance must be met between premature downtimes due to maintenance work and downtimes due to decreased process quality or machine damage. To plan maintenance periods efficiently and detect looming failures at an early stage, methods of condition monitoring can be applied. The objective of condition monitoring is to obtain an estimate of the current machine health status through the automated evaluation of operation data gathered from a machine [1].

In the field of hydraulics, different approaches for condition monitoring have been investigated. Solutions studied and suggested in the literature range from rule-based systems to model-based approaches, to purely data-driven approaches [2,3]. Most recently, the increasing integration of measurement technology and information technology in industrial products [4] turned the focus to another group of methods, the data-driven approaches. These methods, often summarized by the term machine learning, aim to algorithmically extract a model from data, which maps patterns in recorded data to corresponding machine health statuses. In the case of methods of supervised learning, the health state of a machine can be determined based on the algorithmic processing of observed sets of sensor and control signals [5]. Moreover, given the labeling of the data, the cause of deviating machine behavior can even be traced down to faults in the components of the system.

However, challenges in the application of data-driven approaches for condition monitoring still exist. One major issue is the lack of methods for the general and generic development of data-driven solutions. Another challenge is the requirement of comprehensive data. To obtain robust models, the training data used should cover a broad range of possible operating scenarios and machine configurations. Additionally, machine learning-based approaches are regarded as black-box models, which hardly provide human inspectors an insight into the process of decision making learned by such models [6].

In this contribution, a method for developing data-driven condition monitoring solutions for industrial hydraulics is presented. For this, a feature-based approach of supervised learning is applied, which allows the generic extraction of features from multivariate, cyclic and non-stationary time series data. Moreover, the presented procedure includes a feature selection step, which helps in identifying relevant data characteristics and gives an insight into the effect of faults on observable measures of the system. For the setup and training of classification models, a flexible approach is suggested, which allows the modular setup of classification models for individual machine and component faults. Investigations are carried out on a simulation model of a hydraulic press test bench, such that the issue of data availability is not a limiting factor in this study.

2. Related Work and State of the Art

In this section, an overview of previous work on condition monitoring in hydraulic applications is given. Subsequently, the general concept of neural networks is described with respect to relevant design aspects for the setup and use in a fault classification solution.

2.1. Condition Monitoring in Hydraulics

In general, applying machine learning for the condition monitoring of hydraulic applications has proven to yield satisfactory results [2,7,8]. Most publications in this field focus on the use of data-driven techniques for monitoring single hydraulic components such as pumps and valves [9–12]. The algorithms mainly used in these studies are based on variations of neural networks, which are used to process raw time series for the inference of the health states of the components. In [13], features are extracted from the raw time series in order to detect anomalies in the operation of a speed-variable pump by the use of methods of unsupervised learning. A similar feature-based approach is pursued in [14] in the framework of a systematic investigation of techniques of supervised learning for monitoring an axial piston pump. As typically only a few sensors are applied at the component level, data are scarce in these use cases, such that mostly only one single sensor recording can be considered. On the contrary, applying condition monitoring on a system level enables to include further sensor data available in the system. Hence, a component fault cannot only be inferred directly from data acquired on that component, but can also be inferred from indirect effects on other components and the overall system behavior. This concept is successfully applied in [15,16] for hydraulic system test benches by also using a feature-based approach. The features mostly used are measures from descriptive statistics, such as the mean, median and variance of a signal, while, for rotating machinery, features from the frequency domain are often considered [15]. Alternatively, features can be manually defined, as in [17]. In [18], the authors have investigated refinement strategies of generic feature generation, where including process knowledge into the feature generation was shown to improve the detection of faults. Furthermore, in [14] and [16], a feature reduction is performed by applying filter methods and analytical methods of principal component analysis, respectively. While a significant reduction in features can be achieved with a principal component analysis or a linear discriminant analysis, the resulting feature sets are not interpretable by a human inspector after the transformation [19].

2.2. Design Considerations for Neural Networks

In contrast to conventional approaches, data-driven methods depend less on domain-specific knowledge but extract the knowledge implicitly or explicitly from data. In the field of machine learning, this process of knowledge extraction is referred to as model training, where the model is a mathematical structure that is fit to reproduce a desired input–output behavior. The type of model mainly considered in this study is feedforward neural networks in the form of multilayer perceptrons, as shown in Figure 1.

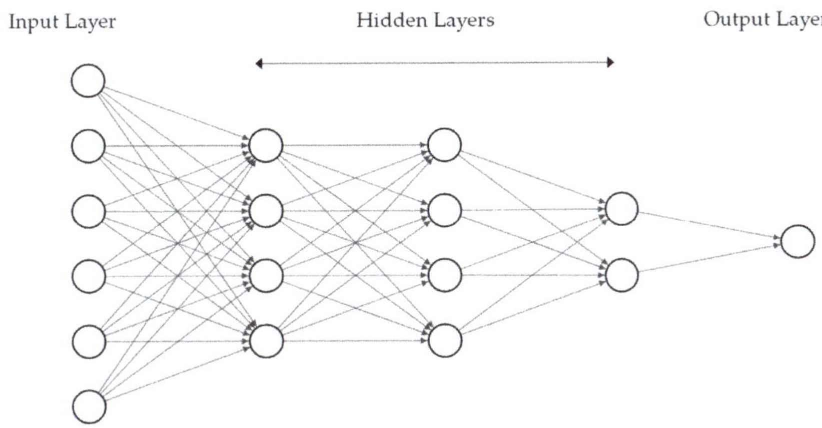

Figure 1. Feedforward neural network with sic nodes in the input layer, three hidden layers and one node in the output layer.

Feedforward neural networks consist of neural units sequentially stacked in layers, where the outputs of a preceding layer form the inputs for the subsequent layer. In regular fully connected neural networks, the outputs of all units of a preceding layer are connected to each neural unit of the next layer, where each connection is attributed a weight parameter. Within the units, the respective input values are multiplied with the corresponding weight parameter, summed up and passed through an activation function. The main purpose of the activation function is applying a transformation to the data between the layers and using nonlinear activation functions to enable nonlinear mappings between input samples and the output of the network [20]. Accordingly, typically used activation functions are the sigmoid activation function, tanh activation function or a rectified linear activation function (ReLU).

The layers located between the input layer and the output layer of the network are referred to as hidden layers. Theoretically, using one hidden layer with a nonlinear activation and an arbitrary high number of nodes in the hidden layer allows to approximate any given input–output relationship [21]. Accordingly, combining multiple layers linearly, as done in a multilayer perceptron, preserves this characteristic of a universal approximator. Moreover, combining multiple layers adds to the representational capacity of a neural network, which, in turn, can allow to use less nodes in the network in total compared to the network with a single hidden layer [20].

In order to obtain a desired input–output mapping from a neural network, the weight parameters of the model are adjusted algorithmically during the model training. For this, each training sample is first passed to the input of the model in a forward pass, where the model outputs a value. After this, the deviation between the model output from the forward pass and the desired output is computed according to a predefined loss function. Subsequently, in the reverse pass, the gradient of the loss function is computed with respect to the model weights, so that it can be inferred which model weights to adjust in order to minimize the loss. While the reverse pass is performed by generic gradient-descent-based algorithms, the selection of the loss function is linked to the format of the model output of the model, i.e., whether continuous values or discrete class labels are predicted [22]. In contrast to the weight parameters of the model, the structural parameters (number of layers, number of nodes and activation functions) and fitting parameters (loss function and optimization algorithm) remain fixed during training. Therefore, these parameters are referred to as the hyperparameters of a model. Applying neural networks hence requires to define these hyperparameters for the specific problem at hand. As no general rule exists for setting the hyperparameters of neural networks, especially for the structural hyperparameters, they are often determined based on experience or by implementing and comparing several setups. The latter approach is also referred to as grid search.

Furthermore, methods of supervised learning are based on fitting models to training data, for which the expected output of the model is known. In the case of condition monitoring, this can be any indicator of the machine health status, which, in turn, can be represented by a single value or be composed of multiple values. The former, the single-value case, is referred to as single-target prediction, while the latter case is referred to as multi-target prediction. Furthermore, in the case of classification, where the task is to predict discrete classes for a set of inputs, a distinction can be made based on the number of possible classes that each target can represent. In the most basic case, targets are binary values, but cases of multiple classes are possible as well.

2.3. Feature Representation of Time Series Data

As the number of input nodes of a neural network is fixed, the dimensionality of the input data must be uniform among all data instances, too. However, time series data, as typically acquired from sensors in industrial machines, can have varying lengths, depending on variations in the process or the sampling rates of different sensors. Moreover, the longer the time series, the more input nodes are required, and the number of parameters of the neural network unfavorably increases. Therefore, describing a time series by a fixed number of representative characteristics is one common means to obtain the compatibility of time series and a fixed number of model inputs [22]. These characteristics are referred to as features and can be either defined manually [19], by the use of generic descriptive measures [23], or derived by transformation of the data learned by another machine learning model [24].

3. Materials and Methods

In this section, the considered hydraulic system is described and the developed approach for generating fault classification models is presented. This includes the steps of data generation, feature extraction and selection, as well as the process of model setup, training and evaluation.

3.1. Hydraulic Press

Representative of an industrial hydraulic system, a hydraulic press is considered in this study. Figure 2 shows the physical machine, along with the corresponding hydraulic circuit diagram. The system mainly comprises a position-controlled cylinder drive. The hydraulic power is supplied by a pump and conducted to the double-acting hydraulic cylinder (CY) through a proportional 4/3-way directional control valve (PV). Two switching valves (SV1 and SV2) allow to set the system to idle circulation and to lock the ram in position, respectively. The pump provides a constant volumetric flow of approximately 19 L/min at a maximal system pressure of 200 bar. The hydraulic cylinder has a nominal piston diameter of 80 mm and is coupled to a guided ram, which finally acts against a load force. The load force represents the reaction force of a workpiece and is provided by a secondary circuit in the physical test bench. During operation of the press, pressures are measured at the main cylinder and the power supply. Additionally, the position of the cylinder is recorded, as well as the volumetric flow entering the 4/3-way directional control valve and the control signals fed to the valves.

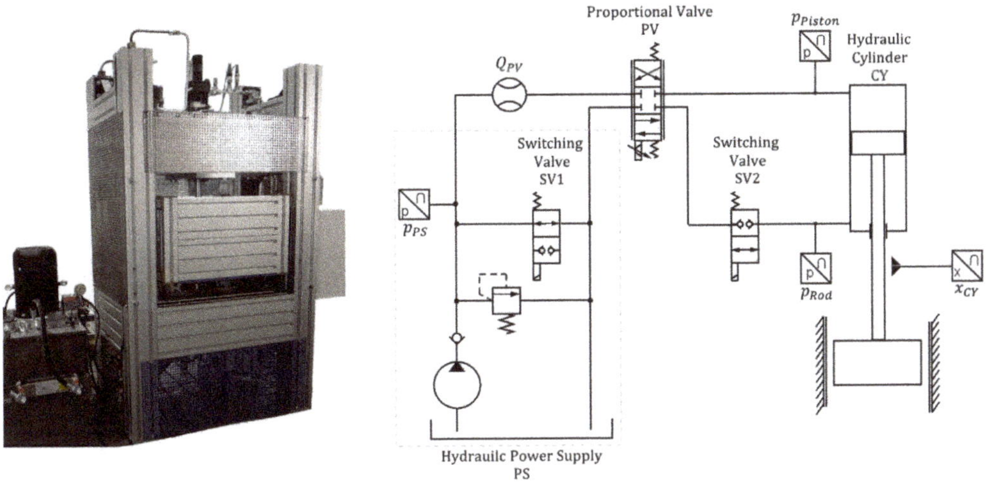

Figure 2. Hydraulic press test bench and its hydraulic circuit diagram.

The function of the system is to perform working cycles, in which the ram is moved to a defined position against the load force. At the beginning of the working cycle, the system is in idle mode, where the switching valve SV1 activates idle circulation of the pump and the switching valve SV2 locks the ram in the top position. Then, starting from the top position, SV1 and SV2 are switched and the main cylinder extends towards the workpiece in rapid traverse. During this phase, neither an external load nor a limitation of the velocity is active. As the ram approaches the workpiece and reaches a defined distance from it, the velocity is restricted through limiting the control signal of the valve PV and the volumetric flow to the cylinder, respectively. As the ram is in contact with the workpiece, an external load acts against the cylinder and increases linearly over the stroke of the cylinder. In this load phase, the system is given a fixed time to reach the set value, which depends on the desired target position of the ram. After this time, the controller initiates the backstroke in open-loop control and the cycle ends with another idle phase. Figure 3 shows the progression of cylinder position, velocity and load force over one working cycle of the press.

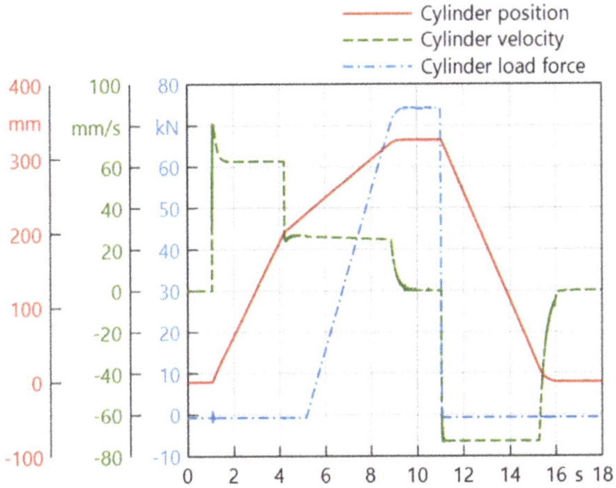

Figure 3. Simulated working cycle of the hydraulic press in regular operation.

For the generation of the large amounts of data required in this study, a corresponding simulation model of the system is set up as a lumped parameter model using the simulation software *SimulationX* (software version 4.2) [25]. Hydraulic components in the simulation model are parametrized according to nominal data sheet specifications. The pump is modeled with a pressure-dependent volumetric loss, while no information on hydro-mechanical parameters could be derived from the pump data sheet. Furthermore, the dynamic characteristics of the pump and the driving electric motor are neglected. A linear opening characteristic is defined for the proportional valve, while the relationship between the spool position and control signal is modeled by a second-order differential equation. Pressure losses in the piping are neglected. However, the capacitance of conductors is considered where hoses or long pipes are installed in the real system. This is especially the case for the hosing and piping from the hydraulic power supply to other parts of the system.

To enable the effective training of a fault classification model from data, the training data ideally comprise all possible cases of operating scenarios and fault conditions. Hence, besides the regular operation of the press, operation under fault conditions must be emulated in the simulation model as well. Here, faults are emulated by introducing fault models into the lumped parameter simulation, which then are adjusted in intensity by varying the parameters of the fault models. The considered component faults are listed in Table 1. Leakages at the cylinder are modeled as pressure- and velocity-dependent flow through circular gaps, while the leakage at a connector of the hydraulic power supply is modeled by a pressure-dependent leakage factor. Increased friction at the main cylinder is modeled as an increased friction level, which is adjusted by setting the static friction force, whereas the dynamic friction force is lower than the static friction force by a constant factor. A worn valve spool is modeled by the valve spool overlap, while a sensor offset fault is introduced by adding an offset to the true sensor value. More detailed descriptions of the fault models used are given in [26].

Table 1. Faults of the hydraulic system considered in this study, along with the corresponding model parameters of the simulation model and their ranges of variation.

Fault	Symbol	Model Parameter	Parameter Range	Fault Threshold
Cylinder, Internal Leakage	$Q_{CY,Li}$	Gap height piston sealing	1–100 μm	>25 μm
Cylinder, External Leakage	$Q_{CY,Le}$	Gap height rod sealing	0.5–75 μm	>20 μm
Cylinder, Increased Friction	$F_{CY,Fr}$	Static friction force	500–5000 N	>1500 N
Cylinder, Position Sensor Offset	$x_{CY,off}$	Offset value	0–410 mm	>1 mm
Proportional Valve, Worn Control Edges	$y_{0,\ PV}$	Valve spool overlap	−(2–10)%	<−4%
Hydraulic Power Supply, External Leakage	$Q_{PS,Le}$	Leakage factor	0–0.1 L/(min·bar)	>0.0005 L/(min·bar)

3.2. Data Generation and Labeling

A dataset is generated from the simulation model by means of a variational study, where model parameters are varied. For this, a sampling scheme must be defined for the variable parameters of the simulation model. Here, the basis of the sampling scheme is a Latin Hypercube design. It is generated by creating ten random sampling schemes and choosing the design that best fulfills the maximin criterion. Accordingly, the scheme with the largest minimal distance between samples in the parameter space is selected. The eight-dimensional parameter space is formed by the six fault-indicating parameters and two process parameters, namely the target position of the press and the load force. Both process parameters are varied in three different levels of 290 mm, 310 mm and 340 mm and 40 kN/mm, 60 kN/mm and 80 kN/mm, respectively. Hence, in the obtained dataset, different fault intensities, fault combinations and operating conditions are represented. The temperature, however, is kept constant at 40 °C in all simulation runs.

While a Latin Hypercube design generally aims at sampling uniformly from the parameter space, different regions in the design space can be of different relevance for a fault classification. On the one hand, some fault cases and fault combinations can be of low practical relevance, such as the simultaneous occurrence of a high fault intensity for all considered faults. On the other hand, the severity of some of the considered faults is non-linearly related to the fault-indicating parameter. In this case, placing the samples uniformly over the range of the fault-indicating parameters, a non-uniform representation of fault severities would be obtained in the dataset. This can be illustrated by the example of the internal leakage at the main cylinder of the press, which is modeled as flow through a circular gap and adjusted by setting the height of the gap. As Figure 4 shows, the leakage flow non-linearly increases with increasing gap height. As a result, the range of the gap height settings, which are considered to not represent a fault, is smaller than the range of settings that represents a fault. Consequently, in this case, uniform sampling leads to a higher representation of settings that represent a fault. Accordingly, an adjustment of the distribution of sampled points is required, to obtain a balanced representation of different fault states. For the given reasons, the density of samples is manually reduced in practically less relevant regions of the parameter space, while, in regions of low fault settings, the density of samples is increased by removing or adding random samples in these regions. Finally, by following the described sampling scheme, a variational study is performed with the simulation model and a dataset of 200,000 data instances is generated.

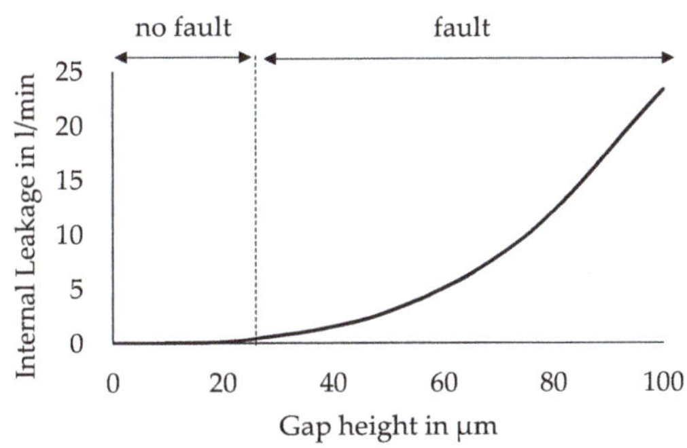

Figure 4. Example of non-uniform class distribution over the range of fault-indicating parameters.

For the application of methods of supervised learning, the generated data instances must be provided along with their class labels, which, in the framework of condition monitoring, indicate the fault state of the considered system. Here, the labels are defined for the single-target classification of binary classes. In the process of simulative data generation, continuous model parameters are used to indicate and adjust the fault intensities. Now, these fault-indicating parameters are transformed into categorical class labels by binning parameter ranges to classes of "fault" and "no fault". The threshold values, which define whether a certain degradation level is considered a fault or not, depend on the specific application and the tolerances allowed therein. In this study, the threshold values listed in Table 1 are used. For the three different leakage faults, thresholds are set under consideration of the leakage flows resulting at the respective fault intensities. The defined thresholds of the gap heights, which mark the beginning of a fault for the internal and external leakage at the cylinder, produce volumetric leakage flows of 0.36 L/min and 0.1 L/min, respectively, at 40 °C and a differential pressure of 150 bar. The threshold value for the leakage factor, which marks the beginning of a fault of external leakage at the hydraulic power supply, is equivalent to a leakage flow of 0.1 L/min at 40 °C and the maximal pressure of 200 bar. For the other three faults, the thresholds are determined heuristically.

Another issue that arises during the variational study is the setting of the initial conditions for the simulation runs. Because model parameters are constantly varied between simulation runs, a single initial system condition, which is valid for all runs, cannot be determined. In order to circumvent this issue, the simulated press cycle is considered to represent one cycle within a series of multiple press cycles. As a result, the initial state of the system is determined by the state at the end of the previous cycle. To achieve this, every simulation run is started with the final second of the control sequence of the working cycle, instead of only simulating the actual working cycle. Afterwards, the first second of the simulated cycle is removed after the data are generated. Consequently, setting the initial conditions of the simulation is no longer critical while performing the variational study. The described approach and its effects are shown in Figure 5 by the example of the pressure at the rod side of the hydraulic cylinder. When only simulating the actual working cycle, the progression of the rod-side pressure depends on the selection of the initial value, as Figure 5a shows. In contrast, Figure 5b shows that starting the simulation run with the ending sequence of a cycle allows to obtain results independent of the selected initial condition, when the first simulated second is removed afterwards.

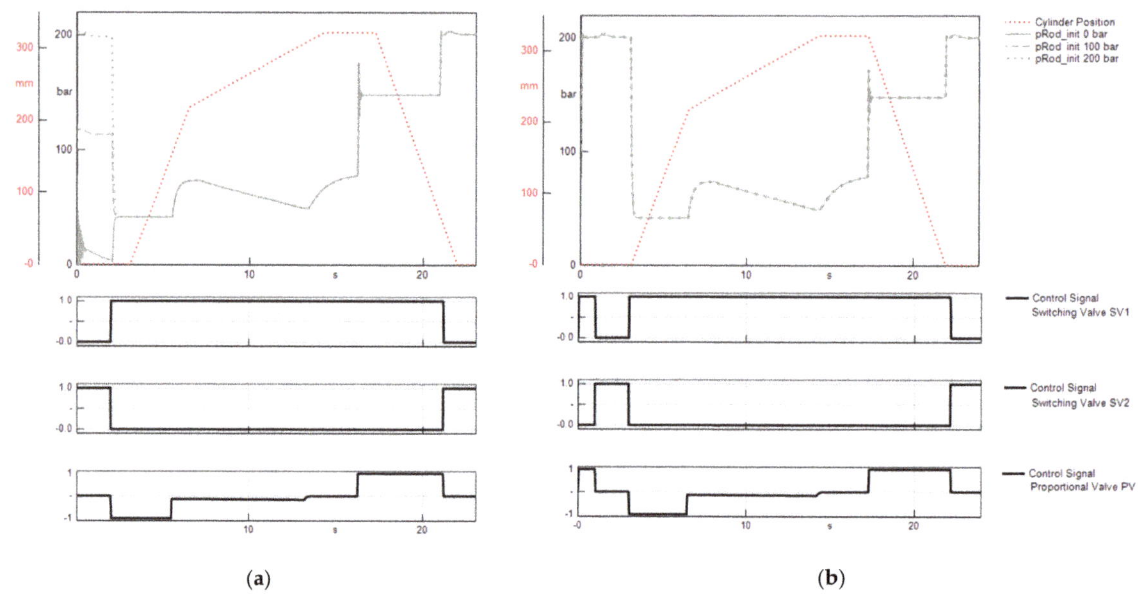

Figure 5. Cylinder position, cylinder pressures and valve control signals simulated under fixed initial conditions while (**a**) starting the simulation with the actual control sequence of the press cycle and (**b**) starting the simulation with the ending sequence of a press cycle prior to the actual control sequence.

3.3. Feature Extraction and Selection

Before using the generated data for training a fault classification model, the raw time series are transformed into a fixed number of features. For this, the python-based open-source package *tsfresh* is used, which provides the calculation of a large variety of features for time series [23]. As the features must be calculated for each of the eight subsignals of the 200,000 multivariate time series, the following base set of 16 features with low computational costs is extracted:

- minimum;
- maximum;
- mean;
- median;
- standard deviation;
- variance;
- skewness;
- kurtosis;
- first location of minimum;

- first location of maximum;
- last location of minimum;
- last location of maximum;
- sum of values;
- absolute energy (sum over squared values);
- absolute sum of changes (sum over absolute values of consecutive changes);
- mean absolute change (mean over absolute values of consecutive changes).

Extracting this base set of features for a time series yields a feature vector of 128 features for one data instance. Generally, the resulting feature vectors only comprise subsignals recorded during data generation. However, derivations and transformations of the raw data often can reveal additional information about the characteristics of a signal. Moreover, the extracted features only represent characteristics over complete working cycles and thus only carry information on the aggregated effect of a fault on the observed data. In contrast, a conventional, visual inspection of sensor data by a human expert would typically involve a detailed inspection of anomalies in specific segments of the working cycle, such as checking for leakages in idle mode or the maximal cylinder velocity in rapid traverse. Such short-term effects can even be attenuated, when computing features over complete working cycles. Consequently, in this study, the state-of-the-art feature extraction is extended by incorporating further information into the feature extraction, which is typically available or can be acquired easily.

The first step of the extension consists in adding information about the progression of the recorded subsignals over time. This information is obtained by computing the first time derivatives of the subsignals and adding them to the time series as further subsignals. As a result, the number of subsignals is doubled and the number of extracted features per time series increases accordingly. The second step of the extension focusses on accentuating the characteristics of the subsignals in specific segments of the working cycle. For this, each time series is divided into segments of constant operation phases. These are the idle phases at the beginning and end of the cycle, rapid traverse, load phase and the backstroke. The transitions between operation phases are inferred from control signals given by the machine control. Subsequently, the same set of features that is extracted for the complete time series is again extracted for each subsignal in each of the defined segments, as Figure 6 illustrates for the features "maximum" and "minimum". In the framework of this contribution, the originally generated feature set is referred to as global features, while the newly added features are referred to as segment features. The length l of the feature vector that represents a single data instance is again increased and now given by Equation (1), where $n_{subsignals}$, $n_{features}$ and $n_{segments}$ denote the number of subsignals, features and segments, respectively. Altogether, extracting features from the full time series and its five segments results in a set of 1536 features.

$$l = n_{subsignals} \cdot n_{features} \cdot (1 + n_{segments}) \tag{1}$$

As a result of the extended feature extraction, the number of features is increased, with the aim to also increase the amount of extracted information. However, not all the additionally generated characteristics can be assumed to be useful for fault classification. Concurrently, a concise set of features without irrelevant and redundant information is preferable for the efficient and robust training of most machine learning models and algorithms [19]. Moreover, a small set of only relevant features enables a human inspection and comprehension of the information fed to a classification model and consequently leads to higher transparency of the inference process. Therefore, feature selection follows, to reduce the set of features to those useful for the respective fault classification. This is achieved by using a wrapper method with decision tree classifiers. This means that a classification model is trained with the primary purpose of accessing the features implicitly selected within the model.

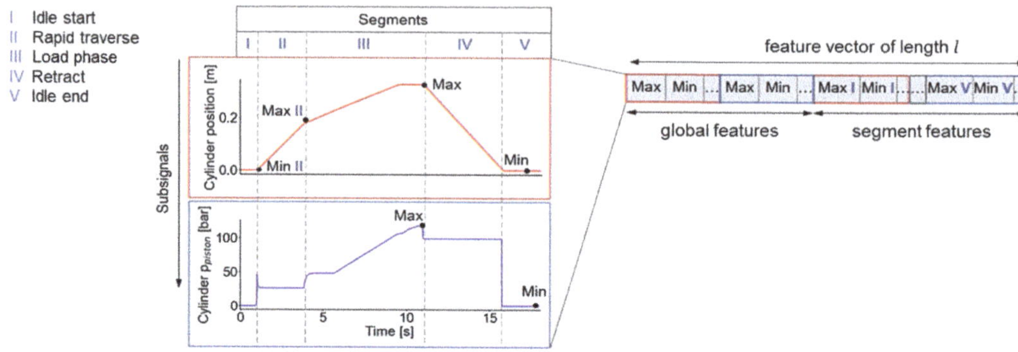

Figure 6. Concept of adding segment features to the base set of global features shown for two subsignals and the features "maximum" and "minimum".

A decision tree classifier is a machine learning model that aims at hierarchically dividing a given dataset into groups of homogenous classes [27], which, in this case, are the classes "fault" and "no fault". The decision rules derived during training are based on defining threshold values for certain features, which then allow to classify a data instance. After training, the features and the corresponding threshold values of the different decision nodes can be inspected. Subsequently, it can be comprehended which features enable the effective separation of the data instances. Figure 7 shows the first two levels of a decision tree for the classification of internal cylinder leakage. Moreover, decision trees, and their underlying algorithm, show low sensitivity of classification performance to the size of the feature sets that they are trained with [19], which makes them well suited for the task of processing the initially large feature set. To mitigate the tendency of decision trees to overfit, the maximum number of levels of the trees is limited. A maximum of five levels is heuristically found to provide a low variance of classification accuracy with no indication of overfitting.

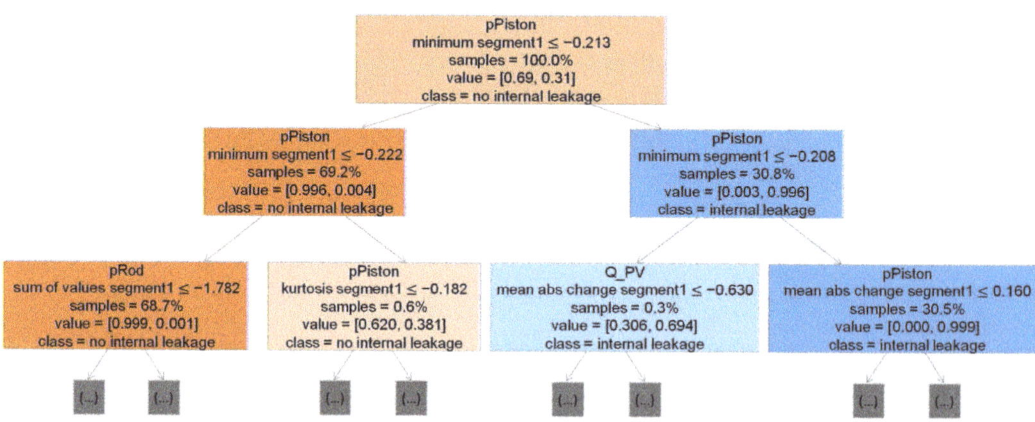

Figure 7. First two levels of a decision tree for the classification of internal leakage at the cylinder. Threshold values relate to features scaled by means of standardization.

As each of the considered component faults of the hydraulic press has a different impact on the machine behavior, the feature selection procedure is repeated for each of the considered faults. Furthermore, the resulting sets of selected features can depend on the specific set of training data used during selection. Random elements included in the

CART algorithm, which is here used for training the decision trees, further add to the stochastic variation of the training results. Therefore, the feature selection is performed in the form of a 5-fold cross-validation. For this, the training and testing of each decision tree is repeated five times, using different excerpts of the data for training and testing in each repetition. Finally, the union set of features selected in all repetitions is formed and used for the training the final classification models. Figure 8 shows the procedure of feature extraction and cross-validated feature selection. The data scaling is not illustrated therein, but standardization is performed within the cross-validation for each training and testing split individually.

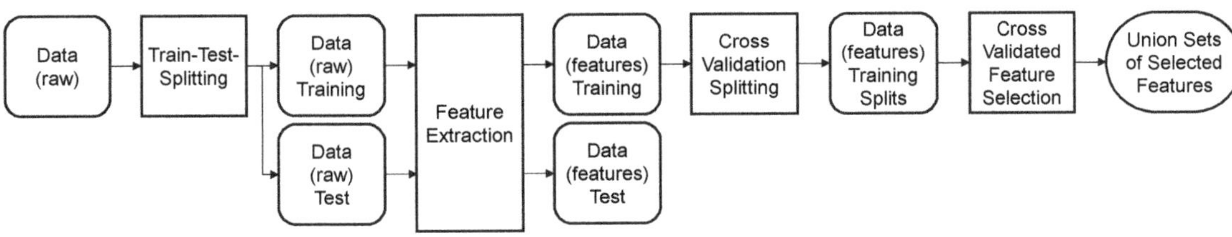

Figure 8. Process of feature extraction and cross-validated feature selection.

3.4. Model Setup, Training and Evaluation

Apart from the decision tree classifiers used during feature selection, the final fault classification is realized using feedforward neural networks. The problem of classifying the machine state is set up as a composition of single-target classifications with binary targets. This means that a model for binary classification is set up and trained for each of the considered component faults individually. This allows us to build smaller models, which potentially can be trained more efficiently and transparently. Moreover, modular fault classification models can be generated independently and on demand, while multi-target classification would require the a priori definition of all possible fault cases that can be expected in a system. Furthermore, using a separate classifier for each fault allows to obtain a diagnosis without the issue of an exponential increase in the number of class labels with an increasing number of faults, as would be the case in multi-target, multi-class classification. However, one disadvantage of the considered setup is that interdependencies between fault occurrences cannot be considered by the models.

The hyperparameters of the feedforward neural networks comprise the structural hyperparameters of the network, such as the number of hidden layers, number of nodes per layer and activation function of each layer, as well as fitting hyperparameters such as the loss function and the learning rate. As with the definition of the labels described previously in this section, the definition of the fitting hyperparameters depends on the definition of the problem at hand. Given that a binary classification is the targeted task, options for selecting the loss function are narrowed down to loss metrics, which are suitable for evaluating the outputs of the model against binary labels. Here, the binary cross-entropy is used as a loss metric. Furthermore, the selection of a learning rate can be simplified to only defining a starting learning rate when a state-of-the-art optimization algorithm with an adaptive learning rate is used. The *Adam* algorithm is used in this study, with a default setting of the starting leaning rate of 0.001.

Similarly, the numbers of nodes on the input layers and output layers of the neural networks are predetermined by the number of features fed to the models and the number of labels to predict, respectively. While the input layer is set to have a ReLU activation function, a sigmoid activation function is used in all hidden layers and the output layer. Further structural hyperparameters of the neural networks, which are the number of hidden layers and the number of nodes therein, are determined by means of a grid search. Different model structures are set up and trained so that the best model structure of the considered configurations can be identified for each fault case. The scheme of the grid search and the concept of varying the model structure are shown in Figure 9. The scheme shows that three different numbers of hidden layers and four different settings of the maximal number of nodes are varied. For a given number of hidden layers, the first hidden layer is constructed with the maximal number of nodes defined. Then, in each subsequent layer, the number of nodes is halved.

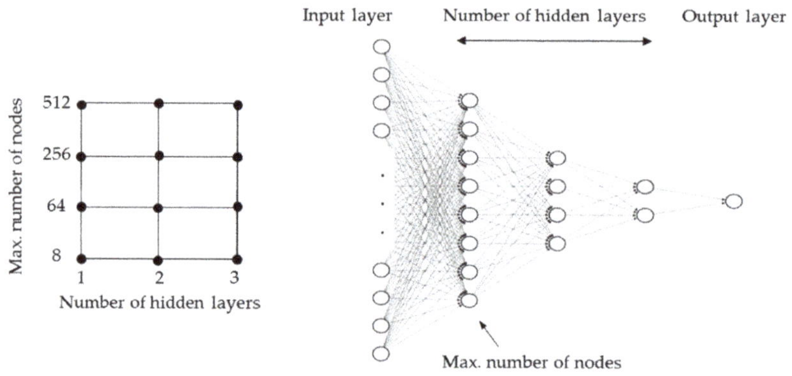

Figure 9. Applied scheme of grid search and corresponding model hyperparameters.

As with the feature selection step, each model configuration considered in the grid search is trained and tested with different excerpts of the data within a 5-fold cross-validation. A further test set, which is held out from the whole process of feature selection and model tuning, is finally used to evaluate the performance of the models selected in the grid search. The test set comprises around 40,000 working cycles, which is 20% of the available data. This form of validation procedure is referred to as the three-way holdout method and is required to obtain an unbiased, final estimate of the model performance, as the test data used in the cross-validation loop were already involved during feature selection and grid search [28]. The classification performance of the models is evaluated using the balanced accuracy. The balanced accuracy is defined as the mean value of the true-positive rate and the true-negative rate and thus gives a balanced estimate independent of the class distribution of the data [29]. The applied procedure of the cross-validated grid search is illustrated in Figure 10.

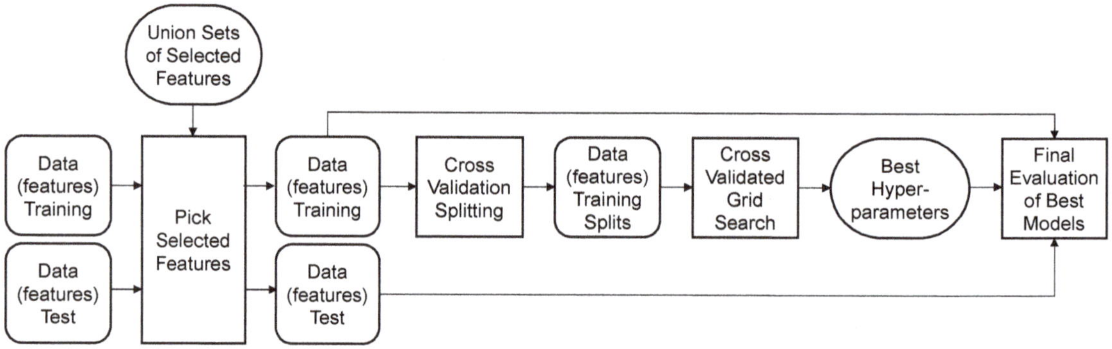

Figure 10. Procedure of cross-validated grid search and final evaluation of models following the three-way hold-out method.

4. Results

4.1. Results of the Grid Search

The mean classification accuracies of the best models identified in the grid search are shown in Table 2 for the different faults, for different feature sets, as well as with and without feature selection. Table 3 lists the corresponding structures of these models. For the faults of internal cylinder leakage $Q_{CY,Li}$, the position sensor offset $x_{CY,off}$ and the external leakage at the hydraulic power supply $Q_{PS,Le}$, high classification accuracies around 99% are already obtained with the base set of global features. This range of accuracies is obtained for these faults with all tested structural configurations of the neural networks, with no clear dependence on model size and a maximal deviation between the best- and worst-performing models of less than 1% across all feature sets. Therefore, the smallest possible model structure (one hidden layer with eight nodes) is also considered the best-performing model for these faults. A marginal increase in the model performance can be observed for these three faults when extending the feature sets and applying feature selection. For the fault of worn control edges on the proportional valve $y_{0,PV}$, a similar trend is observed when increasing the feature set, while the feature selection does not always lead to improved maximal performance. Regarding the model size, however, the feature selection leads to the same effect as stated above for other faults, where the difference in performance between model sizes becomes marginal and the smallest models can be regarded as the best-performing models for a specific configuration. Using an increased set of features clearly improves the performance of the classifiers for the faults of increased friction at the cylinder $F_{CY,Fr}$ and an external leakage at the cylinder $Q_{CY,Le}$. The prediction of increased friction appears to benefit from added derivative and segment features likewise. In contrast, for the external leakage, the information added by segment features appears to mainly induce the improvement, especially when no feature selection is applied. Moreover, for both latter faults, the best-performing models have larger network structures when feature selection is considered. While the achieved maximal accuracies can be considered high performance scores for the increased friction and the external cylinder leakage, the models cannot fully resolve the targeted input–output relationship with the given data and considered model structures.

Table 2. Mean classification accuracies obtained with the best model during the grid search with and without feature selection, for different feature sets and for different faults. Best results for each fault are highlighted in bold.

	Increased Cylinder Friction $F_{CY,Fr}$		Internal Cylinder Leakage $Q_{CY,Li}$		External Cylinder Leakage $Q_{CY,Le}$	
Feature Set	**Without Selection**	**With Selection**	**Without Selection**	**With Selection**	**Without Selection**	**With Selection**
Global	93.60%	93.57%	98.65%	99.29%	78.72%	78.81%
Global + derivative	95.51%	94.97%	98.44%	99.13%	78.68%	82.32%
Global + segment	96.87%	95.91%	99.26%	99.45%	88.69%	86.52%
Global + segment + deriv.	96.81%	**97.09%**	99.21%	**99.51%**	88.38%	**89.69%**
	Position Sensor Offset $x_{CY,off}$		**Worn Control Edges Prop. Valve $y_{0,PV}$**		**External Leakage at Hydr. Power Supply $Q_{PS,Le}$**	
Feature Set	**Without Selection**	**With Selection**	**Without Selection**	**With Selection**	**Without Selection**	**With Selection**
Global	98.51%	99.68%	96.62%	96.42%	99.69%	99.78%
Global + derivative	98.06%	99.64%	96.69%	96.44%	99.71%	99.82%
Global + segment	99.28%	**99.83%**	98.57%	98.93%	99.70%	99.83%
Global + segment + deriv.	99.10%	99.67%	98.38%	**98.94%**	99.70%	**99.85%**

Table 3. Structure of the hidden layers of the best models obtained from the grid search with and without feature selection, for different feature sets and for different faults.

	Increased Cylinder Friction $F_{CY,Fr}$		Internal Cylinder Leakage $Q_{CY,Li}$		External Cylinder Leakage $Q_{CY,Le}$	
Feature Set	**Without Selection**	**With Selection**	**Without Selection**	**With Selection**	**Without Selection**	**With Selection**
Global	(256)	(512, 256)	(8)	(8)	(256)	(256, 128)
Global + derivative	(64)	(256)	(8)	(8)	(64, 32, 16)	(512, 256)
Global + segment	(64, 32)	(512, 256)	(8)	(8)	(64, 32)	(512, 256)
Global + segment + deriv.	(256, 128)	(256, 128)	(8)	(8)	(64, 32, 16)	(512, 256, 128)
	Position Sensor Offset $x_{CY,off}$		**Worn Control Edges Prop. Valve $y_{0,PV}$**		**External Leakage at Hydr. Power Supply $Q_{PS,Le}$**	
Feature Set	**Without Selection**	**With Selection**	**Without Selection**	**With Selection**	**Without Selection**	**With Selection**
Global	(8)	(8)	(512, 256, 128)	(512, 256)	(8)	(8)
Global + derivative	(8)	(8)	(256)	(256)	(8)	(8)
Global + segment	(8)	(8)	(64, 32, 16)	(8)	(8)	(8)
Global + segment + deriv.	(8)	(8)	(8)	(8)	(8)	(8)

A final evaluation of the classification performance of the identified model structures is performed on a held-out test set, which was not involved during the cross-validated feature selection and grid search. This is carried out to validate that the trained models are not overfitted to the training data and that the features selected during feature selection are also important for classifying new, unseen data instances. For this, all the data, which were partitioned into five subsets during cross-validation, are jointly used to train the models that were identified as the best model structures during the grid search. The results are given in Table 4 and show that, generally, the same scores are achieved as given in Table 2 and estimated during the cross-validated grid search. Moreover, the scores obtained on the test set are mostly slightly higher, which can be attributed to the larger set of training data used in this final evaluation.

Table 4. Classification accuracies obtained with the best model structure on the held-out test set with and without feature selection, for different feature sets and for different faults. Best results for each fault are highlighted in bold.

	Increased Cylinder Friction $F_{CY,Fr}$		Internal Cylinder Leakage $Q_{CY,Li}$		External Cylinder Leakage $Q_{CY,Le}$	
Feature Set	**Without Selection**	**With Selection**	**Without Selection**	**With Selection**	**Without Selection**	**With Selection**
Global	94.51%	93.37%	99.08%	99.17%	81.30%	72.94%
Global + derivative	96.38%	94.73%	98.94%	99.07%	77.73%	74.04%
Global + segment	97.62%	96.59%	99.46%	99.47%	**91.89%**	87.11%
Global + segment + deriv.	**97.72%**	97.64%	99.43%	**99.47%**	90.63%	90.17%
	Position Sensor Offset $x_{CY,off}$		**Worn Control Edges Prop. Valve $y_{0,PV}$**		**External Leakage at Hydr. Power Supply $Q_{PS,Le}$**	
Feature Set	**Without Selection**	**With Selection**	**Without Selection**	**With Selection**	**Without Selection**	**With Selection**
Global	97.86%	98.98%	97.12%	97.13%	99.61%	99.66%
Global + derivative	98.53%	98.80%	98.00%	97.01%	99.68%	99.69%
Global + segment	**99.49%**	98.82%	**99.13%**	98.55%	99.75%	**99.86%**
Global + segment + deriv.	98.53%	98.74%	98.25%	98.79%	99.59%	99.76%

4.2. Results of the Feature Selection

Apart from the effects discussed in the previous subsection, further effects of the feature selection are described in the following. Table 5 shows the number of features selected from the different feature sets for the different faults. Compared to the case of no feature selection, the number of features is significantly reduced after selection in all the listed cases. Moreover, considering the number of selected features across feature sets, for most faults, the number of selected features steadily increases when increasing the feature set before selection. The main increase in the number of selected features compared to the base set of global features is obtained by adding segment features, which is in alignment with the increase in performance induced by the segment features observed in the previous subsection. The exception is the fault of increased friction $F_{CY,Fr}$, where a higher number of features is selected that represent trend information of the time series.

Table 5. Number of features before and after feature selection for the different feature sets and different faults.

Feature Set	All Faults Before Selection	Number of Selected Features					
		$F_{CY,Fr}$	$Q_{CY,Li}$	$Q_{CY,Le}$	$x_{CY,off}$	$y_{0,PV}$	$Q_{PS,Le}$
Global	128	40	29	36	14	21	28
Global + derivative	256	38	36	39	14	26	39
Global + segment	738	25	57	46	20	46	42
Global + segment + deriv.	1536	39	59	52	22	51	42

A benefit of a reduced set of features is that it enables to inspect features manually. This allows us to gain an insight into the process of fault inference and into the general system behavior under faults. Moreover, using decision trees for feature selection, the selected features are provided with relative importance, which indicates which fraction of the data instances could be separated into classes of "fault" and "no fault" by using this feature. Accordingly, a ranking of the features is obtained, in addition to the reduction. However, as the feature selection is performed in a cross-validation loop and union sets of selected features are formed afterwards, the ranking information is not maintained after this step. Therefore, the ranking information is acquired from a single run of the cross-validation loop. In Table 6, the three most important features selected for each of the considered faults are listed. The subsignals can be found in the hydraulic circuit diagram in Figure 2.

Table 6. Three most important features identified through feature selection for each fault.

Fault	Subsignal	Features
F_{Fr}	p_{Piston}	median—segment IV
	p_{Rod}	minimum of derivative—segment IV
	p_{Rod}	standard deviation of derivative—segment III
$Q_{CY,Li}$	p_{Piston}	minimum—segment I
	Q_{PV}	maximum of derivative—segment I
	Q_{PV}	absolute sum of changes—segment V
$Q_{CY,Le}$	p_{Piston}	minimum—segment III
	p_{Rod}	maximum—segment III
	p_{Piston}	standard deviation—segment II
$x_{CY,off}$	x_{CY}	minimum—segment I
	x_{CY}	median—segment I
	x_{CY}	minimum

Table 6. *Cont.*

Fault	Subsignal	Features
$y_{0,PV}$	Q_{PV}	maximum—segment III
	Q_{PV}	maximum of derivative—segment III
	Q_{PV}	median
$Q_{PS,Le}$	Q_{PV}	minimum—segment I
	p_{PS}	absolute sum of changes—segment I
	Q_{PV}	mean absolute change—segment V

To check the plausibility of the selected features, an exemplary visual inspection of the respective subsignals is performed. Figure 11 shows the progression of the pressures recorded at the hydraulic cylinder in regular operation, as well as under the influence of an internal cylinder leakage $Q_{CY,Li}$. For the detection of this fault, the minimum of the pressure on the piston side of the cylinder in the idle phase at the beginning of the press cycle (segment I) is selected as the most important feature. This is in accordance with the observed subsignals shown in Figure 11. An internal leakage clearly alters the pressure progression in the idle phases at the beginning and end of the press cycle. Moreover, an internal leakage leads to an increased pressure level at the piston side and thus to an altered minimum value of this pressure. This increase is induced by the pressure equalization between the two cylinder chambers, which is enabled by the existence of a leakage gap. An example for the effect of an external cylinder leakage $Q_{CY,Le}$ on both cylinder pressures is shown in Figure 12. For this fault, the minimum value of the pressure on the piston side during the load phase (segment III) is selected as the most important feature. Again, the selection of this feature appears plausible, as this characteristic of the rod-side pressure is uniquely impacted by the occurrence of the external leakage. Additionally, the drop in the rod-side pressure in segments I and V is strongly affected by the external leakage. However, the fact that this symptom is also observed with an internal cylinder leakage explains why this feature was not selected as important indicator of an external cylinder leakage.

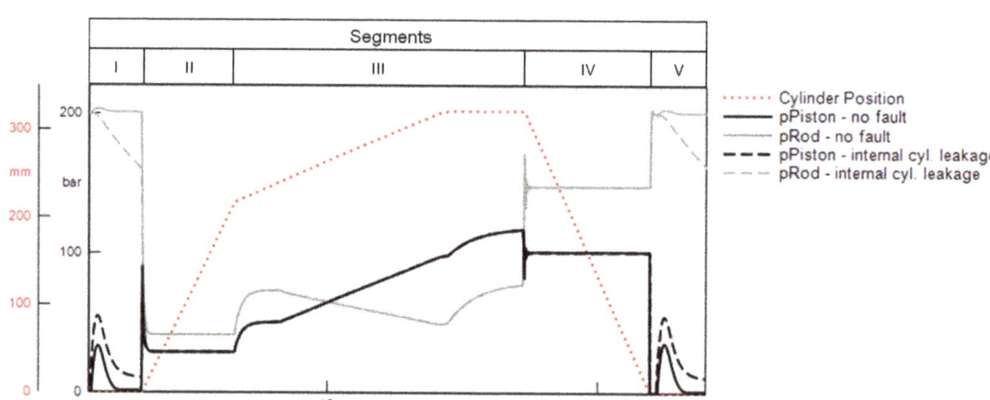

Figure 11. Simulated effect of an internal cylinder leakage on the cylinder pressures.

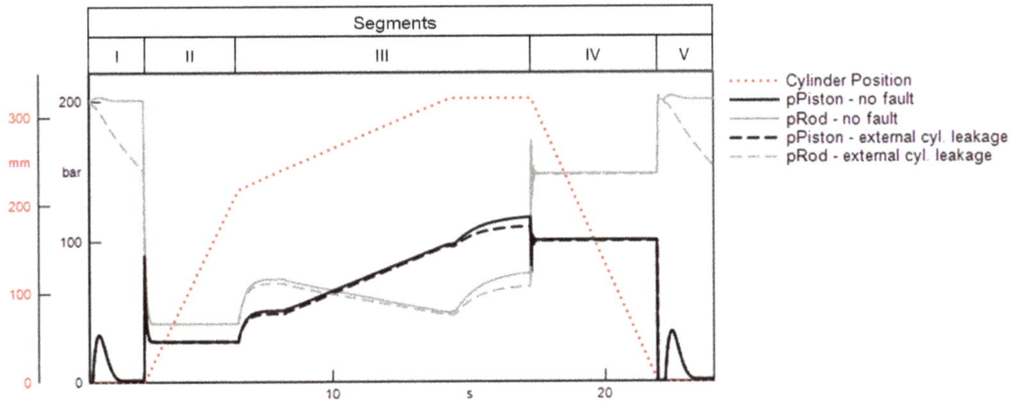

Figure 12. Simulated effect of an external cylinder leakage on the cylinder pressures.

5. Discussion

Overall, the results show that the generic development procedure applied in this study allows to effectively develop fault classification models for the condition monitoring of a hydraulic press. In general, the results show that different faults in the system require different model structures and information for their effective classification. Especially for the faults of the internal leakage at the cylinder, the offset at the position sensor and the external leakage at the hydraulic power supply could be inferred with a base set of generic features and small neural network structures. This indicates that the occurrence of these faults uniquely impacts a small set of features, independently of other faults and operating conditions. For the other faults, the presented approach of extended feature extraction was shown to clearly impact the performance of the respective classification models. The difference in required features and obtained model sizes indicates the sensibility of the applied approach in performing single-target classifications for each fault individually, instead of defining a multi-target task to predict all faults concurrently.

In contrast to extending the feature set, the feature selection does not clearly affect the maximal model performance, regardless of the feature set and fault considered. The feature selection, however, impacts the sizes of the model structures of the best-performing models. Except for the fault cases, where the smallest considered model structure was sufficient, the best-performing models had larger network structures when feature selection was applied. The results indicate that the feature selection yields larger models, when the relationship between features and the existence of a fault cannot be fully resolved. This is particularly indicated by the results for the fault of worn control edges of the proportional valve, shown in Tables 2 and 3. As the feature set is increased and the classification accuracy approaches a level of 99%, the difference in performance between different model structures diminishes and smaller models achieve the same performance as larger model structures. Another aspect to consider is that, in the grid search, only the structure of hidden layers was varied, while the number of input layers depended on the size of the feature vector fed to the model. Consequently, a larger set of features at the input also adds a larger number of model parameters to start with.

Finally, the feature selection provides an insight into the development workflow and increases the transparency of the data-driven development. This is especially the case when human-interpretable features are extracted from the raw time series data. For the examples of an internal and external cylinder leakage, it was shown that the feature selection provided plausible information on symptoms uniquely caused by certain faults.

Author Contributions: Conceptualization, F.M. and K.S.; methodology, F.M.; software, F.M.; validation, F.M.; investigation, F.M.; data curation, F.M.; writing—original draft preparation, F.M.; writing—review and editing, K.S. and F.M.; supervision, K.S.; project administration, F.M. and K.S.; funding acquisition, K.S. All authors have read and agreed to the published version of the manuscript.

Funding: The authors thank the Research Association for Fluid Power of the German Engineering Federation VDMA for its financial support. Special gratitude is expressed to the participating companies and their representatives in the accompanying industrial committee for their advisory and technical support.

Institutional Review Board Statement: Not applicable.

Informed Consent Statement: Not applicable.

Data Availability Statement: The datasets generated during and/or analyzed during the current study are available from the corresponding author on reasonable request.

Conflicts of Interest: The authors declare no conflict of interest.

References

1. *ISO 17359:2018*; Condition Monitoring and Diagnostics of Machines—General Guidelines (ISO 17359:2018). International Organization for Standardization: Geneva, Switzerland, 2018.
2. Watton, J. *Modelling, Monitoring and Diagnostic Techniques for Fluid Power Systems*; Springer: London, UK, 2007; ISBN 9781846283734.
3. Isermann, R. *Fault-Diagnosis Systems*; Springer: Berlin/Heidelberg, Germany, 2006; ISBN 978-3-540-24112-6.
4. Alt, R.; Murrenhoff, H.; Schmitz, K. A Survey of "Industrie 4.0" in the Field of Fluid Power—Challenges and Opportunities by the Example of Field Device Integration. In Proceedings of the 11th International Fluid Power Conference, Aachen, Germany, 19–21 March 2018; Murrenhoff, H., Ed.; RWTH Aachen University: Aachen, Germany, 2018; ISBN 3958862152.
5. Kuhn, M.; Johnson, K. *Applied Predictive Modeling, Corrected at 5th Printing*; Springer: New York, NY, USA, 2016; ISBN 978-1-4614-6848-6.
6. Döbel, I.; Leis, M.; Vogelsang, M.M.; Neustroev, D.; Petzka, H.; Rüping, S.; Voss, A.; Wegele, M.; Welz, J. Maschinelles Lernen—Kompetenzen, Anwendungen und Forschungsbedarf: Projektbericht. 2018. Available online: https://www.bigdata-ai.fraunhofer.de/content/dam/bigdata/de/documents/Publikationen/BMBF_Fraunhofer_ML-Ergebnisbericht_Gesamt.pdf (accessed on 28 July 2022).
7. Paulmann, G.; Mkadara, G. Condition monitoring of hydraulic pumps—Lessons learnt. In Proceedings of the 11th International Fluid Power Conference, Aachen, Germany, 19–21 March 2018; Murrenhoff, H., Ed.; RWTH Aachen University: Aachen, Germany, 2018; ISBN 3958862152.
8. Torikka, T.; Haack, S. Predictive Maintenance Service Powered by Machine Learning and Big Data. In Proceedings of the 11th International Fluid Power Conference, Aachen, Germany, 19–21 March 2018; Murrenhoff, H., Ed.; RWTH Aachen University: Aachen, Germany, 2018; ISBN 3958862152.
9. Schraft, J.P.; Becher, D.; Weber, J. Condition Monitoring Strategy for Pump-Driven Hydraulic Axes. In Proceedings of the ASME/BATH Symposium on Fluid Power and Motion Control, Online, 9–11 September 2020; The American Society of Mechanical Engineers: New York, NY, USA, 2020; ISBN 9780791883754.
10. Azeez, A.A.; Vuorinen, E.; Minav, T.; Casoli, P. AI-based condition monitoring of a variable displacement axial piston pump. In Proceedings of the 13th International Fluid Power Conference, Aachen, Germany, 13–15 June 2022; Schmitz, K., Ed.; RWTH Aachen University: Aachen, Germany, 2022.
11. Wiens, T.; Fernandes, J. Application of Data Reduction Techniques to Dynamic Condition Monitoring of an Axial Piston Pump. In Proceedings of the ASME/BATH Symposium on Fluid Power and Motion Control, Longboat Key, FL, USA, 7–9 October 2019; The American Society of Mechanical Engineers: New York, NY, USA, 2020; ISBN 9780791859339.
12. Yao, J.; Li, X.; Wang, P.; Cheng, Y.; Yang, S. Data-driven Fault Diagnosis Based on Deep Learning for Multiple Failures of High Speed On/off Valve. In Proceedings of the 13th International Fluid Power Conference, Aachen, Germany, 13–15 June 2022; Schmitz, K., Ed.; RWTH Aachen University: Aachen, Germany, 2022.
13. Ali, E. Self-Learning Condition Monitoring for Smart Electrohydraulic Drives. Ph.D. Thesis, Technische Universität Dresden, Dresden, Germany, 2019.
14. Torikka, T. Bewertung von Analyseverfahren zur Zustandsüberwachung einer Axialkolbenpumpe. Ph.D. Thesis, Shaker, Aachen, Germany, 2011.
15. Duensing, Y.; Rodas Rivas, A.; Schmitz, K. Machine Learning for Failure Mode Detection in Mobile Machinery. In Proceedings of the 11th Kolloquium Mobilhydraulik, Karlsruhe, Germany, 10 September 2020; Geimer, M., Synek, P.-M., Eds.; KIT Scientific Publishing: Karlsruhe, Germany, 2020; pp. 1–25.
16. Helwig, N.J. Zustandsbewertung Industrieller Prozesse Mittels Multivariater Sensordatenanalyse am Beispiel Hydraulischer und Elektromechanischer Antriebssysteme. Ph.D. Thesis, Shaker, Aachen, Germany, 2019.
17. Adams, S.; Beling, P.; Farinholt, K.; Brown, N.; Polter, S.; Dong, Q. Condition Based Monitoring for a Hydraulic Actuator. In Proceedings of the Annual Conference of the PHM Society, Bilbao, Spain, 5–8 July 2016.

18. Makansi, F.; Schmitz, K. Feature Generation and Evaluation for Data-Based Condition Monitoring of a Hydraulic Press. In Proceedings of the 13th International Fluid Power Conference, Aachen, Germany, 13–15 June 2022; Schmitz, K., Ed.; RWTH Aachen University: Aachen, Germany, 2022.
19. Kuhn, M.; Johnson, K. *Feature Engineering and Selection: A Practical Approach for Predictive Models*; Chapman & Hall/CRC: Boca Raton, FL, USA, 2021; ISBN 978-1-03-209085-6.
20. Aggarwal, C.C. *Neural Networks and Deep Learning: A Textbook*; Springer International Publishing: Cham, Switzerland, 2018; ISBN 978 3 319 94462 3.
21. Hornik, K.; Stinchcombe, M.; White, H. Multilayer feedforward networks are universal approximators. *Neural Netw.* **1989**, *2*, 359–366. [CrossRef]
22. Géron, A. *Hands-on Machine Learning with Scikit-Learn, Keras, and TensorFlow*; O'Reilly Media, Inc.: Sebastopol, CA, USA, 2019.
23. Christ, M.; Kempa-Liehr, A.W.; Feindt, M. Distributed and Parallel Time Series Feature Extraction for Industrial Big Data Applications. 2016. Available online: http://arxiv.org/pdf/1610.07717v3 (accessed on 28 July 2022).
24. Radford, A.; Metz, L.; Chintala, S. Unsupervised Representation Learning with Deep Convolutional Generative Adversarial Networks. 2015. Available online: http://arxiv.org/pdf/1511.06434v2 (accessed on 28 July 2022).
25. ESI Group. SimulationX. Available online: https://www.esi-group.com/products/system-simulation (accessed on 28 July 2022).
26. Makansi, F.; Schmitz, K. Simulation-Based Data Sampling for Condition Monitoring of Fluid Power Drives. *IOP Conf. Ser. Mater. Sci. Eng.* **2021**, *1097*, 012018. [CrossRef]
27. Fürnkranz, J. Decision Tree. In *Encyclopedia of Machine Learning and Data Mining*; Sammut, C., Webb, G.I., Eds.; Springer: Boston, MA, USA, 2017; pp. 330–335, ISBN 978-1-4899-7685-7.
28. Raschka, S. Model Evaluation, Model Selection, and Algorithm Selection in Machine Learning. 2018. Available online: http://arxiv.org/pdf/1811.12808v3 (accessed on 28 July 2022).
29. Brodersen, K.H.; Ong, C.S.; Stephan, K.E.; Buhmann, J.M. The Balanced Accuracy and Its Posterior Distribution. In Proceedings of the 2010 20th International Conference on Pattern Recognition (ICPR), Istanbul, Turkey, 23–26 August 2010; IEEE: Piscataway, NJ, USA, 2010; pp. 3121–3124, ISBN 978-1-4244-7542-1.

energies

Article

LQG Control of an Open Circuit Axial Piston Pump

Alexander Mitov [1,*], Jordan Kralev [2] and Tsonyo Slavov [2]

1 Department of Hydroaerodynamics and Hydraulic Machines, Technical University of Sofia, Kliment Ohridski 8 Boulevard, 1000 Sofia, Bulgaria
2 Department of Systems and Control, Technical University of Sofia, Kliment Ohridski 8 Boulevard, 1000 Sofia, Bulgaria
* Correspondence: a_mitov@tu-sofia.bg; Tel.: +359-886208937

Abstract: In recent years, the development of hydraulic variable displacement axial piston machines has been focused in two main directions: improvement of their construction and improvement of their displacement control methods. The goal of both directions is to increase the efficiency of the machines. Increasing their efficiency is key to improving the efficiency of the entire hydraulic system, whether they are used as pumps or hydraulic motors. This motivates the present work, which essentially contains a developed embedded control system designed for a known type of open circuit axial piston pump. The developed solution is implemented on a laboratory test rig. A detailed description of the hydraulic system in the context of pump displacement control is presented, as well as the developed system architecture for its control. The control system is based on a linear-quadratic Gaussian (LQG) controller. The controller is synthesized on the basis of a model obtained by means of identification based on experimental data. The designed controller is validated through experimental studies, enabling the analysis of its performance.

Keywords: linear-quadratic Gaussian controller; axial piston pump; embedded control

Citation: Mitov, A.; Kralev, J.; Slavov, T. LQG Control of an Open Circuit Axial Piston Pump. *Energies* **2022**, *15*, 6800. https://doi.org/10.3390/en15186800

Academic Editors: Paolo Casoli and Massimo Rundo

Received: 31 July 2022
Accepted: 10 September 2022
Published: 17 September 2022

1. Introduction

Axial piston displacement machines are widely used in hydraulic drive systems [1]. The reason for this is a number of advantages compared to other types of hydraulic displacement machines. These advantages are mainly expressed in the following aspects: high density of transmitted power per unit of weight with relatively compact overall dimensions, low level of flow rate fluctuations, workability with relatively higher pressure, the possibility of variable displacement control according to different control laws, possibility to work in both open and closed-loop circulation, etc. The last of the listed advantages proves their applicability for both mobile and industrial applications [2]. Of course, their relatively high cost is often pointed out as a significant drawback, which is justified by the need for greater power of the hydraulic drive.

These advantages, in addition to expanding the range of applicability of axial piston machines, have motivated scientific research in the last few decades [3–6]. This makes variable displacement axial piston machines an object not only of application but also of scientific interest.

It is pertinent to note that the development of software products for performing simulations in various types of hydraulic elements, devices and machines facilitates scientific research for the purpose of constructive improvements [3]. A number of scientists from various institutes and development units, scientific–educational, have determined constructive qualitative improvements on the basis of in-depth simulation studies, making a connection between the structural parameters and the flow parameters in axial piston machines [1,7–9]. Less research exists on the aspect of improving the regulation of the displacement volume, especially in pumps of this type [10]. In retrospect, a significant impetus was first provided by the development of high response proportional electro-hydraulic spool valves, which replaced conventional hydraulic controllers [11,12]. Then, along with the development of proportional hydraulic technology, so-called secondary control systems were also developed [13,14]. As is known, axial piston machines are mainly used in these systems, and their efficiency depends on the strategy by which they are controlled [15]. In the last two decades, there has been a growing number of applications of non-conventional controllers for axial piston pump control [16–18].

Looking at the development of control systems applied in various fields of technology, it can be seen that research is mainly focused on embedded systems and the implementation of complex control laws through programmable platforms [19,20]. The authors have accumulated experience in the implementation of this type of control systems for plants of different natures, particularly hydraulic drive systems.

This motivates the present work, which essentially contains a developed embedded control system designed for a known type of open circuit axial piston pump. The developed solution is implemented on a laboratory test rig. A detailed description of the hydraulic system in the context of pump displacement control is presented, as well as the developed system architecture for its control. The main goal is to design a linear-quadratic Gaussian (LQG) controller. The controller is synthesized on the basis of a model obtained by the "black box" system identification approach based on experimental data. The designed controller is validated through experimental studies enabling the analysis of its performance.

2. System Description

For the purpose of system identification and the subsequent development of embedded controllers, a laboratory test rig for the study of open circuit axial piston pumps with proportional valve control is developed and implemented [4]. The hydraulic circuit diagram of the test rig is shown in Figure 1, and its photo is presented in Figure 2. The system is based on a well-known model pump, A10VSO (Bosch Rexroth company). The regulation of the displacement volume of this type of pump is well known in the theory and practice of hydraulic drives [2]. It is based on two main approaches. The first consists of using a conventional hydraulic controller of type DR, DFR, DFLR, etc. These controllers are directly coupled to the pump casing and perform their regulating function (of pressure and flow rate) according to a certain law depending on their schematic and constructive solution. The other approach for their regulation consists of the use of a hydraulic valve that is specially developed for this type of pump with proportional electric control, which is mounted on the same plate of connection intended for the listed conventional controllers [21]. In this way, through integrated electronics or an external card type VT-5041, control of the spool of this valve is performed, and, in turn, the displacement volume of the pump is changed via the main energy parameters, pressure and flow rate [21,22]. In practice, this approach is designed to implement secondary control; however, it requires the use of a variant of the same pump that has a swash plate swivel angle sensor and the ability to change the angle of the swash plate in the opposite direction to work in the motor mode of operation. Inevitably, this modernization approach has a significantly higher cost but with an expanded range of applications. In this aspect, the developed control solution incorporates both regulation approaches. However, a pump is used without a swivel angle sensor and without the possibility of operation in motor mode. On the other hand, both types of controllers are used, connected in parallel by means of an additional developed mounting hydraulic block with a built-in check valve. It must be noted that this cannot be used for secondary control. Its advantage is that it can implement a hydraulic system with speed and load feedback based on a variable displacement axial piston pump of the most common type. A detailed description of the system was carried out in a previous publication [4]. Here, only a brief presentation is made, emphasizing the mentioned advantages and disadvantages compared to solutions imposed in practice.

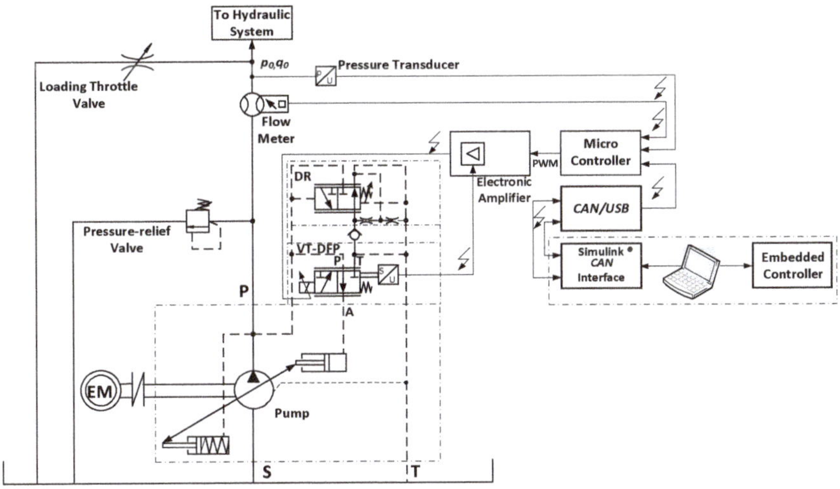

Figure 1. Hydraulic circuit diagram of a test rig for the axial piston pump investigation.

Figure 2. Photo of the test rig.

As mentioned above, the system consists of a variable displacement axial piston type A10VSO designed for open-circulation hydraulic drive systems. The pump has an 18 cubic centimeters displacement volume. It is equipped with a hydraulic pressure controller whose designation is better known as DR. Parallel to it is a proportional hydraulic valve VT-DFP (Bosch Rexroth Company) designed only for this type of pump and having a built-in spool position sensor. Control of this type of valve is realized by a selective external electronic amplifier type VT-5041, of which only the valve spool position feedback function and the power stage are used to power its proportional solenoid [23]. Classically, the system structure is presented in Figure 3. In the developed solution, an electronic amplifier is used to receive a reference signal from a PLC MC012 [24]. The system software is distributed between the industrial controller and the desktop workstation to allow rapid prototyping of the different types of identification experiments and embedded controllers. This is another advantage of the system as it makes it possible to realize advanced control laws, not just conventional PID controllers and their varieties established in hydraulic drives. The distributed system is based on a real-time communication channel between the PLC and the workstation running a Simulink® model (Mathworks). Communication is carried out via a USB/CAN network in blocking mode with a synchronization packet emitted from the industrial controller with a sample time of 10 milliseconds. This sampling period is fast enough to allow precise control of the pressure and flow rate.

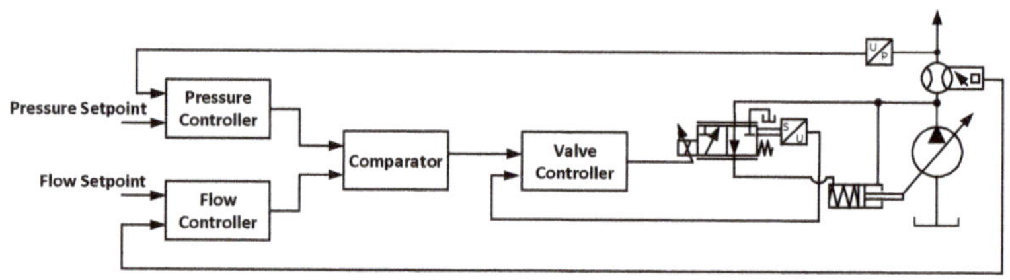

Figure 3. Hydraulic circuit diagram of a test rig for investigation of the axial piston pump.

3. System Identification

To employ the linear quadratic control method, a linear time-invariant state space representation of the axial piston pump is required. Generally, two approaches are available to build a state space model, either through the application of first principle physical laws [25] and the consecutive estimation of model parameters ("gray box" model), or alternatively, the application of "black box" system identification techniques [26,27]. The "gray box" approach requires a detailed understanding of the internal construction of the axial piston pump, as well as the exact measurement of geometrical and physical variables. Even though that approach is beneficial for advancing construction aspects of the axial piston pump machine by reaching a high-fidelity simulation model, it does not guarantee successful control system design. As we know, the purpose of control-oriented models is to focus on the numerical signal transformations between manipulated and performance variables because the controller will aim to invert these transformations in order to achieve the desired behavior of the system. Certainly, input–output transformation can be obtained from a high-fidelity physical model of the system; however, a lot of non-control-related information will also be incorporated into the physical model. Hence, considering the amount of effort and experimental work necessary to achieve precise physical modelling, in most situations, control engineers prefer the estimation of black box functional models. Additionally, digital controllers require discrete-time representation of the system dynamics, which can be naturally achieved by selecting a discrete-time structure for the representation of the black box model. The key elements in black box system identification are the model structure, identification experiment and estimation method.

For the purpose of axial piston pump control, the selected model structure is a discrete-time state space model in controllability canonical form (Cauchy representation):

$$x(k+1) = Fx(k) + Gu(k) + Ke(k), \tag{1}$$

$$y(k) = Cx(k) + Du(k) + e(k), \tag{2}$$

where $x = (x_1, x_2, \ldots, x_n)^T$ is a state vector, $u(k)$ is the PWM signal applied to a proportional spool valve, $y = \left(y_q, y_p\right)^T$ is the output signal, y_q is the pump flow rate, y_p is the pump pressure and $e = \left(e_q, e_p\right)^T$ is the residual error with respect to pump flow rate (e_q) and e_p pump pressure. F, G, K, C, D are model parameters which are matrices with appropriate dimensions. To obtain model parameters, open loop identification experiments are performed for variable loading conditions according to the scheme presented in Figure 4. The discrete-time experiment is carried out with sample time $T_s = 0.01$ s, which is sufficiently small for this type of dynamic system. The identification input signal is a white Gaussian noise that is applied as a PWM voltage signal to a proportional valve.

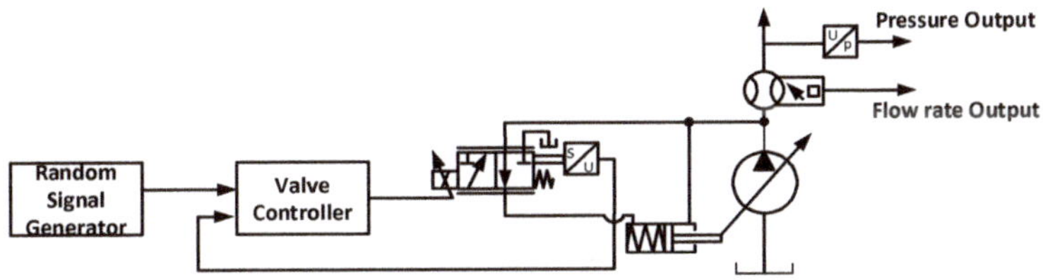

Figure 4. Structure scheme of open loop identification experiment.

The measured output signals are pump flow rate and pump pressure. The date set obtained after the experiment in the presence of full loading (the loading throttle valve depicted in Figure 1 is approximately close) is depicted in Figures 5–7.

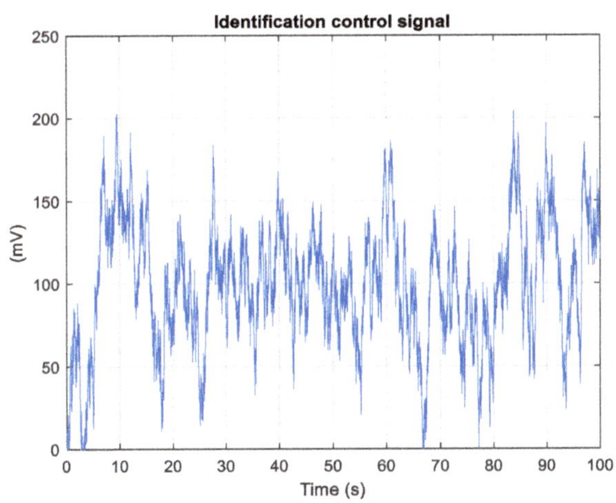

Figure 5. PWM voltage signal proportional to valve used for identification experiment.

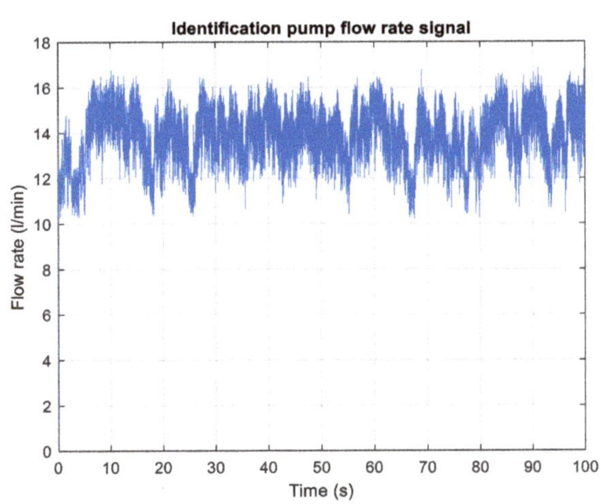

Figure 6. Pump flow rate signal used for identification experiment.

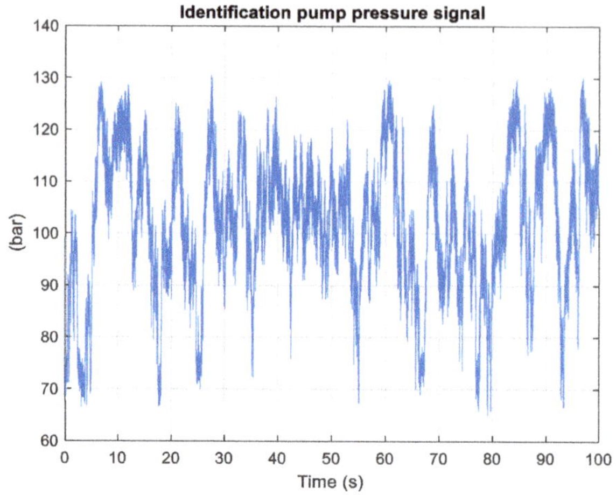

Figure 7. Pump pressure signal used for identification experiment.

The first 1000 points are removed from the data set in order to exclude transient responses. The remaining data are centered and divided into two data sets, the first is used for parameter estimation, and the second is used for model validation. Due to the absence of accurate information for the order of models (1) and (2), the models of the 1st to 10th order are estimated.

The Hankel singular values of models from the 1st to 10th order are depicted in Figure 8. As can be seen from the figure, the appropriate model order is the 5th, but the model of the 6th order has singular values close to the ones of the 5th order model. The Hankel singular values test provides only initial information about the supposed model order and cannot be used alone to determine the appropriate model. Thus, the models of the 4th, 5th, 6th and 8th orders are estimated. The comparison of measured outputs and those simulated by estimated model outputs are presented in Figure 9. As can be seen from the pump flow rate, the FIT between the measured and simulated signals is almost the same for all models. However, for the pump pressure, the FIT between the measured and simulated signal for the models of the 5th, 6th and 8th orders is significantly large compared to the one for the 4th order model.

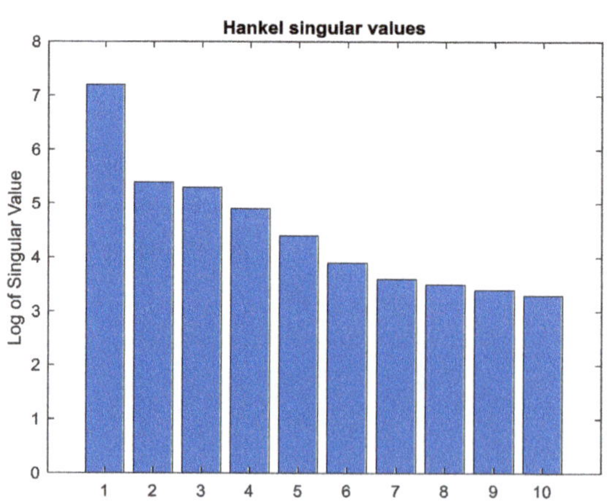

Figure 8. Hankel singular values of estimated models.

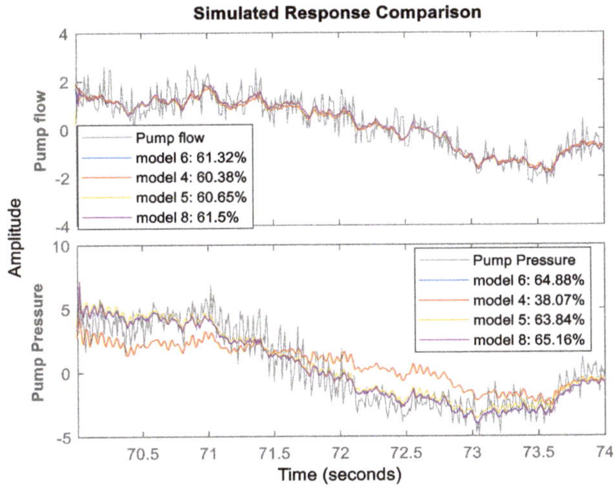

Figure 9. Comparison of model outputs and measured outputs.

The models of the 5th, 6th and 8th orders passed residual correlation tests (Figure 10). As can be seen, the residual error for both outputs is white Gaussian noise. This indicates that the obtained parameter's estimates are unbiased. Additionally, the residual error is uncorrelated with input signals which shows that the model sufficiently describes the pump dynamics for both channels.

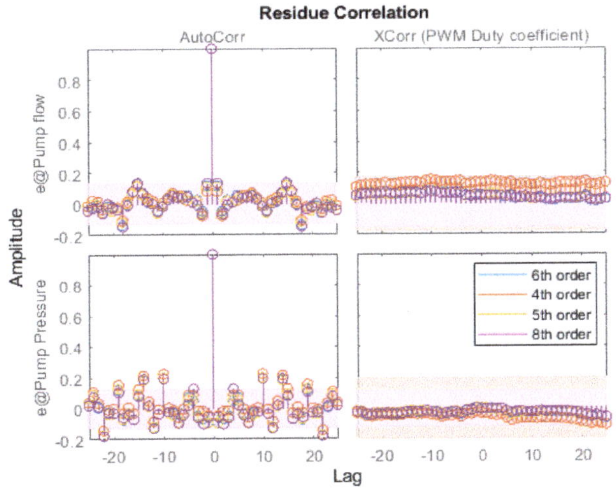

Figure 10. Residual correlation test of 6th order model.

The 6th order is most appropriate due to the fact that it also has good performance in the case of non-loading conditions (the loading throttle valve is fully open). This can be seen from the frequency responses of estimated models in non-loading conditions, which are depicted in Figure 11. It is seen that in a high-frequency range, the responses of the 6th and 8th model orders are close, whereas the response of the 5th order model is approximately the same as the 4th order model.

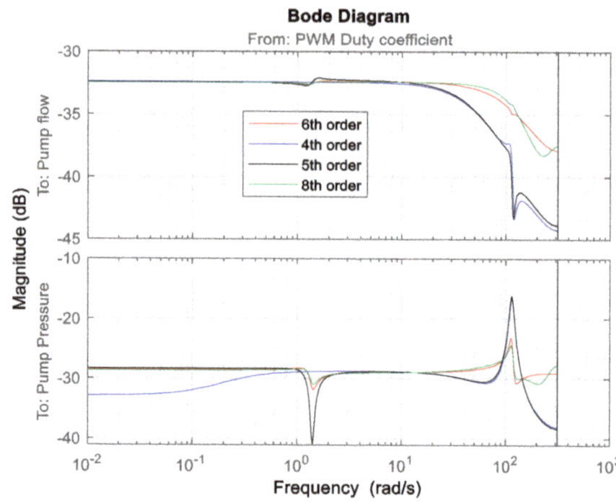

Figure 11. Comparison of frequency responses of 4th, 5th, 6th and 8th order model in case of non-loading.

The step responses along their confidence intervals with respect to pump flow rate and pump pressure for the 4th and 6th model orders are depicted in Figure 12. It is seen that the confidence intervals are small.

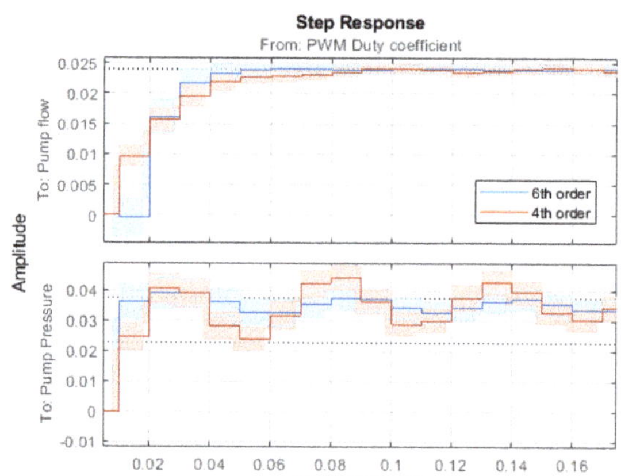

Figure 12. Comparison of magnitude frequency of 6th order model.

The matrices of the estimated 6th order model take the values:

$$\mathrm{A} = \begin{pmatrix} 0 & 1 & 0 & 0 & 0 & 0 \\ 0 & 0 & 1 & 0 & 0 & 0 \\ -0.3 & 0.43 & 0.84 & -0.01 & 0.04 & -0.03 \\ 0 & 0 & 0 & 0 & 1 & 0 \\ 0 & 0 & 0 & 0 & 0 & 1 \\ 0.57 & -11.17 & 10.83 & 0.72 & -1.62 & 1.9 \end{pmatrix}, \mathrm{B} = \begin{pmatrix} 0.005 \\ 0.161 \\ 0 \\ 0.209 \\ 0.046 \\ 0.058 \end{pmatrix},$$
$$\mathrm{K} = \begin{pmatrix} 0 & 0.075 \\ 0.082 & -0.015 \\ 0.031 & 0.031 \\ 0.423 & 0.099 \\ -0.033 & 0.259 \\ 0.141 & -0.029 \end{pmatrix}, \mathrm{C} = \begin{pmatrix} 1 & 0 & 0 & 0 & 0 & 0 \\ 0 & 0 & 0 & 1 & 0 & 0 \end{pmatrix} \text{ and } \mathrm{D} = \begin{pmatrix} 0 \\ 0 \end{pmatrix}. \quad (3)$$

4. LQG Control Design

The structure of the designed control system is depicted in Figure 13. The purpose of the controller design in this research is to regulate the flow rate y_q of the axial piston pump to the desired reference value r_q with maximal steady-state accuracy and minimal settling time. Several problems arise in the control of axial piston pumps. First, the pump is based on a single input multiple output (SIMO) model. It can be established through the controllability test that we can regulate both pressure y_p and the flow rate y_q to desired values, if and only if they lie in a common equilibrium manifold $y_p = f(y_q)$. However, due to the uncertainty in the system dynamics, the exact analytical form of the manifold is not established; hence, we can either regulate the pressure or the flow rate alternately. The second problem arising from the control design is the inherent uncertainty in the identified pump model. This uncertainty can be characterized by the limited size of the identification data, the presence of uncontrollable disturbances acting on the plant and the non-stationarity of the pump dynamics with the selected operating point. The operating point is selected in our experiment by the level of closing of the precise resistance throttle valve bypassing the pump. Hence, the third problem with the present system is that due to uncertainty, it requires a feedback controller. However, due to the static nature of the plant model, we need to extend the feedback loop with an additional integrator element to guarantee reference tracking in the presence of output loading disturbances.

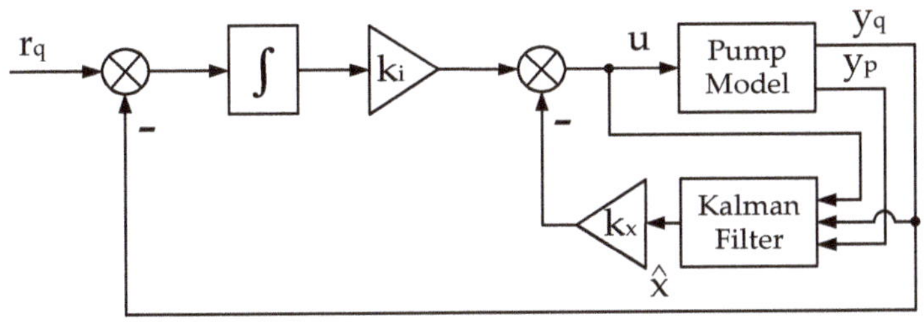

Figure 13. Structure of closed-loop system with LQG controller and Kalman filter.

The linear quadratic control approach is derived from the key results of optimal control theory, where a quadratic cost functional is minimized with respect to the realization of the control signal. The benefit of the LQR optimal solution is that it can be represented as a function of the Lagrange multiplier costate vector which is coupled to the system state vector through the solution of the differential Ricatti equation [28]. The quadratic cost function for the closed-loop pump model is selected as

$$J(u) = \sum_{k=0}^{\infty} \boldsymbol{x}^T[k] Q_x \boldsymbol{x}[k] + r u[k]^2 + w_{I_y} I_y[k], \tag{4}$$

where $\boldsymbol{x}[k] = \begin{pmatrix} x_q[k] & x_2[k] & x_3[k] & x_p[k] & x_5[k] & x_6[k] \end{pmatrix}^T$ is the state vector of the identified pump model, $Q_x \in \mathbb{R}^{6\times 6}$ is the positive definite real matrix weight of the state vector components and their correlations and $r \in \mathbb{R}$ is the weight gain for the amplitude of the control signal. In the cost function, we also include the signal $I_y[k]$ weighted with $Q_y \in \mathbb{R}$, which represents the integrated output error with respect to the flow rate:

$$I_y[k] = \sum_{i=0}^{k} T_S (r_q[k] - y_q[k]) = \sum_{i=0}^{k} T_S e_q[k]. \tag{5}$$

For calculation purposes, the integrated signal is represented as a finite difference equation for a first-order Euler approximation of the integration operation in continuous time as

$$I_y[k] = I_y[k-1] + T_S e_q[k]. \tag{6}$$

The weighting matrix Q_x, together with r and w_{I_y}, are tunable parameters reflecting performance requirements for the closed-loop system. However, due to the implicit representation of the physical state variables in the state vector, it is not obvious how to select the components of the matrix Q_x. Hence, we employ mapping with matrix C between the output and the state variables to translate the tuning of Q_x into the output space as

$$\boldsymbol{y}^T[k]Q_y\boldsymbol{y}[k] = \boldsymbol{x}^T[k]\begin{pmatrix} C_q{}^T & 0 \\ 0 & C_p{}^T \end{pmatrix} Q_y \begin{pmatrix} C_q & 0 \\ 0 & C_p \end{pmatrix}\boldsymbol{x}[k] = \boldsymbol{x}^T[k]Q_x\boldsymbol{x}[k], \quad (7)$$

where the matrix $Q_y \in \mathbb{R}^{2\times 2}$ is a corresponding weighting matrix in the output space of the model. Since the physical meaning of the output signals is established as $y = \begin{pmatrix} y_q & y_p \end{pmatrix}^T$, we can specify the matrix $Q_y = diag(w_q, w_p)$, where w_q represents the cost weight for the flow rate channel and w_p represents the cost weight for the pump pressure channel. In this mapping, the matrix

$$Q_x = \begin{pmatrix} C_q{}^T w_q C_q & 0 \\ 0 & C_p{}^T w_p C_p \end{pmatrix}. \quad (8)$$

Considering that in the identified model, matrices C_q and C_p contain just one non-zero element, i.e., a unit corresponding to the state for the output signal, we see that with such a weighting matrix, we put weight on only two of the states and the rest are left outside of the cost function, or we do not care for their amplitude, since they do not carry an explicit physical meaning.

Equivalently, we can extend the state vector of the plant model, including the integrator state $I_y[k]$, as

$$x_{ext} = \begin{pmatrix} I_y \\ x \end{pmatrix}, \quad (9)$$

and then the plant model matrices F and G will become

$$F = \begin{pmatrix} 1 & -T_S C_q & 0 \\ 0 & F_q & F_{qp} \\ 0 & F_{pq} & F_p \end{pmatrix}, G = \begin{pmatrix} 0 \\ G_q \\ G_p \end{pmatrix}. \quad (10)$$

Additionally, the state weighing matrix for the extended state vector will become

$$Q_{ext} = \begin{pmatrix} w_{I_y} & 0 \\ 0 & Q_x \end{pmatrix}. \quad (11)$$

Following the methodology for LQR design, we look for a linear state feedback controller gain such that the control signal is

$$u[k] = -K_p\boldsymbol{x}[k] + k_i e_q[k] = -K_{ext}\boldsymbol{x_{ext}}[k], \quad (12)$$

where $K_p \in \mathbb{R}^{1\times 6}$ is a row vector with proportional gains multiplying the state vector, and $k_i \in \mathbb{R}$ is the integral gain multiplying the reference tracking error $e_q[k]$ for the flow rate, then $K_{ext} = \begin{pmatrix} k_i & K_p \end{pmatrix}$. According to ref. [28], these gains can be obtained as

$$K_{ext} = \frac{G^T PF}{G^T PG + r}, \quad (13)$$

where the denominator is scalar because the system has a single input. The matrix P is a positive definite solution of the infinite time algebraic Ricatti equation

$$F^T PF - P - F^T PG\left(G^T PB + r\right)^{-1} G^T PF + Q_{ext} = 0. \quad (14)$$

The numerical values we select for the integrated error $w_{I_y} = 500$, the flow rate channel $w_q = 0.001$, the pump pressure $w_p = 0.001$ and the control signal amplitude $r = 0.001$ lead to the following solution for the state feedback

$$k_i = 637.7478, \tag{15}$$

$$K_p = (-1.27, 19.9, 28.91, -0.88, 1.77, -0.92). \tag{16}$$

As can be seen, the integral gain of 637 dominates the controller response, followed by the gain of 28.9 for the third state variable, representing the delayed flow rate output $y_q[k-2]$, followed by the gain of 19.9 for the $y_q[k-1]$.

The key problem with the practical implementation of the LQR controller is that the internal state vector of the system $\boldsymbol{x}[k]$ is generally inaccessible for direct measurement. Hence, it must be estimated by a calculated value $\hat{\boldsymbol{x}}[k]$ from the previous values of the measurable system signals (the input and output plant channels). To produce the state vector estimate, we employ the stationary Kalman filter algorithm, which minimizes the variance $E\left((\boldsymbol{x}[k] - \hat{\boldsymbol{x}}[k])^2\right)$ of the estimation error over all initial conditions and realizations of the model residuals $e_q[k]$ and $e_p[k]$, which are assumed to behave as Gaussian white-noise processes with unit variance. In addition, an unbiased estimate is usually required, meaning

$$E\{\hat{\boldsymbol{x}}[k]\} = E\{\boldsymbol{x}[k]\}. \tag{17}$$

The Kalman filter algorithm has many modifications [29] depending on the model of the process and the assumptions about the statistical distribution of the noise and disturbances. The model of the axial piston pump is a linear time-invariant system with stationary noise action on the state and output. Hence, we design the state observer algorithm as

$$\hat{\boldsymbol{x}}[k+1] = F\hat{\boldsymbol{x}}[k] + Gu[k] + L(\boldsymbol{y}[k+1] - CF\hat{\boldsymbol{x}}[k] - CG\hat{\boldsymbol{x}}[k]), \tag{18}$$

where $L \in \mathbb{R}^{6\times 2}$ is the filter gain matrix calculated to minimize the state error variance. The filter gain is calculated as

$$L = DC^T\left(CDC^T + 0.01I_2\right)^{-1}, \tag{19}$$

where $V_e = E\left(ee^T\right) \in \mathbb{R}^{2\times 2}$ is the variance of models' (1) and (2) residuals, and D is the error covariance matrix calculated as a solution of the algebraic Ricatti equation,

$$D = FD_eF^T + KV_eK^T. \tag{20}$$

The resultant filter gain matrix obtained from the identified model becomes

$$L = \begin{pmatrix} 0 & 0.07 \\ 0.08 & -0.02 \\ 0.03 & 0.03 \\ 0.42 & 0.1 \\ -0.03 & 0.26 \\ 0.14 & -0.03 \end{pmatrix}. \tag{21}$$

The first column from the filter gain matrix corresponds to the flow rate channel and the second column corresponds to the pressure output signal. The filter gain depends on how much information is carried from each state to the output signal and how much we can trust the output signal given the present level of external noise variance. Therefore, the higher gains in matrix L indicate that the respective state variable will be more strongly compensated by the output estimation error. The optimal error covariance matrix obtained from the solution of the Ricatti equation is

$$D = 10^{-4} \times \begin{pmatrix} 2 & 0 & 1 & -2 & 2 & -1 \\ 0 & 2 & 1 & 3 & -1 & 0 \\ 1 & 1 & 1 & 0 & 0 & 0 \\ -2 & 3 & 0 & 40 & 11 & 13 \\ 2 & -1 & 0 & 11 & 24 & 12 \\ -1 & 0 & 0 & 13 & 21 & 21 \end{pmatrix}. \quad (22)$$

As can be seen from the diagonal elements, the variance of the first three states determining the flow rate output signal of the model is 10 times lower than the variances of the pressure determining states. The same is true for the cross-correlations between the errors of the estimated states. The fundamental mechanical constraints can explain that in the pump, when provided with a fixed input signal to the proportional valve, the swash plate swivel angle will determine a fixed displacement volume leading to a fixed flow rate in dependence on the shaft rotation speed. On the other hand, the pressure in the system will be more strongly dependent on the externally attached loading system to the pump and not so much on the input signal $u(t)$. Hence, the variance in the estimated pressure states will become higher. Another observation from the state covariance matrix is that the 3rd and 6th states have lower variances than the 1st and 2nd or 4th and 5th states. This can be related to the casual relationships between those states if one observes the structure of matrix F.

Figure 14 presents the closed loop system frequency responses with the designed linear quadratic regulator and Kalman filter. The output sensitivity function characterizes the ability of the closed loop to attenuate the loading disturbances. We see that the designed regulator dampens the amplitudes of the output disturbances up to 100 times in the low-frequency domain and around 10 times in the middle frequency up to 2 rad/s. From the complementary sensitivity function, we can say that the bandwidth of the closed loop is around 10 rad/s which is acceptable for a regulated pump, considering the capabilities of the employed actuator and flow meter. Figure 14 also shows the control signal sensitivity. As expected, it is higher in the low-frequency range and lowers down in the high frequencies to prevent amplification of measurement noise in the actuator command.

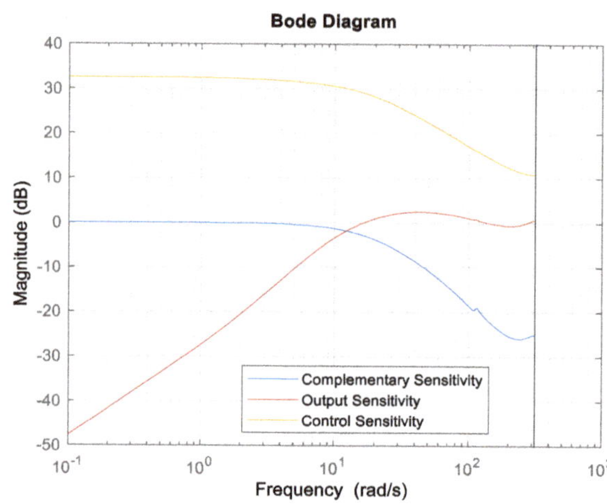

Figure 14. Sensitivities of closed-loop system with LQG controller and Kalman filter.

5. Experimental Validation of Designed Controller

This section presents the experimental results measured with the designed controller in a hardware in the loop (HIL) experiment with the MC012 microcontroller. The loading system for the pump is represented by a variable throttle valve bypassing the output of the pump to the tank, which is presented in the hydraulic circuit diagram in Figure 1. The opening of the loading throttle valve can be manually selected by a precision micrometer scale from 0 to 5 mm. We test the performance of the LQG controller for loading throttle valve openings of 2 mm, 3 mm, 4 mm and 5 mm. The recorded signals are the output pump pressure, output flow rate, LVDT feedback from the pump proportional valve and control signal sent to the electronic amplifier. The experimental results are summarized in Figures 15–18.

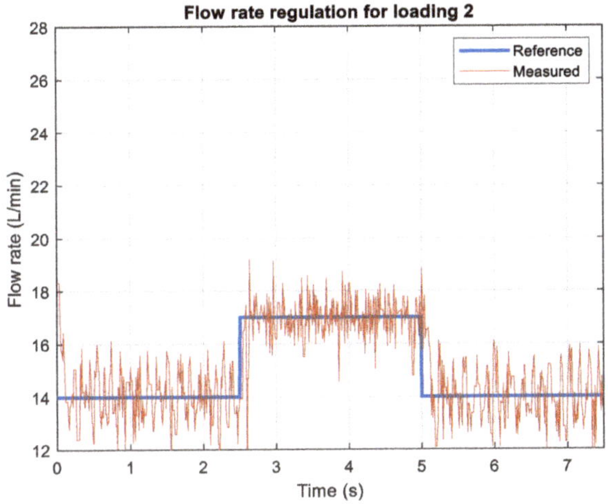

Figure 15. Experimental results for loading condition 2.

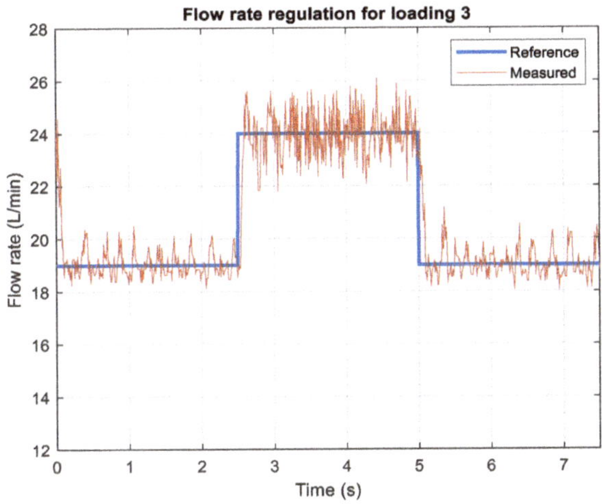

Figure 16. Experimental results for loading condition 3.

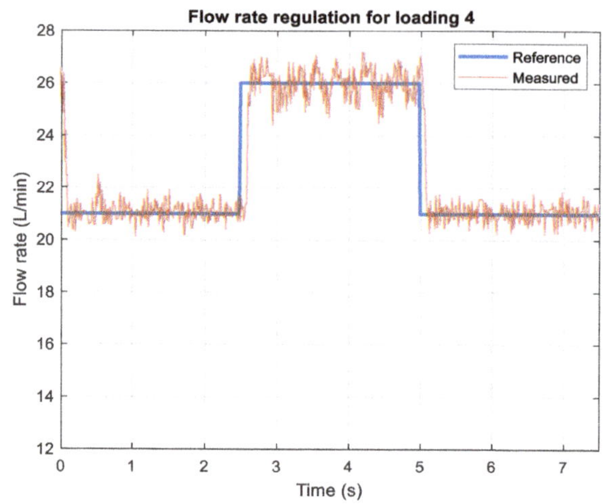

Figure 17. Experimental results for loading condition 4.

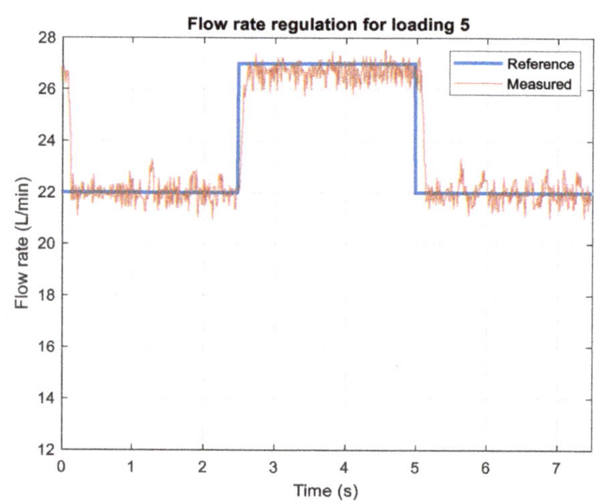

Figure 18. Experimental results for loading condition 5.

For the minimal opening of the loading valve of 2 mm, the closed loop pump system will experience maximal loading as pressure. In such conditions, the reaction force on the swash plate from the output pressure will be maximal, and the control effort from the actual will be required to compensate for the variations in the flow rate. To examine the system during such conditions, we apply a reference flow rate of 14 L per minute and a step reference jump to 17 L per minute for a period of 2.5 s. As can be seen in the figure, we note a very fast system response below 100 ms. The flow rate is held around the reference value with random deviations of +/− 0.5 L per min. Part of these observed variations can be attributed to the sensor noise and imprecision. As noted in the introduction, we employ a low-cost flow rate meter based on an orbital hydraulic motor and a hall sensor. Measurement is based on the capturing of the impulses period from the Hall sensor with a precision of 100 ns. Hence, this measurement technique will introduce some limitations to the maximal accuracy we can obtain for the flow rate. Another component of the observed flow rate variations can be attributed to the actual irregularities in the flow through the gear flow meter. During these high loading experiments, we observe very demanding conditions for the system with pressure in the range of 80 to 120 bar, providing a power output of around 2 kW. Since the electrical motor driving the pump shaft is an unregulated three-phase induction machine, such a high load will lead to variation in the rotational speed of the motor, reflecting directly on the generated flow rate by the pump. Even in such high loading conditions, the designed control system is able to keep the mean value of the flow rate signal around the reference value without much amplified noise in the control signal.

Figure 16 examines the transient behavior for lower loading corresponding to the opening of the loading throttle valve of 3 mm. The pressure during this experiment is around 50 bars, which corresponds to the output power of the pump of around 1.5 kW. The reference signal of the controller is a step from 19 L per minute up to 24 L per minute. The transient response and accuracy of the regulations are comparable to the case of a 1.5 kW operating point with a settling time below 100 ms. It is interesting to note that with an increase in flow rate through the pump, increasing the pressure and the output power, the amplitude of flow rate variations increases. This can be taken as a confirmation of the hypothesis that flow rate variations are due to increased load on the induction motor causing the instability of its rotational speed.

Figure 17 examines the reference tracking performance for the case of a 4 mm opening of the loading throttle valve. In this case, we set the reference step levels between 21 L/min and 26 L/min. The operating pressure in these conditions is between 20 to 30 bars, providing a power output for the pump of around 1 kW. In these conditions, the settling time is increased to around 100 ms. Due to decreased pressure in the system, the reaction of the swash plate to change in the actuating piston becomes slower; hence, the transient responses are delayed. However, such a scale of the response is within sufficient limits for the practical application of the controller. The flow rate variations, in this case, are comparable to the case with the opening of the loading valve up to 3 mm. Similar to the previous case, the amplitude of the variations increases with the increasing flow rate, hence the increased power output of the pump.

The last examined loading condition (Figure 18) corresponds to a 5 mm opening of the loading valve, which can be characterized as minimal loading. During the experiment, the reference step is from 22 L per minute to 27 L per minute. The settling time is around 200 ms, which again is increased because of the decrease in the pressure in the system, reducing the forces acting on the action valve to move the swash plate. It is worth noting that, in these cases of loading, minimized variations of the stationary flow rate around the reference are observed. The pressure in the system is around 20 bar which corresponds to power output below 1 kW. In this case, the loading of the induction motor driving the pump is reduced, and its rotation speed is more stable, which minimizes the variations in the flow rate.

Experimental results for the flow rate regulation prove the experimental stability and performance of the designed LQG controller. The controller is examined for four operating points in the full range of loading conditions for the pump between 14 to 26 L per minute. Even though the identified model is for a single operating point, the closed loop system is able to maintain the reference tracking performance, which is an indication of its robustness.

Figure 19 compares the resultant pressures in the system for the four operating points selected by opening the loading throttle valve. As can be expected with the increase in flow rate, the pressure in the system is increased too. It is notable that for the highest loading conditions, the oscillations in pressure are much more pronounced. Hence, it is not recommended to operate the closed loop system at such high pressure for prolonged intervals. The working envelope of the designed closed loop system can be characterized by a flow rate range between 14 and 26 L/min and a pressure range between 20 to 80 bar. Another thing which can be noted in the figure is that there is a nonlinear relationship between the opening of the loading valve and the mean value of the pressure in the system.

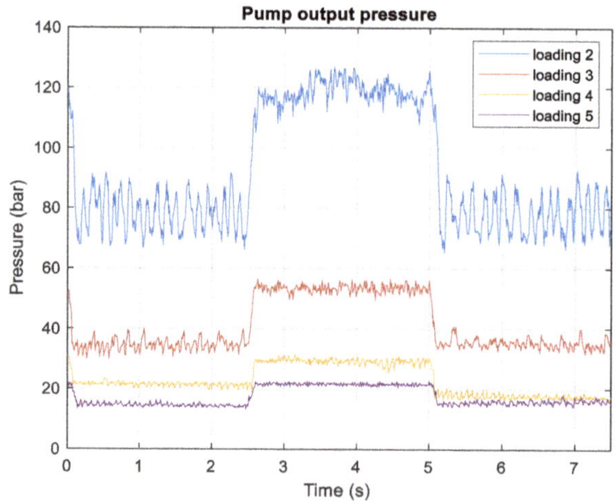

Figure 19. Comparison of pump output pressure for different loading conditions.

Figure 20 shows the control system from the LQG controller during the experiments. The control signal sent to the valve is from 0 to 10V. Therefore, the negative values in the figures are saturated at 0. The most important feature of the examined control signals is that they do not amplify the sensor noise present in the output. This is because of the application of the noise model from the system identification and because of the design of the Kalman filter. Such output noises can be easily amplified by some classical control algorithms, such as PID, leading to the overloading of the actuator valve. It is interesting to note that the average value of the control signals for the selected operating points is not much different, which is an indication that the identified model is able to carry information not only for its operating point but for others too.

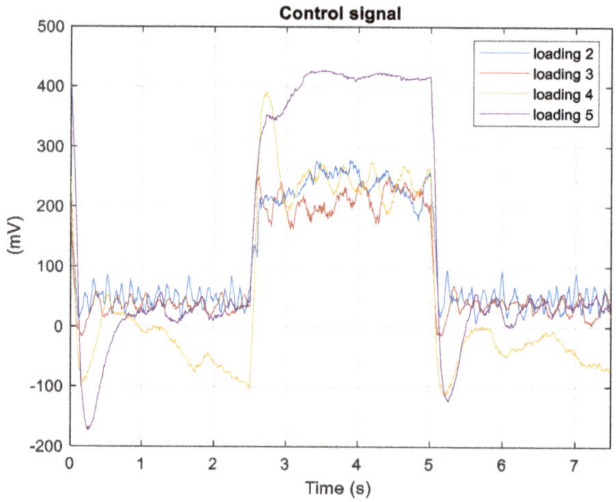

Figure 20. Comparison of control signals for different loading conditions.

Additionally, in Figure 21, we present the LVDT signal for the proportional valve position during the examined loading conditions. The valve we use is bidirectional; hence, we limit its range from 0 to 50% because in the pump we test, the swash plate can work only on positive swivel angles. Alternatively, if the pump can be inverted to work as a motor, the proportional valve operating range will be from 0 to 100%. In the figure, it is important to note that the proportional valve is fast enough to follow the dynamics of the generated control signal from the controller.

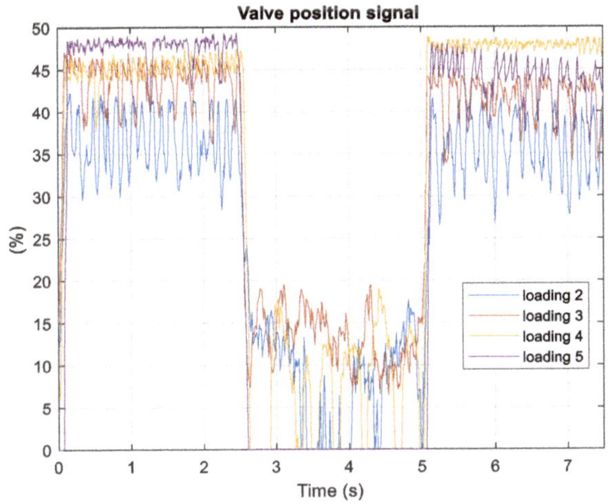

Figure 21. Comparison of LVDT signals for different loading conditions.

As an additional experiment, we examine the performance of the pump for a fixed reference flow of 22 L per minute and with variable (dynamic) loading conditions simulating a real application operation. The loading is modeled by manually decreasing the opening of the loading valve from 5 mm to 1 mm. Figures 22 and 23 present the results of this experiment. As can be seen in Figure 22, the flow rate is held steady by the control system at 22 L/min for all loading pressures in Figure 23 from 20 to 100 bar. With the increase in the loading of the pump and the required output power from the induction motor, the oscillations in the flow rate begin to increase too.

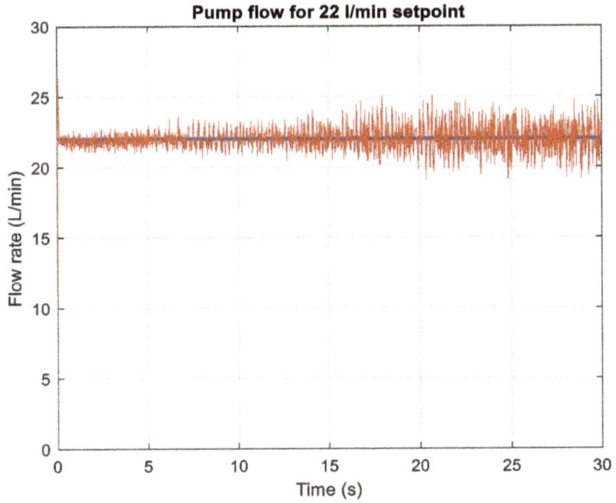

Figure 22. Pump flow rate for 22 L/min setpoints.

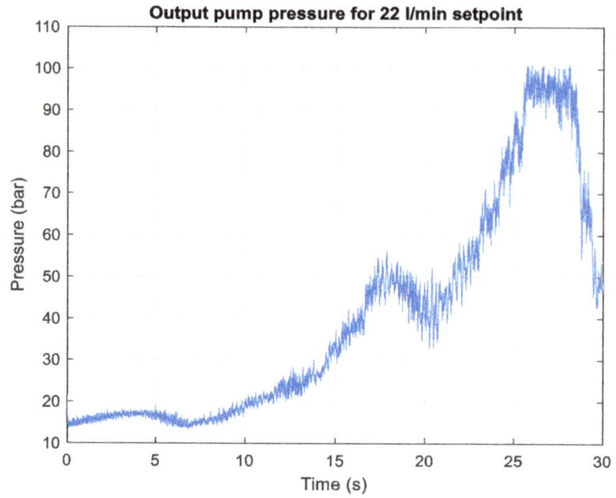

Figure 23. Output pump pressure for 22 L/min setpoints.

6. Conclusions

The main contribution of the article is the developed embedded system for control of an axial piston pump with a proportional valve. In contrast with existing solutions, in this work, the developed embedded system uses a pump that is not equipped with a swash plate swivel angle sensor and rotational speed sensor. This solution is not appropriate for secondary control but is appropriate for flow and pressure control of open-circuit hydraulic drive systems. The proposed control solutions provide possibilities to implement not only existing analogue PID controllers but also advanced control algorithms. First, the "black box" model of the axial piston pump with proportional valve control is estimated from experimental data. The results of validation tests show the model accuracy and its applicability for controller design. The designed control system is based on a linear-quadratic Gaussian (LQG) controller with a time-invariant Kalman filter that is implemented on a laboratory test rig. A detailed description of the hydraulic system in the context of pump displacement control is presented, as well as the developed system architecture for its control. The designed control system is validated through experimental studies for various loading conditions that enable the analysis of its performance.

Author Contributions: Conceptualization, A.M., T.S. and J.K.; methodology, J.K., A.M. and T.S.; software, J.K.; validation, A.M.; formal analysis, T.S., J.K. and A.M.; investigation, A.M., T.S. and J.K.; resources, A.M.; data curation, A.M.; writing—original draft preparation, T.S., A.M. and J.K.; writing—review and editing, J.K. and A.M.; visualization, A.M.; supervision, T.S.; project administration, A.M.; funding acquisition, A.M. All authors have read and agreed to the published version of the manuscript.

Funding: The APC was funded by the Research and Development Sector at the Technical University of Sofia.

Institutional Review Board Statement: Not applicable.

Informed Consent Statement: Not applicable.

Data Availability Statement: When contacted, authors can provide a particular data set from the present article.

Acknowledgments: The authors would like to thank the Research and Development Sector at the Technical University of Sofia for the financial support.

Conflicts of Interest: The authors declare no conflict of interest.

References

1. Manring, N. *Fluid Power Pumps and Motors: Analysis, Design, and Control*; McGraw-Hill Education: New York, NY, USA, 2013.
2. Frankenfield, T. Using Industrial Hydraulics. In *Rexroth Worldwide Hydraulics*; Penton Publishing Inc.: Belfast, UK, 1984.
3. Fresia, P.; Rundo, M. Lumped parameter model and experimental tests on a pressure limiter for variable displacement pumps. *E3S Web. Conf.* **2020**, *197*, 07005. [CrossRef]
4. Mitov, A.; Kralev, J.; Slavov, T.; Angelov, I. Design of Embedded Control System for Open Circuit Axial Piston Pump. In Proceedings of the 22st International Symposium on Electrical Apparatus and Technologies (SIELA 2022), Bourgas, Bulgaria, 1–4 June 2022, *in press*.
5. Casoli, P.; Pastori, M.; Scolari, F. Swash plate design for pressure ripple reduction—A theoretical analysis. *AIP Conf. Proc.* **2019**, *2191*, 020038.
6. Karpenko, M.; Marijonas, B. Investigation of hydrodynamic processes in the system—Axial piston pumps–pipeline–fittings. In *Transport Problems 2018, X International Scientific Conference, VII International Symposium of Young Researchers: Proceedings, Katowice, Wisła, 27–29 June 2018*; Silesian University of Technology: Gliwice, Poland, 2018; pp. 832–843.
7. Ivantysyn, J.; Ivantysynova, M. *Hydrostatic Pumps and Drives*; Vogel: Wurzburg, Germany, 1993.
8. Zeiger, G.; Akers, A. Dynamic Analysis of an Axial Piston Pump Swashplate Control. *Proc. Inst. Mech. Eng. Part C J. Mech. Eng. Sci.* **1986**, *200*, 49–58. [CrossRef]
9. Manring, N.; Mehta, V. Physical limitations for the bandwidth frequency of a pressure controlled, axial-piston pump. *J. Dyn. Syst. Meas. Control. Trans. ASME* **2011**, *133*, 1–12. [CrossRef]
10. Murrenhoff, H. Regelung von Verstellbaren Verdrängereinheiten am Konstantdrucknetz. Ph.D. Thesis, RWTH Aachen, Aachen, Germany, 1983.
11. Lin, S.S.; Akers, A. Optimal control theory applied to pressure-controlled axial piston pump design. *J. Dyn. Syst. Meas. Control. Trans. ASME* **1990**, *112*, 475–481. [CrossRef]
12. Akers, A.; Lin, S. Optimal control theory applied to a pump with single-stage electrohydraulic servovalve. *J. Dyn. Syst. Meas. Control. Trans. ASME* **1988**, *110*, 120–125. [CrossRef]
13. Kordak, R. Hydrostatic Drives with Control of the Secondary Unit. In *The Hydraulic Trainer*; Mannesmann Rexroth GmbH: Lohr a. Main, Germany, 1996; Volume 6.
14. Belan, H.C.; Locateli, C.C.; Lantto, B.; Krus, P.; de Negri, V.J. Digital secondary control architecture for aircraft application. In Proceedings of the Seventh Workshop on Digital Fluid Power, Linz, Austria, 26–27 February 2015.
15. Berg, H.; Ivantysynova, M. Design and testing of a robust linear controller for secondary controlled hydraulic drive. *Proc. Inst. Mech. Eng. Part I J. Syst. Control. Eng.* **1999**, *213*, 375–385. [CrossRef]
16. Kemmetmüller, W.; Fuchshumer, F.; Kugi, A. Nonlinear pressure control of self-supplied variable displacement axial piston pumps. *Control. Eng. Pract.* **2010**, *18*, 84–93. [CrossRef]
17. Park, S.; Lee, J.; Kim, J. Robust control of the pressure in a control-cylinder with direct drive valve for the variable displacement axial piston pump. *Proc. Inst. Mech. Eng. Part I J. Syst. Control. Eng.* **2009**, *223*, 455–465. [CrossRef]
18. Wei, J.; Guo, K.; Fang, J.; Tian, Q. Nonlinear supply pressure control for a variable displacement axial piston pump. *Proc. Inst. Mech. Eng. Part I J. Syst. Control. Eng.* **2015**, *229*, 614–624. [CrossRef]
19. Kralev, J.K.; Petkov, P.H.; Slavov, T.N. *Design of Embedded Robust Control Systems Using MATLAB®/Simulink®*; IET Control: London, UK, 2018.
20. Forrai, A. *Embedded Control System Design: A Model Based Approach*; Springer: Warsaw, Poland, 2013.
21. Rexroth Bosch Group. *Pressure and Flow Control System*; Technical Data Sheet, RE 30630; Rexroth Bosch Group: Lohr a. Main, Germany, 2015.
22. Rexroth Bosch Group. *Proportional Directional Valves, Direct Operated, with Electrical Position Feedback as Pilot Control Valve for Control Systems SY(H)DFE*; Technical Data Sheet, RE 29016; Rexroth Bosch Group: Lohr a. Main, Germany, 2019.
23. Tonyan, M. *Electronically Controlled Proportional Valves*; Marcel Dekker, Inc.: New York, NY, USA, 1985.
24. Danfoss. *Plus+1 Controllers MC012-020 and 022*; Data Sheet, 11077167; Danfoss: Nordborg, Denmark, 2013.
25. Merrit, H. *Hydraulic Control Systems*; John Wiley & Sons Inc.: Hoboken, NJ, USA, 1967.
26. Ljung, L. *System Identification: Theory for the User*, 2nd ed.; Prentice Hall: Hoboken, NJ, USA, 1999.
27. Slavov, T.; Mitov, A.; Kralev, J. Advanced Embedded Control of Electrohydraulic Power Steering System. *Cybern. Inf. Technol.* **2020**, *20*, 105–121. [CrossRef]
28. Zhou, K.; Doyle, J. *Robust and Optimal Control*; Prentice Hall International: Upper Saddle River, NJ, USA, 1996.
29. Grewal, M.; Andrews, A. *Kalman Filtering: Theory and Practice with MATLAB®: Fourth Edition*; Wiley-IEEE Press: Hoboken, NJ, USA, 2014.

energies

MDPI

Article

Taguchi Techniques as an Effective Simulation-Based Strategy in the Design of Numerical Simulations to Assess Contact Stress in Gerotor Pumps

Pedro Javier Gamez-Montero [1,*] and Ernest Bernat-Maso [2]

1 Department of Fluid Mechanics, Universitat Politecnica de Catalunya, Campus Terrassa, Colom 11, 08222 Terrassa, Spain
2 Strength of Materials and Structural Engineering Department, Universitat Politecnica de Catalunya, Campus Terrassa, Colom 11, 08222 Terrassa, Spain
* Correspondence: pedro.javier.gamez@upc.edu

Abstract: The contact problem of a trochoidal gear is a drawback and a well-known performance indicator of a gerotor pump. Although numerical simulations aid in the evaluation of contact stress, the difficult task of determining geometrical parameters, operating conditions, and the number of simulations to run falls to the designer. This paper presents the Taguchi techniques as an effective simulation-based strategy to narrow down the geometrical parameter combinations, reducing the solution space and optimizing the number of simulations. The work is first focused on the validation of the proposed numerical model by means of published contact stress results of recognized researchers in the field, as well as the unification of nomenclature and notation. Then, the Taguchi approach is based on a sequence of four experiments, ranging from the screening case with two levels and seven parameters to multiple levels and four parameters with three software input operating conditions (temperature, torque, and friction coefficient) emulating noise effects. The contact stresses of 128 gear sets, having common volumetric capacity and dimensional constraints to detach mechanical performance from flow rate and casing, were analyzed. Results prove the feasibility of the proposed methodology by identifying the most suitable gear set configuration and predicting the quantifiable performances of a real-working gerotor pump.

Keywords: trochoidal profile; gear pump; gerotor technology; contact stress; finite element method; design of experiments; Taguchi method; simulation-based design; process planning; fluid power

Citation: Gamez-Montero, P.J.; Bernat-Maso, E. Taguchi Techniques as an Effective Simulation-Based Strategy in the Design of Numerical Simulations to Assess Contact Stress in Gerotor Pumps. *Energies* **2022**, *15*, 7138. https://doi.org/10.3390/en15197138

Academic Editors: Paolo Casoli and Massimo Rundo

Received: 5 September 2022
Accepted: 23 September 2022
Published: 28 September 2022

1. Introduction

Fluid technology, as a part of human activities, has become one of the important issues in connection with environmental concerns. Hydraulic machines play an essential role in fluid technology and will further accelerate their green transition in the coming years with the improvement of models and simulations [1]. The gerotor pump is intended to be a part of the solution in leading research technologies in industrial and aeronautical areas, mechanical engineering, pharmacy, medicine and, more recently, in areas such as the design of a virtual prototype to be used in the context of a Digital Twin tool [2].

Nowadays, attention has been focused on reducing energy loss and improving efficiency, since any relatively small proportion of them is becoming crucial in any technology. Improvement of electric consumption, heralded as the critical step in the reduction of transportation's impact on climate change, can be used as an example in the adapted Figure 1 from the work of Kawamata et al. [3]. Loss due to friction in the motor-based drive system of an electric vehicle (EV) system is still responsible for approximately 10% of the total global energy consumption. Loss due to aerodynamics is the most important factor of an EV during operation compared with an internal combustion engine (ICE), but there is still room for improving the current levels of energy consumption by paying attention to these slight proportions as loss due to friction.

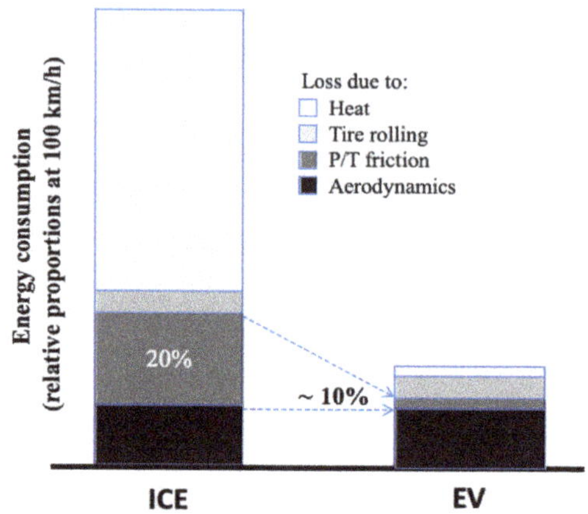

Figure 1. Comparison of energy consumption loss between ICE vehicle and EV (adapted from Kawamata et al. [3]).

Contact stresses and friction are key performance indicators of a trochoidal gear set working as a gerotor pump. Over the years, from the work on the reduction in contact stress in internal gear pumps by Colbourne [4] in 1976 up to recent works such as gap estimation in ORBIT motors by Roy et al. [5,6] and the in-depth design concept of hydraulic displacement units by Sliwinski [7], an important body of literature has been published related to contact, deformation, clearance, and stress in the teeth of gerotor pumps. In fact, "contact" and "stress" appeared among the top ten keywords in the list of the thirty most frequently used keywords in the publications from 2009 to 2019 [8].

It should be pointed out that two researchers, Ivanović and Biernacki, have conducted remarkable contact stress studies on gerotor pumps over the last twelve years. Ivanović and the colleagues have explored this challenging area in analytical methods [9–11], wear rate, friction, and stress index evaluation [12–16]. Biernacki and the colleagues have investigated the field with an emphasis on new materials [17–21] and new concepts of design [22,23]. Both research groups have stood out and exceled in the area.

When the question of how to choose the gear set to improve performance indexes arises, optimization has come to assist. Focusing on works of the past five past years, Robinson and Vacca [24,25] and De Martin et al. [26] have published complete studies in multi-objective optimization to traditional circular-toothed profiles. As a result, the procedure pinpointed that the industry was able to find optimal solutions at almost the same level as the state-of-the-art pump designs. A thorough study of commercial gear sets of a company carried out by Tessari et al. [27] and Pulliti et al. [28] allowed them to extrapolate an empirical formulation for the eccentricity, the initial parameter of the optimization strategy to maximize volumetric efficiency.

Whenever the question of how to choose the geometrical parameters to design a new-born trochoidal gear pump comes up, the answer is not straightforward and an intermediate step amid the proceedings from academia and industry, between commercial gear sets and optimization procedures, would be especially welcome. In addition, the designer's goal is not only to reduce the time and effort to set up the analysis, but also to diminish and even remove the various costs associated with conducting any trial. Hence, numerical simulation comes as a design aid. A new numerical simulation may confront a new set of issues, such as how to evaluate the parameters, how many numerical simulations to carry out, how to estimate parameter combinations, or how to link mechanical and fluid-dynamic characteristics. The purpose of this paper is to attempt to address the above researcher's questions from a practitioner's perspective and the Taguchi method is chosen to blend knowledge from commercial gear sets to numerical simulations. Another important advantage of the Taguchi method is its accessibility to any end-user, becoming an open tool not restricted by software.

The Design of Experiments (DoE) is a well-known and powerful tool [29]. The DoE process is divided into three main phases as follows: (*1*) the planning phase: factors and levels are selected, (*2*) the conducting phase: test results are collected, and (*3*) the analysis phase: positive or negative information concerning the selected factors and levels is generated based on phase (*1*) and (*2*). A full-factorial experiment, having an equal number of test data points under each level of each factor, is acceptable when only a few factors are to be investigated, but not very advisable when there are many factors [30].

The Taguchi approach is a holistic view based on the use of orthogonal arrays (OA) to conduct small, highly fractional factorial experiments up to larger, full-factorial experiments [31]. An alternative approach could be the use of Monte Carlo simulation, but it requires a large number of testing conditions for an accurate estimation of the mean and variance, which is expensive and time-consuming [29,32]. Each OA has a determined maximum possible resolution [33]. The selection of which OA to use predominantly depends on (in order of priority): (*i*) the number of factors and interactions of interest, (*ii*) the number of levels for the factors, and (*iii*) the desired experimental resolution or cost limitations. This strategy will minimize the total number of tests to be conducted yet will yield meaningful information at the same time. A lack of communication between the academics and specialists in industrial worlds is a cause that has restricted the application of the Taguchi method on a larger scale [34]. To analyze the data and determine the optimum levels of selected factors, the signal-to-noise (S/N) ratio was developed by Dr. Taguchi as a performance measure to choose control levels that best cope with noise effects. All these uncontrolled and unknown factors that actually vary from one degree to another, causing differences from data point to data point, generate noise effects. Rather than assuming that error variation is an aggregate of all these noise effects and equally distributed in all treatment conditions, Taguchi parameter design utilizes these repetitions to aid in identifying what levels of which control factors might have reduced variation.

To varying degrees of depth, the Taguchi method has been applied to volumetric characteristics in a gerotor pump. Jung et al. [35] implemented the method for the design of lobes with multiple profiles. The optimal design parameters obtained are promising in the experimental results of volumetric performance, but unfortunately, the explanation of the implementation is scarce and noise effects are not mentioned. Ivanović et al. studied the selection of optimal parameters for the volumetric characteristics of a gerotor pump, such as flow rate and volumetric efficiency, through a factorial design and response surface methodology [36] and by using the Taguchi method [37]. Three tested geometries at three different rotation speeds and pressures were established as influential parameters. In these experimental works, a selection of noise factors is not reported. As a result, one important contribution of the presented work is the use of several input parameters as noise factors to assess the contact stress characteristics in a gerotor pump.

The application of the Taguchi method in software and computer design has been reported. As an example, assistance for the management of defects found in software development by quantifying performance has been proven [38]. The multitude of parameters in Computer Aided Design (CAD) and Computer Aided Manufacturing (CAM) are also benefited by the structured generation of a representative subset of them into a solution space, as is shown in bevel gear machining simulations [39]. The mechanical performance can be modeled mathematically by using the finite element methodology, which makes the contact stress a prime candidate for designed experiments using the environment of the software packages. Choudhury et al. [40] demonstrated the feasibility of this approach by working with the mechanical design of an epicyclic gear train to find an optimized set of various input factors for which the stresses are the lowest and, thus, the chances of tooth failure are reduced. The simulated experiments can use the operating conditions to emulate noise effects (prescribed factors in the software) in an outer array, in combination with an inner array for only control factors (controllable parameters in the simulation). Hence, the use of the Taguchi method in numerical simulation to assess the contact stress characteristics of a gerotor pump is the other key contribution in the present research.

The methodology in this work follows three consecutive phases to achieve the goals: the background of a trochoidal gear set, the verification of the contact stress numerical simulation, and the Taguchi approach to a gerotor pump. Establishing the right blueprint is a challenge unique to each phase. The first phase establishes the basic definition of a trochoidal profile, the function of a gerotor pump, the selected gears from the literature under study, the geometry generation, and the integration of nomenclature and notation.

The second phase aims to build a solid foundation on which the implementation of the Finite Element Model (FEM) to evaluate contact stresses is supported by verification. Hence, a set of simulations, along with their geometries, have been selected and extracted from the literature, and the fundamental criterion was to be able to reproduce the results since not all published articles provide the necessary technical data. The objective of this approach is to reconstruct those simulations and corroborate the validity of the results within reasonable bounds.

Once the outcomes are verified, the last phase involves a sequential simulation-based design strategy of four Taguchi experiments turning out 128 numerical simulations: from the screening case with two levels and seven parameters up to multiple levels and four parameters with three noise factors. This innovative approach to the Taguchi method is the study of the contact stress by using a software package that is feasible to emulate noise effects as input operating data for a set of gerotor geometries that have in common a constant volumetric capacity and dimensional constraints. Hence, the outer noise (environmental-related, such as temperature) and inner noise (function-related, such as torque and friction coefficient) retain significance with the noise effects of the Taguchi method.

This paper endeavors to prove the reliability of the proposed methodology to predict optimum settings in the geometry of gerotor pumps and be able to find gear sets in industrial designs. This study looks to formulate its own approach that identifies the most suitable gear set configuration and predicts the quantifiable performance of a real-working gerotor pump, taking into account best practices in the design process to guide the designer to make good decisions in the numerical simulation procedure.

2. Trochoidal Gear Set Background

A trochoidal gear set is a pair of gears, inner/internal and outer/external, with trochoidal tooth profiles (see Figure 2 and check the Nomenclature section). The two gears are mated so that each tooth/lobe of the internal gear is always in sliding contact with a tooth of the external gear, and inter-tooth contact occurs. These points are known as contact points. The pumping action is generated since the volume trapped between two consecutive contact points varies, increasing and decreasing with the angular position of the mated gears. This internal gear pump is well-known as the gerotor pump. For the sake of simplicity, and because they can be found thoroughly in the literature, theoretical analysis and formulae are not explicitly presented in this work; as a reference, see work [41].

2.1. Gear Sets under Study

Works from the literature were selected to validate the implemented numerical model. The main criterion of election was that selected works should contain numerical mechanical analyses, including an evaluation of the contact stress level. In addition, the researchers of these selected works ought to have a long and solid career in the mechanical analysis of trochoidal gear sets and gerotor pumps. Hence, the researchers Ivanović and Biernacki fulfil more than enough of the above-mentioned selection criteria, and their studies selected have been [13,14,17,23], respectively. In addition, a reference geometry with theoretical, numerical, and experimental work is also added to the study [42].

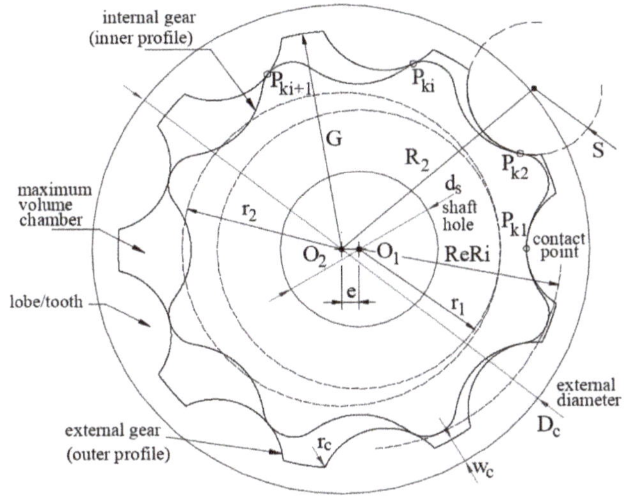

Figure 2. Basic geometric parameters of the trochoidal gear set.

The codification used in the present study to unify all gear sets follows the same sequence as Per___: PZXeYYY. X is a figure coding the number of teeth of the outer gear and YYY are three figures coding the non-decimal eccentricity value of the gear set (the three digits without the decimal dot). This codification is illustrated in Table 1, along with supplementary information.

Table 1. Gear sets under study (PZXeYYY, where X is the number of teeth of the outer gear and YYY are the non-decimal eccentricity value).

Gear Set	Researchers' Work	Analysis–Environment	Observation
PZ7e377	Biernacki [23]	FEM–ABAQUS	The keyway is removed
PZ6e356/1375 PZ6e356/1575	Ivanović et al. [14]	FEM–CATIA/FEMAP	$\lambda' = 1/\lambda = 1.375$ $\lambda' = 1/\lambda = 1.575$
PZ9e285	Gamez-Montero et al. [42]	FEM–GiD/COMET	The shaft hole is included
MZ9e855		Photoelasticity model	Scaled 3:1 of PZ9e285

2.2. Nomenclature and Notation

Nomenclature and notation of the basic geometric parameters are not standardized, and each researcher uses their own. The present study works with the GeroLAB notation [43], then the researchers' geometries under analysis are contextualized in this specific nomenclature and presented in Table 2.

2.3. Geometry Generation

Once basic geometrical parameters are decoded, as presented in Table 2, the full set of trochoidal gears has to be generated. The well-known formulae for calculating the trochoidal profiles of a gear set can be used by introducing these geometrical parameters into specialized free software, such as GeroLAB (Terrassa, Spain) [44], which will calculate and provide the technical drawings in CAD format. By importing the drawing into CAD software (Vélizy-Villacoublay, France), a planar body of the geometry is created and modified with the appropriate operations, such as splitting, adding, and assembling, to be able to be imported into a mechanical simulation software.

Table 2. Geometrical and technical parameters of gear sets (please refer to Nomenclature section and Figure 2).

Input Gear Set Variable/ Parameter	PZ7e377	PZ6e356/1375 PZ6e356/1575	PZ9e285	MZ9e885
GeroLAB				3:1 (PZ9e285)
Z [-]	7	6	9	
e [mm]	3.77	3.56	2.85	
S [mm]	11.11	9.79 (/1375) 14.06 (/1575)	10.85	
$DeRi$ [mm]	53.24	46.28	65.45	
G [mm]	30.50	26.94	35.80	
r_c [mm]	0	0	0	
H [mm]	10.40	16.46	9.25	
Geometrical				3:1 (PZ9e285)
w_c [mm]	7.0	4.1	4.2	
D_c (O_2) [mm]	75	62	80	
d_s (O_1) [mm]	25	16	44	
Material properties	POM *	Steel	Sintered metallic powder	Epoxy
Young's module [GPa]	3	200	115	3
Density [kg/m^3]	1410	-	6800	1160
Poisson's coefficient [-]	0.43	0.30	0.25	0.35
Friction coefficient, μ [-]	0.40	0.40	0.40	0.30
Torque, T [N·m]	7.16	0.621 (/1375) 0.632 (/1575)	18.75	37.5

* Polyoxymethylene.

3. Verification of Contact Stress Numerical Simulation

3.1. Numerical Model Definition

Two main series of numerical simulations are considered. The first one is aimed at validating the implemented Finite Element Model (FEM) and consists of simulating real cases of gerotor systems published in the literature, which include mechanical data. The second series is aimed at analyzing the influence of different parameters according to the Taguchi method.

All simulations are run in ANSYS® 19.2R (Canonsburg, PA, USA). This general-purpose software environment was selected because of its wide availability among practitioners and also because most of the definitions in terms of mechanical description are analogous and comparable to those of other FEM simulation packages. Moreover, the multiphysics capability of ANSYS® would allow future enhancement of the model applicability including fluid or thermal transitive simulations, although these steps are far beyond the aim of the current research. In addition, many other researchers have used this software package to study gear contact stresses before, such as Ram Kumar et al. [45], Karthick et. Al. [46], Benaïcha et al. [47], Rao [48], Lahtivirta and Lehtovaara [49], or Lisle et al. [50]. The most relevant information about the implemented numerical models is described in the following subsections.

3.1.1. Geometry and Material Properties

The geometry in all cases consists of two parts: the internal and the external gear. For every reference case, two angular positions of the gear sets are simulated: 0^0 and 25^0. One additional case at 40^0 is considered for the PZ9e285 case. The keyway is not allocated in any geometry. Hence, since two specific positions in the gear set are only required, the Static Structural module is the most suitable selection.

Despite the fact that all the simulations chosen from the literature share common characteristics, two of them are computed with a 3D geometry. On the one hand, the specific reasons why these researchers decided to implement 3D are unknown. On the other hand, although these simulations were run in 3D, the results are shown in 2D, in the front plane of the gear sets. Then, this fact endorsed the decision to assume a 2D plain stress hypothesis with gear thickness as input data in this study. Two main advantages are added: the remarkable diminution of mesh elements and the contact definition as edge–edge and not surface–surface, which helps the program to be more precise.

For all simulations, isotropic linear elastic material is considered. Thus, the only properties required for the numerical simulation are Young's modulus and Poisson's and friction coefficients (see Table 2).

3.1.2. Loads, Contacts, and Boundary Conditions

The boundary conditions consist of fixing the edge of the external diameter of the external gear and imposing a restrained and fixed displacement on the edge of the shaft hole of the internal gear, while allowing its rotation. The load is performed by a counter-clockwise torque applied on the edge of the shaft hole of the internal gear. For the validation cases, the value of the torque is set according to the corresponding reference in the researchers' works.

Finally, contacts are described as frictional through the friction coefficient definition. All contacts are set between the specific edges of the teeth (internal and external) in interaction. Thus, it is required to divide the continuous internal and external edges in the trochoidal tooth profiles into as many parts as the corresponding number of teeth. This definition is essential to correctly set the contact areas to allow an independent analysis of the several contact points at the same time between gears: the quasi-static condition.

Using frictional contact rather than frictionless is preferred because it more accurately denotes real cases and allows one to study the possible life-long variation of the friction coefficient as an external factor on the contact stress, which is addressed later. Frictional contacts [5,51] and frictionless contacts [49,50] are both previously used by other researchers in gear contact studies.

3.1.3. Elements and Mesh

PLANE183 elements are used to mesh both internal and external gears. CONTA172 and TARGE169 elements are used to define the contacts. The average size of the mesh is 5 mm. This dimension is reduced to an edge sizing of 1 mm near the contact areas by imposing this seeding distance on the edges that define the contacts. Since the goal is to validate the model by obtaining results akin to the researchers' work, an equivalent configuration must be used. The quality of the mesh is checked with the orthogonal quality indicator, with an average value of over 0.9, and also with the skewness indicator, with an average value of below 0.3, for all cases. Those indicators fulfilled the recommended quality mesh requirements in [52].

3.1.4. Analysis

The full Newton–Raphson solution procedure is employed to perform the implicit calculation of the implemented non-linear model. Average solution time is approximately ten minutes in an Intel® Core ™ i7-9700 K CPU @ 3.60 GHz with 16 GB RAM memory.

The use of equivalent von Mises stress is widely accepted and used in several research studies [45–48], although some of the research [51] points out that von Mises stress values could be approximately one half of the theoretical Hertz contact stress. In any case, the proposed model is validated through comparison with other accepted models that used von Mises's stress. Finally, the influence of the analyzed parameters is also studied in terms of maximum von Mises's stress at contacts, being mostly a comparative than a quantitative analysis.

3.2. Simulation Results: Validation and Discussion

The FEM validation aims to assess the recreated authors' results from the researchers' works aforementioned. If the validation is positive, meaning that the values obtained with the simulations match those published within reasonable bounds, the methodology will prove to be valid for future simulations using the same model applied up to here.

The maximum equivalent von Mises stress among all contact zones is the variable for comparison with reference stresses. Figure 3 graphically gathers authors' FEM results of maximum contact stress in each gear set from researchers' works at two angular positions, 0° and 25° (the angular position of each gear set in the researcher's reference). A worthwhile comparison of results could have come from the deformation performance of the gear sets. However, limited information is available in the available literature and, consequently, it was decided to concentrate on contact stress results, which are the ultimate goal.

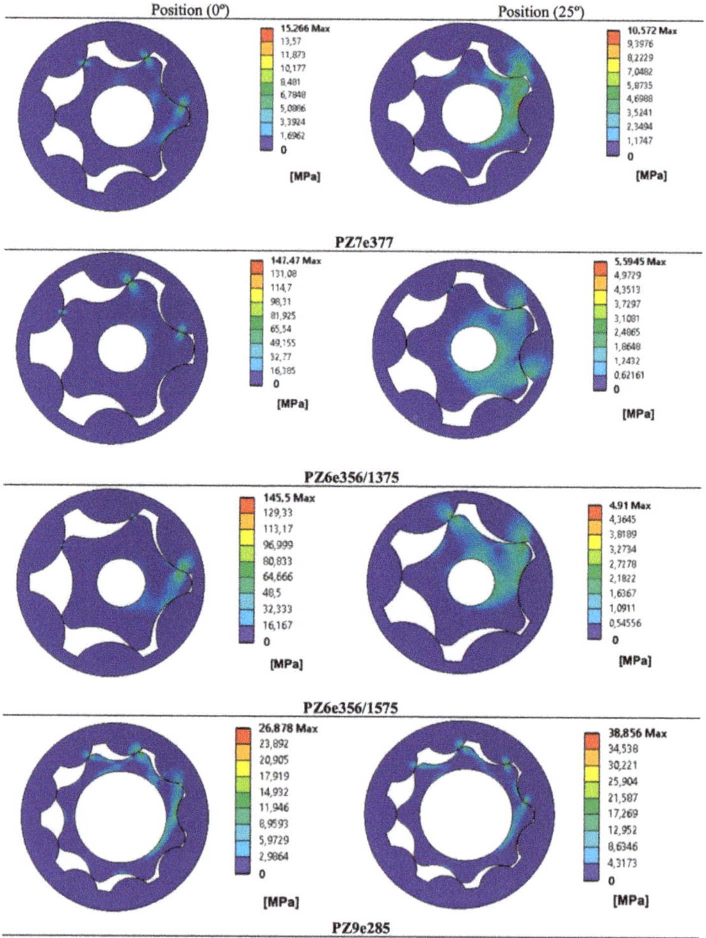

Figure 3. Authors' FEM results of maximum contact stress [MPa] in each gear set from researchers' works at 0° and 25° angular positions. (The commas in this specific figure represents decimal dots. For interpretation of the references to color in this figure legend, the reader is referred to the Web version of this article.).

For the PZ9e285(25°) case, the mesh convergence analysis was performed obtaining the results plotted in Figure 4. The figure shows a clear contact stress value trend when reducing the contact edge sizing from 5 mm (11.5% variation) to 2 mm and from 2 mm (0.4% variation) to 1 mm. However, for smaller mesh sizes (0.50 mm and 0.25 mm), local stress concentration effects were observed to increase the maximum contact stress in a fictitious way. Thus, 2 mm edge size was precise enough to carry out the research, but 1 mm was chosen because it was more precise and was not a significantly greater calculation effort than the 2 mm case. For a 1 mm mesh size case, the PZ9e285 model included 7164 nodes, 2218 2D elements, and 420 contact elements.

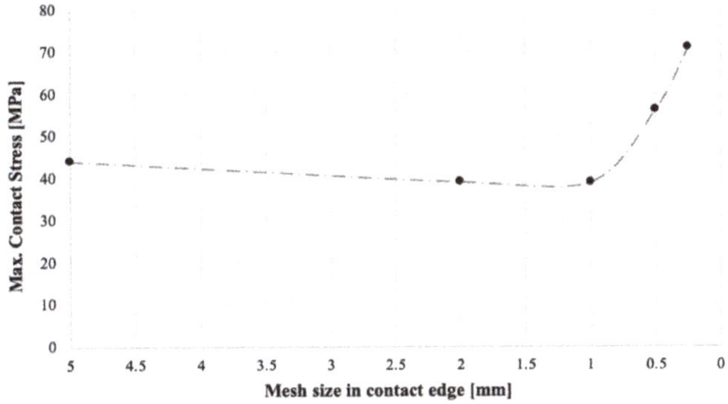

Figure 4. Mesh convergence analysis of PZ9e285(25°) case.

The main results of the nine simulations are summarized in Figure 5. When researchers' and authors' results are contrasted at the maximum contact stress, a 9% average error is obtained when the PZ9e285(0°) is not included (this specific angular position of this gear set shows an unexpected value, and it is considered an outlier because clear stress concentration effects were detected in this case. In fact, a 40° position was also simulated to check the stability of the PZ9e285 case and to overcome this limitation). Since there is a restricted amount of available information about mechanical stresses in gerotors, this accuracy is judged to be adequate to validate the implemented model for describing the mechanical response of gerotor sets.

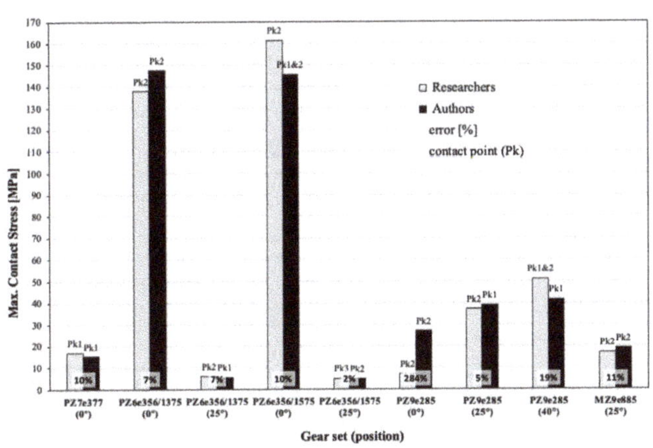

Figure 5. Results comparison Researchers vs. Authors, relative error [%] in maximum contact stress [MPa] and contact point (Pk) of maximum contact stress in each gear set for corresponding angular position in brackets (degree °).

With regard to the location of the contact point at maximum stress, two clear tendencies can be pointed out:

- Maximum volume chamber position corresponding to PZ7e377(25°), PZ6e356/1375(0°), PZ6e356/1575(0°), and PZ9e285(0°). The location of the contact point at maximum stress is Pk2 for all gear sets, according to researchers' and authors' results. (Note: PZ7e377(25°) gear set in reference [17] does not provide an exact value but it can be gathered from the graphical figures);
- Minimum volume chamber position corresponding to PZ7e377(0°), PZ6e356/1375(25°), PZ6e356/1575(25°), PZ9e285(25°), and MZ9e885(25°). The location of the contact point at maximum stress are Pk and Pki + 1 for all results, researchers' and author's, respectively. The exceptions are PZ7e377(0°) and MZ9e885(25°).

Hence, it reveals the importance of having either an odd or even number of external gear teeth in the set-up of the simulation. In addition, the impossibility of simulating exactly the same geometries as the researchers must be remembered, as does how the unnecessary parts of the original geometry were neglected, such as the keyway in the shaft hole.

From these results, the plausibility of the proposed FEM model of the authors is sustained, as well as providing an opening for making progress toward the Taguchi approach. This decision is also justified by the fact that the final purpose of the implemented model is to study the significance and influence of the different design and environmental variables on the contact stress level more than providing a completely accurate result of the stress value.

4. The Taguchi Approach to a Gerotor Pump

The aim of parameter design outlined by Dr. Taguchi is to obtain a gear set that maximises the safety factor (SF) defined as the ratio between the reference yield strength of the material and the maximum contact stress. This gear set will be pursuing a twofold goal of (*i*) detaching the linkage of contact stress with volumetric capacity and dimensional constraints and (*ii*) enhancing the quality of contact stress assessment without controlling or removing the cause of variation contributed to by the numerical simulation's input settings to generate robust FEM results against noise effects. This approach is an iterative process that will be carried out sequentially in four experiments. The means to achieve a certain average contact stress will not be addressed. The Minitab® 19 software package (State College, PA, USA) complements the data processing.

4.1. The Volumetric Capacity Target and the Dimensional Constraints of the Gerotor Pump

Volumetric characteristics and contact stress performance of a gerotor pump are linked by the geometric parameters [53]. Then, in order to soften this link, the contact stress, the volumetric capacity, and several dimensional constraints of the gerotor pump are fixed constants in each gear set for all experiments as follows:

- The PZ9e285 gear set is the chosen prototypical gerotor because it is well-known by the authors regarding, among others, its fluid dynamic performance [54];
- The benchmark gerotor is named D80d40cv1 and the parameters to keep constant are summarized in Table 3 and depicted in Figure 6 based on [55];
- The material properties remain unchangeable in all the experiments: Young's module, density, and Poisson's coefficient of the PZ9e285 gear set (refer to Table 2);
- The shaft keyway is designed with dimensions L_s = 30 mm (parallel 0.75·d_s) shown in Figure 6 as in [55];
- Two specific angular positions as working functions will be under the study: Tip-to-Tip (T2T) and Valley-to-Tip (V2T), both depicted in Figure 6. The V2T corresponds to the maximum volume chamber and the T2T corresponds to the minimum volume chamber. This labeling presumes to enhance the comprehension of the contact points' action;
- The geometries and the analytical values of volumetric capacity and flow irregularity of each experimental gear set are provided by using GeroLAB [43,44].

Table 3. D80d40cv1 benchmark gerotor. (Please refer to Nomenclature section and Figure 2).

Parameter	Value	Significance/Target
$c_{v,H} = c_v/H$ [cc/(rev·mm)]	1	Volumetric capacity/Flow rate
D_c [mm]	80	Housing/Dimensional constraint
d_s [mm]	40	Internal diameter located in the inner gear to accommodate the shaft/Through-shaft application
H [mm]	9.25	Casing/Dimensional constraint

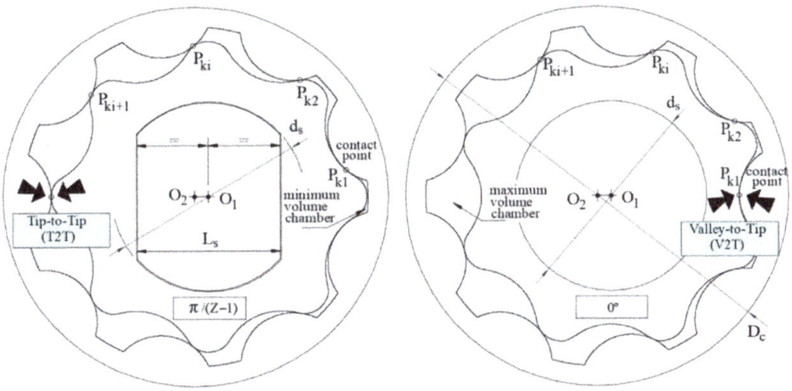

Figure 6. The benchmark gerotor in working function: the Tip-to-Tip (T2T) position (left) and the Valley-to-Tip (V2T) position (right), the shaft keyway ('yes' shaft keyway, left) and the shaft hole without keyway ('no' shaft keyway, right).

4.2. The Taguchi Method: The Designed Experiments

The designed experiments by the Taguchi approach will follow these specific features:

- The number of control factors. The number of parameters is chosen based on the trochoidal gear profile (Z, e, S), external gear (r_c, w_c), and working function (reference position *RP*, shaft keyway *SK*);
- The number of levels. The levels of each parameter are chosen to be significant in the study, from the benchmark gerotor, the literature, the previously presented FEM validation, and the authors' know-how. In addition, all level combinations have to accomplish the feasible geometry of a trochoidal gear set practicable to be generated with GeroLAB, which has become a challenging task. The number of levels is selected as the Taguchi approach is going forward, from two-levels to a mixed-level design;
- The noise factors. The mechanical operating conditions of the gear pump are described by three factors: material temperature (θ), torque (T), and material friction coefficient (μ), which can be tuned by surface modification methods [56,57]. The designed experiments can use these operating conditions as the input of FEM conditions as prescribed by an outer array only for noise factors (T, μ and θ);
- Improve the quality of contact stress assessment. The goal of the experiments is to maximize the safety factor (SF) as the response of the signal-to-noise ratio (S/N) set to Higher-is-Better (HB) where:

$$\text{HB S/N} = -10\log\left(\frac{1}{n}\sum_{i=1}^{n}\frac{1}{y_i^2}\right)$$

where n = the number of tests in a trial (number of repetitions regardless of noise levels) and y_i = each observed value (data points). In all cases, the target is to maximize the HB S/N ratio. The development of a family of matrices based on orthogonal arrays

(OA) and the signal-to-noise (S/N) definition as an indicator of the ratio of the mean to the standard deviation are great contributions of the Taguchi method. As a result, they will be used to evaluate and discuss the results of each Taguchi experiment:

- Statistical treatment. In all experiments, the significance (*alpha*) level was set as 0.05, and means and standard deviation were calculated for all volumetric capacities. The *p*-value inferior to *alpha* concludes that there is a statistically significant association between the response characteristic and the term. The *p*-value between the significance levels of 0.05 and 0.1 can be used for evaluating terms, and it can be considered with practical significance. A higher *p*-value will conclude that there were no statistically significant differences observed.

4.3. First Taguchi Experiment: The Screening Case

The first round, referred to as the screening case, is used to find the few important, influential factors out of the many possible factors involved with a process design. The recommended strategy is to start with the smallest orthogonal array that will accommodate the typically large number of actors under simultaneous evaluation at two levels.

Hence, the $L_8(2^7)$ OA inner array with a resolution number of 1-low (A and B x C are in the same column) without noise factors is selected and presented in Table 4. The resolution number is a measure of the amount of confounding in a column [33]. As the assignment of factors is to all columns, unavoidably, many interactions are confounded (mixed) with the main effects. This is the major compromise of using fractional factorial experiments: by reducing the number of tests, some information must be surrendered.

Table 4. The parameter design table of the screening case of the first Taguchi experiment. (Please refer to Nomenclature section and Figure 2).

8 gear sets in common							
Vol. capacity target:	$\bar{c}_{v,H}$ = 1 cc·rev^{-1}·mm^{-1} (0.093), mean (standard deviation)						
Dimensional constraints:	D_c = 80 mm; d_s = 40 mm; H = 9.25 mm						
FEM conditions:	T = 18.75 N·m; μ = 0.4; θ = 20°						
L8 OA inner array (control factors)	A	B	C	D	E	F	G
Factor parameter (column) Trial gear set::Taguchi experiment (row)	Z	e [mm]	Reference position (RP) (Figure 6)	S [mm]	r_c [mm]	Shaft keyway (SK) (Figure 6)	w_c [mm]
1::1	8	2.50	T2T	5.86	0.0	No	3.0
2::1	8	2.50	T2T	13.96	3.0	Yes	6.0
3::1	8	2.97	V2T	5.86	0.0	Yes	6.0
4::1	8	2.97	V2T	13.96	3.0	No	3.0
5::1	9	2.50	V2T	5.86	3.0	No	6.0
6::1	9	2.50	V2T	13.96	0.0	Yes	3.0
7::1	9	2.97	T2T	5.86	3.0	Yes	3.0
8::1	9	2.97	T2T	13.96	0.0	No	6.0

The meaning of each number in the OA nomenclature $L_8(2^7)$ is: 8 FEM simulations in this screening case (inner array with 8 gear sets trials, without noise factors, corresponding to each row in Table 4), 7 factor parameters (Z, e, RP, S, r_c, SK, and w_c corresponding to each column in Table 4) and 2 levels (low and high values corresponding to the columns of the chosen Taguchi OA family in Table 4). In addition, the table head of Table 4 indicates the volumetric capacity target and the dimensional constraints to remain constant in the gerotor pump, together with the FEM conditions used in the simulation.

The meaning of OA nomenclature and the arrangement of Table 4 will be retained for all incoming Taguchi experiments.

By the inspection of Figure 7, an even number of teeth produces a higher maximum contact stress in all gear sets, with the exception of the 5:1 gear set owing to the selected reference position. The location of the contact point in the second tooth prevails in five out of eight cases. The 8::1 gear set shows the minimum value of contact stress, which is almost half of the 4::1 gear set, indicating the significance of the geometry and the reference position.

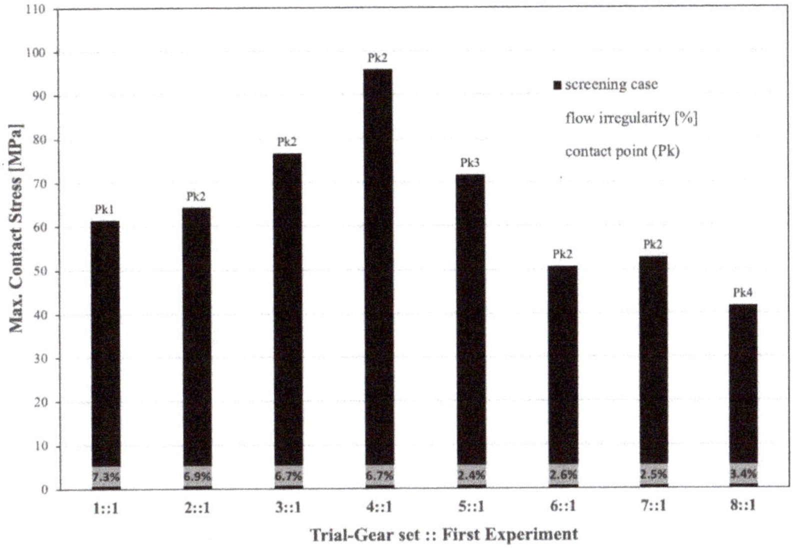

Figure 7. Results of FEM maximum contact stress [MPa], GeroLAB flow irregularity [%], and contact point (Pk) of maximum contact stress in each gear set for the first Taguchi experiment (see Table 4).

Figure 8 illustrates the main effects plot for the screening case. The true critical characteristics (control factors) must be identified and minimized in number. This graphic aids in a better picture of the importance of the chosen control parameters. The larger the vertical increment between the two levels and the higher the slope in each factor, the more emphasis is placed on this parameter: the larger the HB S/N ratio, the better. The figure indicates that the parameters *Z*, *RP*, and r_c are the strength factors. Conversely, the parameter w_c, with a behavior almost horizontal between the two levels, is the first factor to be discarded, together with the parameter *SK*.

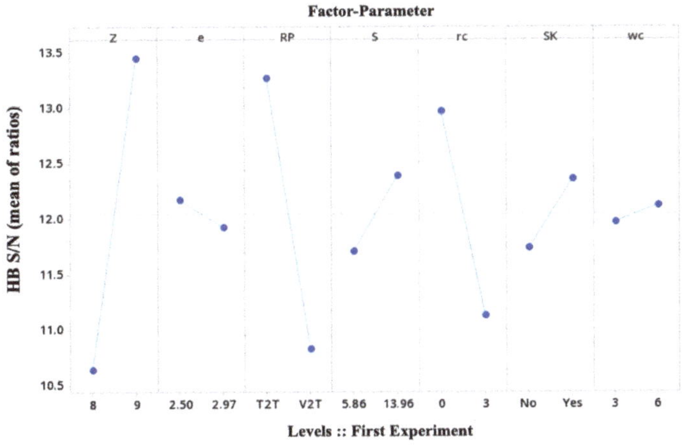

Figure 8. Main effects plot for the first Taguchi experiment (see Table 4).

Obviously, the resolution is low in the first round of experiments (a small fractional factorial) and it will progress to higher-resolution experiments (a large fractional or full factorial) as few factors are going to be identified as influential.

In addition, this screening case has mixed geometrical basic parameters to define the trochoidal gear set, such as Z, e, S, and external gear r_c, with working functions, such as SK and RP, where the former shows less dominance than the latter, and it will be discarded. Owing to the higher effect of the level 'Yes' in the shaft keyway, which is closer to real working function, all the gear sets in the incoming Taguchi experiments have a shaft keyway.

4.4. Second Taguchi Experiment: The Two-Levels-Four-Geometric-Basic-Parameters Case and One Noise Factor

The Taguchi approach to parameter design includes noise factors as related elements of the process that cannot be controlled or removed as a measure of the robust levels of the control factors. Thus, from the screening case, working function related to reference position is going to be used as a non-controlled process noise effect in this second Taguchi experiment.

$L_8(2^4)$ OA inner array with a resolution number of 2 (A and BxCxD, or AxB and CxD are in the same column), with one noise factor at the outer array L1 OA. Here, 16 FEM simulations were carried out in this second Taguchi (8 gear sets trials, 2 working functions noise factors). This second experiment encompasses less factor parameters (Z, e, S, and r_c) corresponding to each column in Table 5. The outer array accommodates the noise factor at two levels, T2T and V2T, corresponding to columns N1 and N2 in Table 5.

Table 5. The parameter design table of the second Taguchi experiment. (Please refer to Nomenclature section and Figure 2).

8 gear sets in common							
Vol. capacity target:	$\bar{c}_{v,H} = 0.997$ cc·rev^{-1}·mm^{-1} (0.023), mean (standard deviation)						
Dimensional constraints:	$D_c = 80$ mm; $d_s = 40$ mm; $H = 9.25$ mm; (SK = yes)						
FEM conditions:	$T = 18.75$ N·m; $\mu = 0.4$; $\theta = 20°$						
L8 OA inner array (control factors)	A	B	D	G	L1 OA outer array (noise factor)	N1	N2
Factor parameter (column) Trial gear set::Taguchi experiment (row)	Z	r_c [mm]	S [mm]	e [mm]	Reference position (Figure 6) (column)	T2T	V2T
1::2	8	0.0	5.86	2.50			
2::2	8	0.0	13.96	2.97			
3::2	8	3.0	5.86	2.97			
4::2	8	3.0	13.96	2.50			
5::2	9	0.0	5.86	2.97			
6::2	9	0.0	13.96	2.50			
7::2	9	3.0	5.86	2.50			
8::2	9	3.0	13.96	2.97			

As in the case of FEM validation, the corresponding concordances reveal the important effect of the reference position in the numerical simulation. Comparing the gear set 8:1 in Figure 7 and the gear set 8:2 in Figure 9, both of them present the lower value of maximum contact stress with an acceptable value of 3.4% of flow irregularity. Since the 8:2 gear set uses the reference position as a noise factor to expose the robust levels of the control factors, these parameters will produce successively better approximations to converge to the sought gear, which is more suitable regarding contact stresses. Considering the location of contact points, it stays the same or within a unit difference.

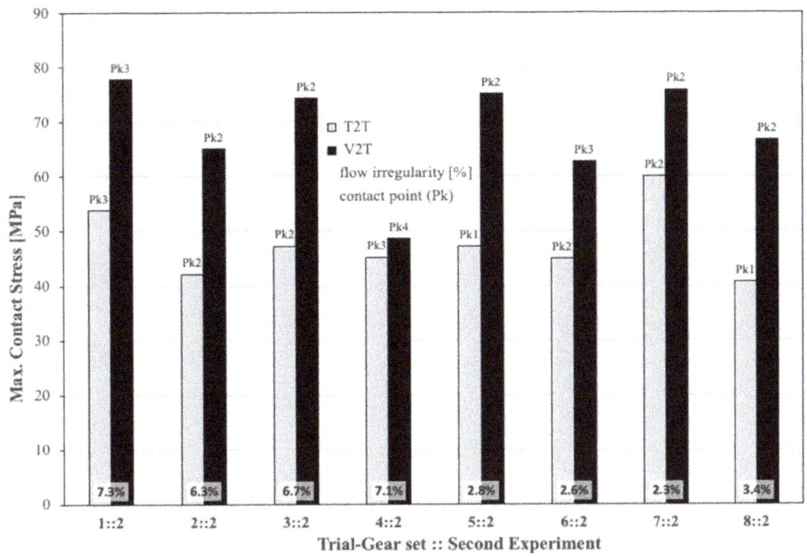

Figure 9. Results of FEM maximum contact stress [MPa], GeroLAB flow irregularity [%], and contact point (Pk) of maximum contact stress in each gear set with the working function noise factor for the second Taguchi experiment (see Table 5).

Since the slope in the graphic is an indicator of the importance of a main effect, Figure 10 shows that the most dominant factor is the arc radius of the external gear tooth, *S*. In addition, this geometrical parameter *S* is considered with practical significance in the analysis of variance for HB S/N ratios ($p = 0.083$).

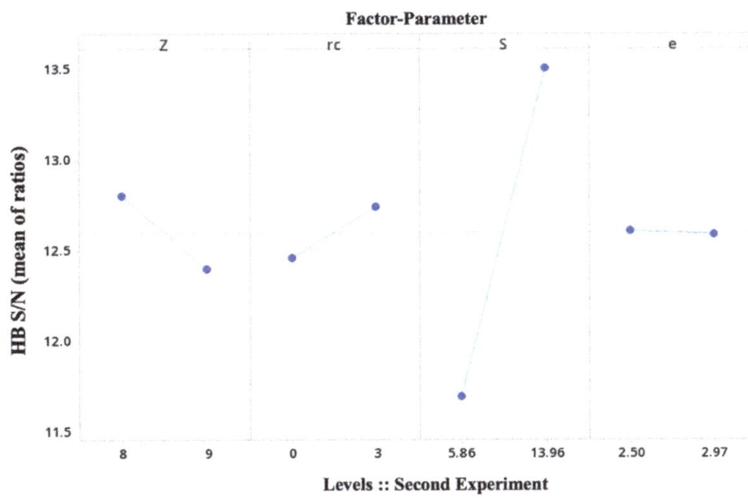

Figure 10. Main effects plot for the second Taguchi experiment (see Table 5).

The mutual dependency between a pair of factors is an interaction of those factors. The interaction plot of the main effect factors *Z* and *S* is constructed by plotting the average response HB S/N values of each factor level combination and is presented in Figure 11. The effect of parameter *S* is estimated to be low since it would depend upon which level of parameter Z is being used. Being almost parallel lines is an indication of the absence of interaction between both factors. In addition, for this experiment, none of the studied interaction effects are statistically significant.

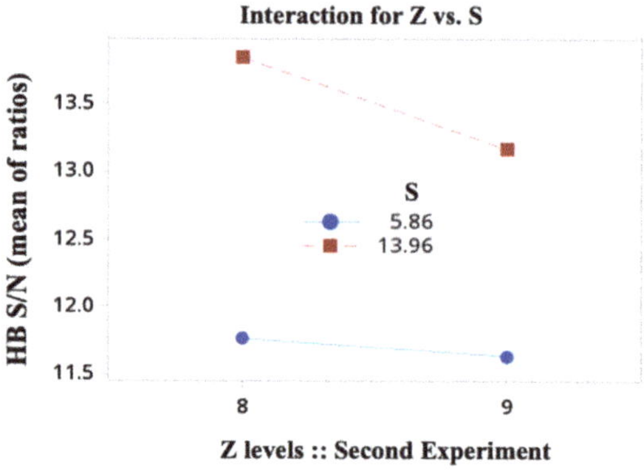

Figure 11. Interaction plot between Z and S for the second Taguchi experiment (see Table 5).

On the other hand, these results are quite confusing, since it is difficult to accept Z as having a similar effect as r_c in the gear set contact stresses as well as no significant interaction effect. One might add the well-known opposite effect on the volumetric characteristics of an even Z value, as it can be seen in the numerals of the flow irregularity in Figure 9. Upon debating this controversial effect, it is decided to go with a full factorial experiment by using the main geometric basic parameters: the number of external teeth, Z, the eccentricity, e, and the arc radius of the external gear tooth, S.

4.5. Third Taguchi Experiment: The Two-Levels-Three-Geometric-Basic-Parameters Case and Three Noise Factors

As previously mentioned, Taguchi parameter design is used to dampen the effect of noise, which includes parameters that are either uncountable or are too expensive to control, by choosing the proper level of control parameters. Then, once more, the question arises: why bother to assign these noise factors when they cannot be controlled or eliminated? The main reason, in this work, is related to the numerical simulations: several settings in the software environment could act as noise since they are associated with input operating conditions in the FEM simulation.

Then, the strategy is to force noise effects into the numerical simulation with an experimental layout based on an outer array only for operating conditions: a combination of outer noise (environmental-related, such as temperature, θ) and inner noise (function-related, such as torque, T, and friction coefficient, μ). If all of these noise factors were mixed together in an inner array for control factors, it would be a traditional cause of detection experiment: this is one of the major contributions of the Taguchi approach to noise emulation in FEM methodology.

Table 6 illustrates the $L_8(2^3)$ OA inner array with a resolution number of 4-high (full factorial, all items are in separate columns), with 3 noise factors. Here, 32 FEM simulations were carried out (8 gear sets trials, 4 FEM conditions noise factors). This third experiment encompasses three factor parameters (Z, e, and S) corresponding to each column. Owing to the low effect of r_c, it was removed from the inner array, avoiding interactions confounded with the main effects and gaining resolution. The outer array accommodates the 3 noise factors at two levels, corresponding to columns N1, N2, N3, and N4 (refer to Table 6).

Table 6. The parameter design table of the third Taguchi experiment. (Please refer to Nomenclature section and Figure 2).

8 gear sets in common Vol. capacity target: Dimensional constraints: FEM conditions:	$\bar{c}_{v,H}$ = 0.997 cc·rev^{-1}·mm^{-1} (0.023), mean (standard deviation) D_c = 80 mm; d_s = 40 mm; H = 9.25 mm; (SK = yes) *Noise factors*							
L8 OA inner array (control factors)	A	B	D	L4 OA outer array (noise factors)	N1	N2	N3	N4
Factor parameter (column) Trial gear set::Taguchi experiment (row)	Z	e [mm]	S [mm]	T [N·m] μ [-] θ [°C] (column)	15.0 0.01 20	22.5 0.50 20	22.5 0.01 40	15.0 0.50 40
1::3	8	2.50	5.86					
2::3	8	2.50	13.96					
3::3	8	2.97	5.86					
4::3	8	2.97	13.96					
5::3	9	2.50	5.86					
6::3	9	2.50	13.96					
7::3	9	2.97	5.86					
8::3	9	2.97	13.96					

Once this new OA with a high-resolution number has been selected, the factors can be assigned to the corresponding columns of the inner array and subsequent interaction columns located. In this new experiment, the interaction of Z and S is again studied since both have shown main effects in the HB S/N ratio of the previous first and second experiments.

In Figures 11 and 12, results referring to the third experiment are reported. Figure 12 illustrates an important effect not influenced by the noise FEM conditions: the location of the maximum contact stress at the same contact point for all operating conditions once the three geometrical basic parameters are taken into account. It is also worth noting that the combination of a higher torque and a lower frictional factor results in higher contact stress, regardless of temperature.

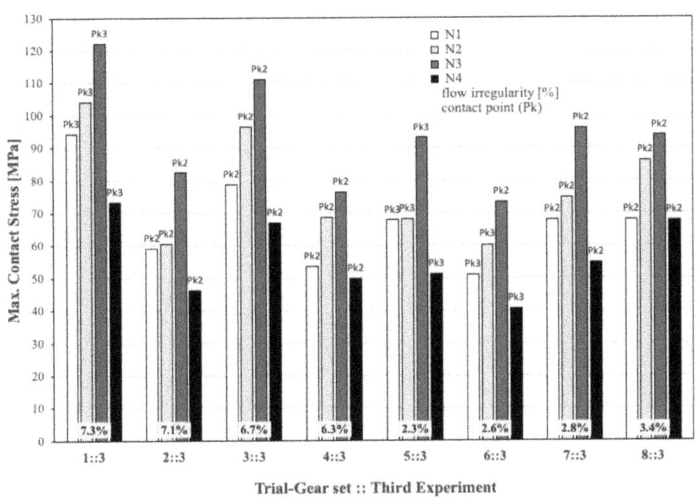

Figure 12. Results of FEM maximum contact stress [MPa], GeroLAB flow irregularity [%], and contact point (Pk) of maximum contact stress in each gear set with 4 FEM conditions noise factors for the third Taguchi experiment (see Table 6).

Figure 13, as previously in Figure 10, shows that the most dominant factor is the arc radius of the external gear tooth, *S*, with practical significance in the analysis of variance for HB S/N ratios ($p = 0.080$). This is an expected result, since removing the parameter cutting radius shows its minimal effect. However, the absence of interactions confounded with the main effects in this full factorial experiment and the inclusion of the operating conditions reinforces this outcome. Now, the odd number of external gear teeth shows a dominant effect, and the eccentricity normalises its intrinsic interest and value.

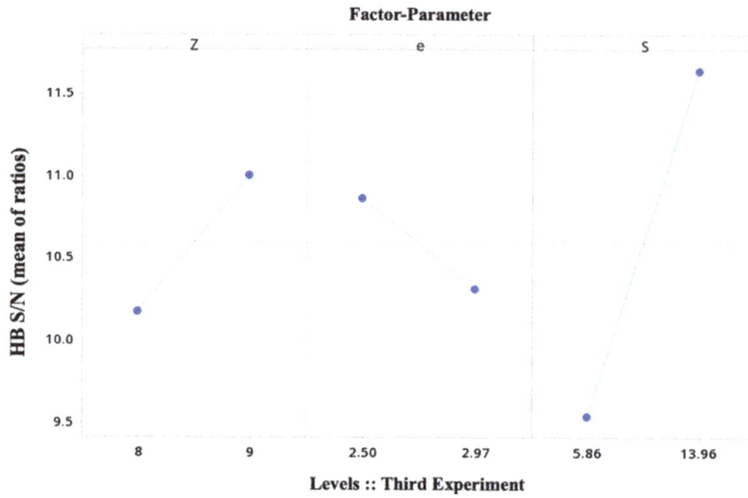

Figure 13. Main effects plot for the third Taguchi experiment (see Table 6).

The interaction plot of the main effect factors *Z* and *S* is again computed and presented in Figure 14. The lines' loss of parallelism and the effect of an odd number value for Z are indicated, although it is not statistically significant yet. The OA size and the interaction potential of interest to the experimenter are the key factors; as a general recommendation, it is preferred to study more factors than to study interactions [33].

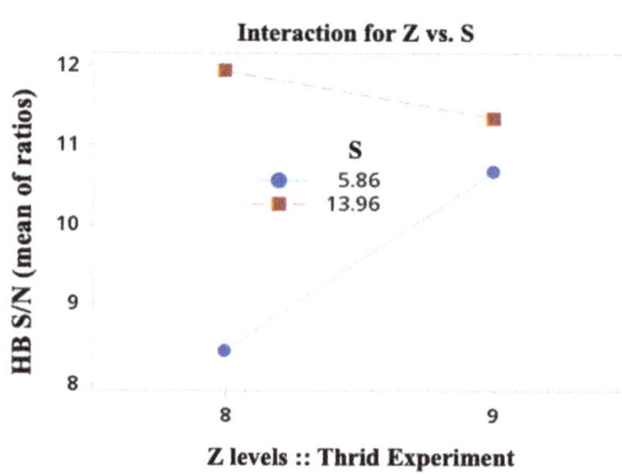

Figure 14. Interaction plot between Z and S for the second Taguchi experiment (see Table 5).

4.6. Fourth Taguchi Experiment: The Mixed-Level-Design-Case with Three Noise Factors

The last experiment in this iterative process is designed to elucidate the ability of the Taguchi method to pinpoint the most suitable gear set configuration in front of FEM contact stress. With the learning from the previous three Taguchi experiments, it is decided to complete the study with an OA inner array with four control factors: the trochoidal gear set parameters (*Z*, *e*, and *S*) and the working function (*RP*). Note that it is decided to include the *RP* in the inner array, since its effects have been shown to be of noticeable importance. The FEM conditions are kept to the three operating conditions (T,μ and θ) in the outer array to emulate noise effects. To complete the experiment, one more level is added to the trochoidal gear set parameters in a mixed-level design case. The Taguchi approach prefers using three or more levels of design parameters to estimate non-linear effects rather than two levels of classical DoE [30].

Table 7 illustrates the $L_{18}(2^1–3^3)$ mixed-level OA inner array with a resolution number of 1-low (no specific interaction columns available, except between AxB by using a specific layout). Here, 72 FEM simulations were carried out (18 gear sets trials, 4 FEM conditions noise factors). The outer array accommodates the three operating conditions (environmental-related, θ and function-related, T and μ) at two levels, corresponding to each column N1, N2, N3, and N4. In this mixed-level design, the parameters of the cutting radius and the wall width of the external gear are kept constant for the 18 gear sets, and factor interactions are not studied based on the results of the previous experiments.

Table 7. The parameter design table of the fourth Taguchi experiment. (Please refer to Nomenclature section and Figure 2).

18 gear sets in common									
Vol. capacity target:	$\bar{c}_{v,H}$ = 0.996 cc·rev^{-1}·mm^{-1} (0.041), mean (standard deviation)								
Dimensional constraints:	D_c = 80 mm; d_s = 40 mm; H = 9.25 mm; (*SK* = yes)								
FEM conditions:	*Noise factors*								
L18 OA inner array (control factors)	A	B	C	D	L2 OA outer array (noise facts)	N1	N2	N3	N4
Factor parameter (column) Trial gear set::Taguchi experiment (row)	Reference position (*RP*) (Figure 6)	*Z*	*S* [mm]	*e* [mm]	*T* [N·m] μ [-] θ [°C] (column)	15.0 0.0 ** 20	22.5 0.4 20	22.5 0.0 ** 40	15.0 0.4 40
1::4	V2T	7	5.8	2.61					
2::4	V2T	7	8.0	2.77					
3::4	V2T	7	10.4	2.93					
4::4	V2T	9	5.8	2.61					
5::4	V2T	9	8.0	2.77					
6::4	V2T	9	10.4	2.93					
7::4	V2T	11	5.8	2.77					
8::4	V2T	11	8.0	2.93					
9::4	V2T	11	10.4	2.61					
10::4	T2T	7	5.8	2.93					
11::4	T2T	7	8.0	2.61					
12::4	T2T	7	10.4	2.77					
13::4	T2T	9	5.8	2.77					
14::4	T2T	9	8.0	2.93					
15::4	T2T	9	10.4	2.61					
16::4	T2T	11	5.8	2.93					
17::4	T2T	11	8.0	2.61					
18::4	T2T	11	10.4	2.77					

** Frictionless.

In Figure 15, the diversity of operating conditions is studied by crossing the inner array of control parameters by the outer array of noise factors. The slope of the HB S/N ratios once more illustrates the dominance of the three trochoidal gear set parameters (*Z*, *e*, and *S*) in front of the working function (*RP*). In addition, the main effects plots for means and standard deviations confirm the same tendency.

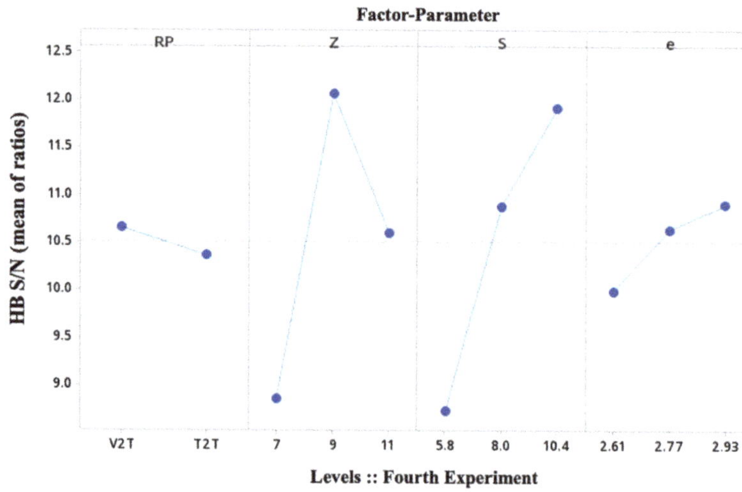

Figure 15. Main effects plot for the fourth Taguchi experiment (see Table 7).

The statistical significance for Z and S supports this result. This fourth experiment shows statistical significance at an *alpha* level of 0.05 for Z and S, in particular, $p = 0.053$ and $p = 0.050$, respectively. Conversely, none of the interaction effects are significant.

For the contact stress experiment, the optimal control factor settings based on the highest HB S/N have been determined by the inspection of Figure 15. In order to decide which level is best to maximize the safety factor, the HB S/N ratios at each level (low, medium, and high) of each factor are compared. Hence, the levels are:

$Z = 9$ (level 2-medium); $S = 10.4$ mm (level 3-high); $e = 2.93$ mm (level 3-high) >>> Gear set 6::4.

Using the Taguchi approach for parametric design, the predicted optimum settings correspond to one of the rows of the matrix experiment: the gear set 6::4 in Table 7. In addition, its flow irregularity is 3%, a good value. (If the predicted optimum settings did not correspond to one of the rows, as a final step, a confirmation trial would be run). The reference position has a poor effect as a control factor. These settings maximize the safety factor for all noise configurations.

Finally, if the three trochoidal gear set parameters of the PZ9e285 gear set in Table 2 are compared with the results of the predicted optimum settings, it shows the reliability of the methodology since it points out the current geometry of a real-working gerotor pump.

5. Conclusions

In this paper, efforts to prove the reliability of the Taguchi techniques as an effective simulation-based strategy in the design of numerical simulations to assess contact stresses in trochoidal gears working in gerotor pumps were presented.

The validity of the Finite Element Model proposed in this paper was first decided to be proven by means of published results in the literature. The works of Ivanović and Biernacki, two researchers with long and solid careers in the mechanical analysis of gerotor pumps, were selected. Nomenclature and notation were integrated and unified for the research presented in this paper.

The average of the results of the nine simulations showed a discrepancy of 9% in the maximum contact stress at contact points, an acceptable error taking into account the available technical data and the limitation of simulating exactly the same geometries as the researchers. In addition, the location of the contact points was linked in all cases. The numerical model was verified to be valid within reasonable bounds and it was applied to the simulations designed by the Taguchi method.

The Taguchi approach was based on a sequential simulation-based design strategy of four experiments enabling a manageable number of gear sets. The resulting 128 numerical simulations accomplished the challenging goal of having the same volumetric capacity and dimensional constraints (gear thickness, external diameter of external gear, and the internal diameter of shaft hole located in the inner gear) to disengage contact stress performance from flow rate and casing. Furthermore, this innovative approach allowed bypassing the inherent multitude of geometric options in FEM numerical simulations and the 'black box' effect in the software input operating conditions, such as environmental-related temperature and function-related torque and friction coefficient. The progressive simulation-based planning narrowed down the number of parameters, reducing the solution space, under operating conditions that emulate noise effects.

Results of contact stress proved the arc radius of the external gear tooth to be the most dominant parameter. The odd number of external gear teeth was shown to be the second dominant effect, and the eccentricity demanded its intrinsic interest and value. The particular values of the statistical significance for both dominant parameters endorsed this conclusion. However, none of the interaction effects were statistically significant.

In particular, the feasibility of the proposed methodology to pinpoint the most suitable gear set configuration in front of FEM contact stress and to predict optimum settings in the geometry of real-working gerotor pumps in industrial designs was demonstrated. This statement was confirmed by the specific value combination of a trochoidal gear set: 9 external gear teeth, an arc radius of 10.4 mm, and an eccentricity of 2.93 mm. This optimized contact stress gear set yields to the targeted volumetric capacity of 1 cc·rev^{-1} per gear thickness and a proper flow irregularity of 3%.

Promising results are obtained in this approach for enhancing the quality of contact stress assessment without controlling or removing the cause of variation contributed to by the numerical simulation's input settings. Taguchi robust design is an easier approach than classical optimization. Its accessibility as an open tool to end-users and its methodology are very much related to real production and products in industry, quality, and costs. This study has formulated its own approach in the field, assisting the numerical simulation procedure in the design process and leading to the gear set configuration of a real-working gerotor pump.

Regarding future work on the research presented in this paper, the next step will be to identify the control factors that move the mean to a pre-established contact stress target and predict the average response for each combination of control factor settings using the model for the response to the signal-to-noise ratio.

Author Contributions: Conceptualization, P.J.G.-M. and E.B.-M.; methodology, P.J.G.-M. and E.B.-M.; software, E.B.-M.; validation, P.J.G.-M. and E.B.-M.; writing—original draft preparation, P.J.G.-M.; writing—review and editing, P.J.G.-M. and E.B.-M.; supervision, P.J.G.-M. and E.B.-M.; project administration, P.J.G.-M. and E.B.-M. All authors have read and agreed to the published version of the manuscript.

Funding: This research received no external funding.

Data Availability Statement: Not applicable.

Acknowledgments: E.B.-M. is a Serra Húnter fellow. The authors would like to acknowledge the Generalitat de Catalunya for providing the necessary support to the research groups IAFARG (SGR 286, https://iafarg.upc.edu, accessed on 1 September 2022), CATMech (Tecnio, https://catmech.upc.edu/home, accessed on 1 September 2022), and Doctorats Industrials (2020 DI 45). The authors would like to thank Byon Zambrano Risco for participating in the research reported in this paper.

Conflicts of Interest: The authors declare no conflict of interest.

Nomenclature

c_v	Volumetric capacity/displacement
$c_{v,H}$	Volumetric capacity/displacement per gear thickness
$DeRi$	External diameter of internal gear ($DeRi = 2 \cdot ReRi$)
d_s	Shaft hole, internal diameter located in the inner gear
D_c	External diameter of external gear
e	Eccentricity (centre distance)
G	Radius of circle to complete external gear
H	Gear thickness
L_s	Shaft keyway length
O_1	Inner/internal gear centre
O_2	Outer/external gear centre
P_k	Contact point
r_c	Cutting radius
r_1	Inner pitch circle
r_2	Outer pitch circle
R_2	Distance O_2P_s
S	Arc radius of the external gear tooth
T	Torque
w_c	Wall width of the external gear
$Z, (Z-1)$	Number external (internal) teeth
λ	Tooth profile height correction coefficient ($\lambda = r_2/R_2$)
μ	Friction coefficient
θ	Temperature
v	Equidistant index ($v = S \cdot Z/R_2$)

References

1. Rundo, M. Models for flow rate simulation in gear pumps: A review. *Energies* **2017**, *10*, 1261. [CrossRef]
2. Pareja-Corcho, J.; Moreno, A.; Simoes, B.; Pedrera-Busselo, A.; San-Jose, E.; Ruiz-Salguero, O.; Posada, J. A Virtual Prototype for Fast Design and Visualization of Gerotor Pumps. *Appl. Sci.* **2021**, *11*, 1190. [CrossRef]
3. Kawamata, H.; Kuroda, S.; Tanaka, S.; Oshima, M. *Improvement of Practical Electric Consumption by Drag Reducing under Cross Wind*; SAE Technical Paper 2016-01-1626; SAE International: Warrendale, PA, USA, 2016; pp. 1–10. [CrossRef]
4. Colbourne, J.R. Reduction of contact stress in internal gear pumps. *J. Eng. Ind.* **1976**, *98*, 1296–1300. [CrossRef]
5. Roy, D. Mechanics and FEM estimation of deformation, gaps and stresses generated in star-ring active contacts of GEROTOR units during operation. *J. Br. Soc. Mech. Sci. Eng.* **2021**, *43*, 429. [CrossRef]
6. Roy, D.; Maiti, R.; Das, P.K. Mechanics and FEM estimation of gaps generated in star-ring active contacts of ORBIT motor during operation. *Int. J. Mech. Mater. Des.* **2020**, *16*, 69–89. [CrossRef]
7. Sliwinski, P. The methodology of design of satellite working mechanism of positive displacement machine. *Sci. Rep.* **2022**, *12*, 1–22. [CrossRef]
8. Gamez-Montero, P.J.; Codina, E.; Castilla, R. A Review of Gerotor Technology in Hydraulic Machines. *Energies* **2019**, *12*, 2423. [CrossRef]
9. Ivanović, L.; Devedžić, G.; Ćuković, S.; Mirić, N. Modeling of the Meshing of Trochoidal Profiles with Clearances. *J. Mech. Des.* **2012**, *134*, 041003. [CrossRef]
10. Ivanović, L.; Josifović, D.; Ilić, A. Modelling of trochoidal gearing at the gerotor pump. *Mech. Mach. Sci.* **2013**, *13*, 553–562. [CrossRef]
11. Ivanović, L.; Ilić, A.; Miloradović, D.; Josifović, D. Modelling and simulation of the load in the epicyclic rotary pump with trochoidal gear profiles. *IOP Conf. Ser. Mater. Sci. Eng.* **2018**, *393*, 1–9. [CrossRef]
12. Ivanović, L.; Devedžić, G.; Mirić, N.; Ćuković, S. Analysis of forces and moments in gerotor pumps. *Proc. Inst. Mech. Eng. C* **2010**, *224*, 2257–2269. [CrossRef]
13. Ivanović, L. Reduction of the maximum contact stresses by changing geometric parameters of the trochoidal gearing teeth profile. *Meccanica* **2016**, *51*, 2243–2257. [CrossRef]
14. Ivanović, L.; Rakić, B.; Stojanović, B.; Matejić, M. Comparative analysis of analytical and numerical calculations of contact stresses at rotational elements of gerotor pumps. *Appl. Eng. Lett.* **2016**, *1*, 1–7.
15. Ivanović, L.; Mackić, T.; Stojanović, B. Analysis of the instantaneous friction coefficient of the trochoidal gear pair. *J. Balk. Tribol. Assoc.* **2016**, *22*, 281–293.
16. Ivanović, L.; Matejić, M. Improving gerotor pump performance through design, modelling and simulation. *Int. J. Fluid Power* **2021**, *21*, 327–346. [CrossRef]

17. Biernacki, K.; Stryczek, J. Analysis of stress and deformation in plastic gears used in gerotor pumps. *J. Strain Anal. Eng. Des.* **2010**, *45*, 465–479. [CrossRef]
18. Biernacki, K. Selection of the optimum tooth profile for plastic cycloidal gears. *Proc. Inst. Mech. Eng. C* **2014**, *228*, 3395–3404. [CrossRef]
19. Stryczek, J.; Bednarczyk, S.; Biernacki, K. Strength analysis of the polyoxymethylene cycloidal gears of the gerotor pump. *Arch. Civ. Mech. Eng.* **2014**, *14*, 647–660. [CrossRef]
20. Stryczek, J.; Bednarczyk, S.; Biernacki, K. Gerotor pump with POM gears: Design, production technology, research. *Arch. Civ. Mech. Eng.* **2014**, *14*, 391–397. [CrossRef]
21. Biernacki, K. Analysis of the material and design modifications influence on strength of the cycloidal gear system. *Int. J. Adv. Manuf. Technol.* **2015**, *16*, 537–546. [CrossRef]
22. Biernacki, K. New concept of power transmission for gerotor hydraulic machines. *Proc. Inst. Mech. Eng. Part C J. Mech. Eng. Sci.* **2021**, *235*, 4873–4883. [CrossRef]
23. Biernacki, K. New construction of cycloidal gear unit made of plastics. *Proc. Inst. Mech. Eng. Part C J. Mech. Eng. Sci.* **2021**, *235*, 800–811. [CrossRef]
24. Robison, A.; Vacca, A. Multi-objective optimization of circular-toothed gerotors for kinematics and wear by genetic algorithm. *Mech. Mach. Theory* **2018**, *128*, 150–168. [CrossRef]
25. Robison, A.; Vacca, A. Performance comparison of epitrochoidal, hypotrochoidal, and cycloidal gerotor gear profiles. *Mech. Mach. Theory* **2021**, *158*, 104228. [CrossRef]
26. De Martin, A.; Jacazio, G.; Sorli, M. Optimization of Gerotor Pumps with Asymmetric Profiles through an Evolutionary Strategy Algorithm. *Machines* **2019**, *7*, 17. [CrossRef]
27. Tessari, F.; Galluzzi, R.; Amati, N. Efficiency-driven design methodology of gerotor hydraulic units. *J. Mech. Des.* **2020**, *142*, 063501. [CrossRef]
28. Puliti, M.; Tessari, F.; Galluzzi, R.; Tonoli, A.; Amati, N. Design methodology of gerotor hydraulic machines for mechatronic applications. In Proceedings of the ASME International Mechanical Engineering Congress and Exposition, IMECE2021, Virtual, Online, 5 November 2021.
29. Phadke, M.S. *Quality Engineering Using Robust Design*; Prentice Hall International: Englewood Cliffs, NJ, USA, 1989.
30. Antony, J. Taguchi or classical design of experiments: A perspective from a practitioner. *Sens. Rev.* **2006**, *26*, 227–230. [CrossRef]
31. Taguchi, G. *Introduction to Quality Engineering: Designing Quality into Products and Processes*; Asian Productivity Organization: Tokyo, Japan, 1986.
32. Unal, R.; Dean, E.B. Taguchi approach to design optimization for quality and cost: An overview. In Proceedings of the 1991 Annual Conference of the International Society of Parametric Analysts, Hampton, VA, USA, 1 January 1991.
33. Ross, R. *Taguchi Techniques for Quality Engineering*, 2nd ed.; McGraw-Hill: New York, NY, USA, 1996.
34. Antony, J.; Jiju Antony, F. Teaching the Taguchi method to industrial engineers. *Work Study* **2001**, *50*, 141–149. [CrossRef]
35. Jung, S.-Y.; Bae, J.-H.; Kim, M.-S.; Kim, C. Development of a new gerotor for oil pumps with multiple profiles. *Int. J. Precis. Eng. Manuf.* **2011**, *12*, 835–841. [CrossRef]
36. Ivanović, L.T.; Veličković, S.N.; Stojanović, B.Ž.; Kandeva, M.; Jakimovska, K. The selection of optimal parameters of gerotor pump by application of factorial experimental design. *FME Trans.* **2017**, *45*, 159–164. [CrossRef]
37. Ivanović, L.; Stojanović, B.; Blagojević, J.; Bogdanović, G.; Marinković, A. Analysis of the flow rate and the volumetric efficiency of the trochoidal pump by application of Taguchi method. *Teh. Vjesn.* **2017**, *24*, 265–270. [CrossRef]
38. Lazić, L.; Milinković, S. Reducing software defects removal cost via design of experiments using Taguchi approach. *Software Qual. J.* **2015**, *23*, 267–295. [CrossRef]
39. Vosniakos, G.C.; Kalattas, A.; Siasos, A. Optimal process planning for helical bevel gears using Taguchi design of simulated machining experiments. *Proc. IMechE Part B J Eng. Manuf.* **2018**, *232*, 2627–2640. [CrossRef]
40. Choudhury, K.; Mandol, S.; Dan, P.K. Optimization of Operations in Epicyclic Gear Train Modelled Through Computer Aided Design. *Asia-Pac. J. Manag. Res. Innov.* **2014**, *10*, 323–335. [CrossRef]
41. Stryczek, J.; Stryczek, P. Synthetic approach to the design, manufacturing and examination of gerotor and orbital hydraulic machines. *Energies* **2021**, *14*, 624. [CrossRef]
42. Gamez-Montero, P.J.; Castilla, R.; Khamashta, M.; Codina, E. Contact problems of a trochoidal-gear pump. *Int. J. Mech. Sci.* **2006**, *48*, 14711480. [CrossRef]
43. Gamez-Montero, P.J.; Castilla, R.; Mujal, R.; Khamashta, M.; Codina, E. GEROLAB package system: Innovative tool to design a trochoidal-gear pump. *J. Mech. Des.* **2009**, *131*, 074502. [CrossRef]
44. GeroLAB Package System. Available online: http://www.gerolab.es (accessed on 1 February 2022).
45. Ram Kumar, A.C.; Mohammed Raffic, N.; Ganesh Babu, K.; Selvakumar, S. Static structural analysis of spur gear using ANSYS 15.0 and material selection by COPRAS, MOORA techniques. *Mater. Today Proc.* **2021**, *47*, 25–36. [CrossRef]
46. Karthick, L.; Shanmugasundaram, A.; Jagadish, C.A.; Prabhu, T.; Vidya Prakash, S. Numerical analysis of Modified tooth in Spur Gear for increasing the performance using fool proofing technology. *Mater. Today Proc.* **2021**, *47*, 5862–5868. [CrossRef]
47. Benaïcha, Y.; Perret-Liaudet, J.; Beley, J.D.; Rigaud, E.; Thouverez, F. On a flexible multibody modelling approach using FE-based contact formulation for describing gear transmission error. *Mech. Mach. Theory* **2022**, *167*, 104505. [CrossRef]

48. Rao, P.S. Contact Stress and Shear Stress Analysis of Spur Gear Using ANSYS and Theoretical. *Int. J. Mod. Stud. Mech. Eng.* **2016**, *2*, 9–14. [CrossRef]
49. Lahtivirta, J.; Lehtovaara, A. Modelling of spur gear contact using a local adaptive finite element mesh. *Tribologia* **2016**, *34*, 41–55.
50. Lisle, T.J.; Shaw, B.A.; Frazer, R.C. External spur gear root bending stress: A comparison of ISO 6336:2006, AGMA 2101-D04, ANSYS finite element analysis and strain gauge techniques. *Mech. Mach. Theory* **2017**, *111*, 1–9. [CrossRef]
51. Puneeth, M.L.; Mallesh, G. Static contact behavior of asymmetric spur gear. *Mater. Today Proc.* **2021**, *47*, 3095–3104. [CrossRef]
52. Munoz, G.A. *Lecture 7: Mesh Quality & Advanced Topics*; ANSYS Inc.: Canonsburg, PA, USA, 2015; p. 37. Available online: https://www.academia.edu/16970000/MESH_QUALITY_AND_ADVENCED_TOPICS_ANSYS_WORKBENCH_16_0 (accessed on 1 February 2022).
53. Sliwinski, P. Determination of the Theoretical and Actual Working Volume of a Hydraulic Motor—Part II (The Method Based on the Characteristics of Effective Absorbency of the Motor). *Energies* **2021**, *14*, 1648. [CrossRef]
54. Garcia-Vilchez, M.; Gamez-Montero, P.J.; Codina, E.; Castilla, R.; Raush, G.; Freire, J.; Rio, C. Computational fluid dynamics and particle image velocimetry assisted design tools for a new generation of trochoidal gear pumps. *Adv. Mech. Eng.* **2015**, *7*, 1–14. [CrossRef]
55. Gamez-Montero, P.J.; Castilla, R.; Codina, E. Methodology based on best practice rules to design a new-born trochoidal gear pump. *Proc. Inst. Mech. Eng. Part C J. Mech. Eng. Sci.* **2018**, *232*, 1057–1068. [CrossRef]
56. Skowrońska, J.; Kosucki, A.; Stawiński, Ł. Overview of Materials Used for the Basic Elements of Hydraulic Actuators and Sealing Systems and Their Surfaces Modification Methods. *Materials* **2021**, *14*, 1422. [CrossRef]
57. Student, M.; Hvozdetskyi, V.; Stupnytskyi, T.; Student, O.; Maruschak, P.; Prentkovskis, O.; Skačkauskas, P. Mechanical Properties of Arc Coatings Sprayed with Cored Wires with Different Charge Compositions. *Coatings* **2022**, *12*, 925. [CrossRef]

energies

Article

Prototyping and Experimental Investigation of Digital Hydraulically Driven Knee Exoskeleton

Rituraj Rituraj [1,*], Rudolf Scheidl [1,*], Peter Ladner [2], Martin Lauber [2] and Andreas Plöckinger [2]

1 Institute of Machine Design and Hydraulic Drives, Johannes Kepler University, 4040 Linz, Austria
2 Linz Center of Mechatronics, 4040 Linz, Austria
* Correspondence: rituraj.rituraj@jku.at (R.R.); rudolf.scheidl@jku.at (R.S.)

Abstract: Digital hydraulic drives are known for their superior efficiency, power density, and robustness. Such advantages make them an attractive alternative (to electric drives) for actuation of exoskeleton devices. This work presents development of a prototype for such a digital hydraulically driven knee exoskeleton and its experimental testing. The device uses two miniature hydraulic cylinders and a novel mechanism to translate the linear motion to rotary motion. The device is controlled via a passive control method in the stance phase and a simplified model predictive control method in the swing phase. In this work, the design of the exoskeleton device is optimized with respect to compactness and weight. Next, the features of the design are further refined to ensure that the device is able to support the operational loads. This design is then realized into a prototype with a mixture of inhouse manufactured parts and procured components. Finally, via experimental tests, the performance of the design and the control strategy are investigated. Certain drawbacks related to valve size and overall weight are observed in the prototype, which will be addressed in future studies.

Keywords: exoskeletons; hydraulic drives; digital hydraulics; miniature applications; control methods

Citation: Rituraj, R.; Scheidl, R.; Ladner, P.; Lauber, M.; Plöckinger, A. Prototyping and Experimental Investigation of Digital Hydraulically Driven Knee Exoskeleton. *Energies* **2022**, *15*, 8695. https://doi.org/10.3390/en15228695

Academic Editors: Paolo Casoli and Massimo Rundo

Received: 31 October 2022
Accepted: 17 November 2022
Published: 19 November 2022

1. Introduction

Exoskeletons are wearable devices that augment the physical strength of their wearers. Such devices are used in a wide variety of applications. The medical application constitutes rehabilitation of the motion and strength of the limbs of patients suffering from limb injuries and post-stroke gait dysfunction [1,2]. Furthermore, exoskeletons are used by industry workers to help them in lifting heavy objects and performing strenuous activities without risking fatigue and injuries [3]. Soldiers, firefighters, and rescue workers can also benefit from exoskeletons for these reasons [4,5].

Exoskeleton devices of different types and capabilities have been developed by both researchers and industries in the past. A systematic review of past exoskeletons can be found in [6]. This reference as well as an internal literature survey by the authors' research group [7] reveal that most of the powered exoskeleton devices being researched and available in the market are electro-mechanically driven.

However, hydraulic drives offer key advantages compared to electro-mechanical drives that can make them a better alternative drive technology for powered exoskeletons. Hydraulic drives are power-dense, which can allow a significant reduction in the mass and space occupied by the exoskeleton device. Moreover, hydraulic drives can be easily configured for energy recuperation functionality, which improves the efficiency of the drive, thus allowing a reduction in the size of the power source. Furthermore, hydraulic drives facilitate motion locking without power supply. Finally, damped motion is easily achievable with hydraulic drives, which allows for smooth and natural motion patterns.

Nevertheless, a number of challenges have prevented hydraulically driven exoskeletons from dominating the exoskeleton market. First, novel hydro-mechanical designs need to be developed that allow integration of the hydraulic drive in the exoskeleton device in a compact fashion. Second, miniature hydraulic components are needed to assemble a compact lightweight hydraulic drive. Third, fast and efficient control strategies appropriate for the hydraulic drives are needed for actuation of the exoskeleton. Last but not least, the image of hydraulics as a big, heavy, and obsolete technology needs revamping, which, until now, has prevented most researchers from pursuing hydraulic drives as an option for exoskeleton actuation.

One of the earliest hydraulically driven exoskeletons was the Berkeley Lower Extremity Exoskeleton (BLEEX), developed at the University of California, Berkeley [8–10]. This exoskeleton augmented the power at the hip, knee, and ankle joints. Further improvements led to development of exoskeleton devices ExoHiker and ExoClimber, which were claimed to be more comfortable to the wearers [11]. Around the same time, the company Sarcos unveiled its full-body exoskeleton XOS for military applications [12]. While BLEEX used linear hydraulic actuators and mechanisms for motion transformation, XOS used rotary actuators directly at the limb joints [13]. Recent works on hydraulically driven exoskeletons have focused on the control methods [14–17], development of compact hydraulic power units [18], and alternate working fluids [19]. However, a key limitation of these exoskeleton devices is that they use traditional servovalve-based resistance control, which is known for its poor efficiency. To overcome this limitation of traditional hydraulic systems, Kaminaga et al. proposed electro-hydraulic systems for actuation of joints in humanoid robots [20] followed by knee exoskeletons [21]. The design and control was focused particularly on achieving enhanced back-drivability [22]. The knee joint power assist exoskeleton developed by them used rotary actuators (vane motors) at the knee joint [23,24] and consisted of additional passive joints to allow secondary knee movements [25]. However, usage of hydraulic motors presents the challenges of internal leakages (leading to efficiency and controllability concerns) and low compactness. Therefore, recently, Lee et al. [26] and Jiang et al. [27] have explored electro-hydraulic systems with hydraulic cylinders instead of hydraulic motors for actuation of exoskeleton devices. In their recent knee exoskeleton design, Lee et al. [28] added a multi-axial structure at the knee joint to minimize misalignments. Last, Sun et al. [29] proposed a lightweight electrohydrostatic actuator (LEHA) for knee exoskeletons where they eliminated some accessory components of traditional EHA and custom-designed components to reduce the weight and volume of their exoskeleton device.

In recent decades, digital hydraulics has emerged as an innovative technology that promises several advantages compared to the traditional hydraulics technologies in terms of high force and power density, reliability, robustness, and inexpensiveness [30]. These advantages make digital hydraulics an ideal candidate for actuation of exoskeleton devices, where compactness and robustness of the drive, lightness of the power source, and precision of motion are key requirements. The leg exoskeleton developed by Cao et al. [31], which used switching control, can be considered as the earliest implementation of digital hydraulic concepts in the exoskeleton devices.

In recent years, the authors' research group has made efforts to leverage the aforementioned advantages of digital hydraulics in development of lower limb exoskeleton devices. In 2017, the first design of a digital hydraulically driven knee exoskeleton was proposed [7,32]. This exoskeleton consisted of a digital hydraulic cylinder driving the knee joint via a four-bar linkage mechanism. A key limitation of this design was the need of multi-chamber cylinders to realize the digital cylinder, which are known to be expensive. In 2021, the authors proposed a novel design of knee exoskeleton that consists of a unique mechanism to enable digital actuation of the exoskeleton with two simple hydraulic cylinders instead of a multi-chamber cylinder [33].

In the work presented in this article, the authors move towards realization of the aforementioned exoskeleton design. The design is first optimized with respect to its size and weight. Next, refined designs of each of the parts are developed, with a focus on their load carrying capabilities. Afterwards, a prototype of the exoskeleton device based on this design is fabricated. Finally, experimental studies are conducted to investigate the performance of the device and the control strategy proposed by the authors elsewhere [34].

Overall, this article showcases successful realization of the concept of digital hydraulically driven exoskeletons. It presents to the research community and industry a power-dense hydraulic alternative to the electromechanically driven exoskeletons. Furthermore, this work brings digital hydraulics technology (and with it, its several advantages) into the field of hydraulically driven exoskeletons.

This article is divided into eight sections including this introductory section. The design of the knee exoskeleton device is presented in Section 2. Section 3 presents the details of the design optimization study. The information related to prototype development is presented in Section 4. The control strategy proposed by the authors is briefly presented in Section 5. The experimental tests and the results are described in Section 6. Section 7 presents a discussion of the work presented in this article in relation to past works. Finally, the key conclusions of this work are provided in Section 8.

2. Knee Exoskeleton Design

The knee exoskeleton design conceived by the authors is shown in Figure 1. The design consists of the thigh and shank parts, which are attachable to the corresponding limb parts of a human. The parts are connected via a revolute joint at the knee. Motion actuation is performed by two mechanisms, each consisting of a hydraulic cylinder and a connecting rod. The joint between the cylinder rod and connecting rod for each mechanism contains a guide pin that is constrained to move along a curved path. This path is present in the form of grooves on the guide plates. There are two guide plates on either side of the mechanisms to support the guide pins (the front guide plate is hidden in Figure 1). The geometric feature of the guide curve (along with the length of the linkages in the mechanisms) dictates the transmission ratios of the mechanisms (Section 2.2).

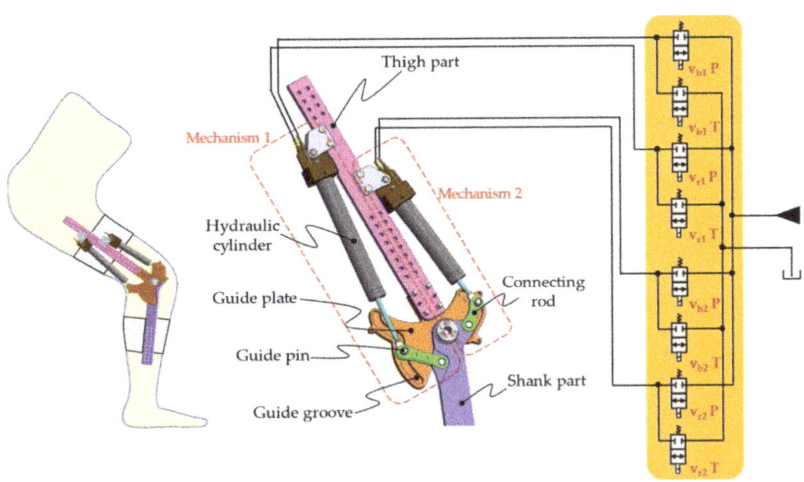

Figure 1. Knee exoskeleton attached to the human leg (**left**); details of the exoskeleton design and the digital hydraulic drive (**right**).

2.1. Hydraulic Drive

As shown in Figure 1, the digital hydraulic system driving the exoskeleton device consists of a pressure source, tank, digital valves (highlighted in yellow), and two hydraulic cylinders. The digital 2/2-way valves connect each cylinder chamber to the pressure source and tank.

There are eight 2/2-way valves in the hydraulic drive, and each valve can have two possible positions (on and off). Thus, the drive has 16 unique valve configurations, as illustrated in Figure 2.

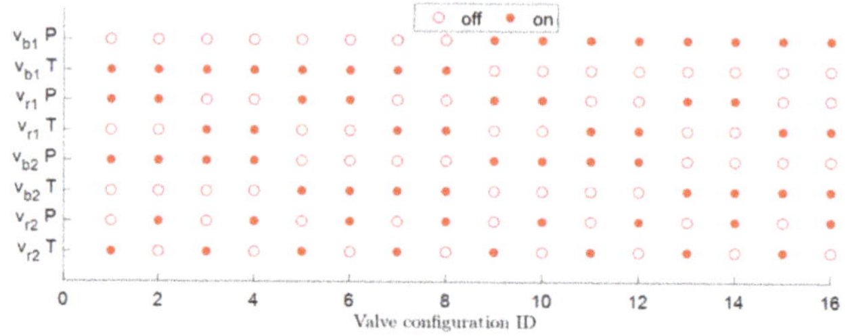

Figure 2. Valve positions for 16 unique valve configurations.

2.2. Transmission Ratio

The transmission ratio of a mechanism (T_R) is defined as the ratio of the linear motion of the pistons in the cylinders and the angular motion of the knee joint:

$$T_{R,1,2} = \frac{dz_{1,2}}{d\psi}, \tag{1}$$

where z is the piston position in the cylinder and ψ is the knee angle, as indicated in Figure 3.

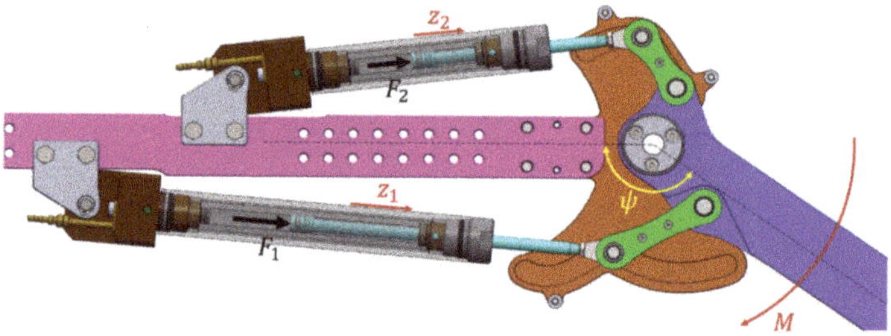

Figure 3. Illustration of the piston positions (z_1, z_2), knee angle (ψ), cylinder forces (F_1, F_2), and knee torque (M).

The transmission ratio establishes the following relationship between the forces at the hydraulic cylinders (F) and the torque delivered at the knee joint (M) [33]:

$$M = T_{R,1}F_1 + T_{R,2}F_2. \tag{2}$$

To determine this knee torque requirement, the limb motion dynamics data from HuMoD database [35] are utilized. The database developed by researchers at TU Darmstadt, Germany [36] contains the biomechanical measurements (motion of the limb joints during typical gait cycles) and anthropomorphic parameters (length and mass of the limb parts) of the test subjects. This information is utilized in a multi-body numerical model of the human leg developed by the authors (a brief description is present in Appendix A) to determine the knee torque requirement in a typical gait cycle.

Figure 4a shows the knee angle and knee torque variation over a gait cycle. In the stance phase (where the foot touches the ground and the leg supports the load of the body), the knee torque requirement is high (up to 100 Nm). In contrast, in the swing phase (where the foot is off the ground), the knee torque remains low.

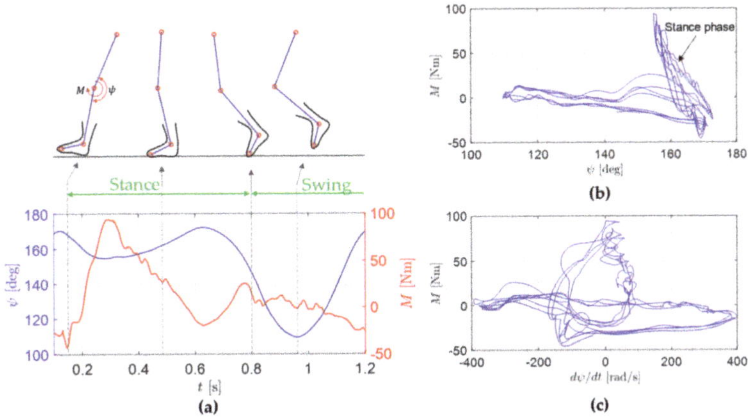

Figure 4. (**a**) Knee angle and torque over one gait cycle; (**b**) knee torque vs. knee angle over multiple gait cycles; (**c**) knee torque vs. knee angular speed over multiple gait cycles.

It is further notable from Figure 4b that the knee torque is high for knee angle $\psi > 155^\circ$ and remains low for lower knee angles. This allows the mechanisms to be designed such that their transmission ratios are low for $\psi < 155^\circ$ and high for $\psi > 155^\circ$ (Figure 5). Low transmission ratio for the majority of the knee angular span allows a smaller piston stroke of the hydraulic cylinders (as per Equation (1)), whereas high transmission ratio at peak torque requirement ensures a low hydraulic force requirement (as per Equation (2)), which enables using small hydraulic cylinders (in terms of bore size) working at medium pressure levels ($\leq$200 bar).

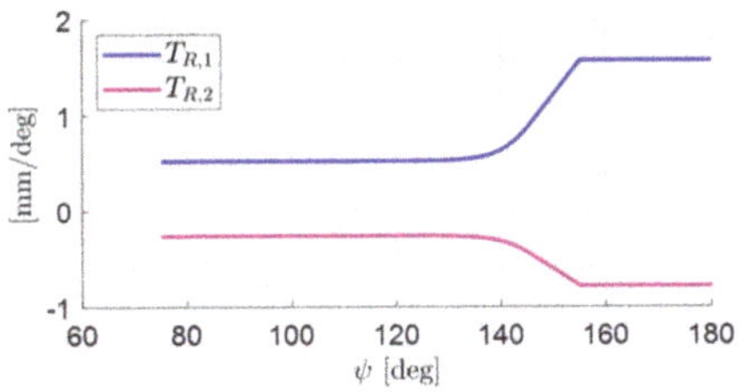

Figure 5. Transmission ratios of the two mechanisms.

The relationship between the transmission ratios of the two mechanisms is chosen as:

$$T_{R,2} = -0.5T_{R,1}, \tag{3}$$

with the negative sign indicating the opposite motion of the pistons for a given knee angular motion. Consequently, the knee torque expression from Equation (2) simplifies to:

$$M = T_{R,1}F_1 - 0.5T_{R,1}F_2. \tag{4}$$

Here, the cylinder force is:

$$F = p_bA_b - p_rA_r. \tag{5}$$

If the area ratio of the cylinders is chosen as:

$$A_b = 4A_r, \tag{6}$$

Equation (4) simplifies to:

$$M = T_{R,1}A_b(p_{b1} - 0.25p_{r1} - 0.5p_{b2} + 0.125p_{r2}). \quad (7)$$

From this expression, the cylinder bore diameter requirement can be determined. In particular, for the source pressure of 200 bar and the transmission ratio characteristic from Figure 5, the maximum torque (100 Nm) can be achieved with cylinders of bore diameter 8 mm.

Next, defining the effective force as,

$$F_{eff} = \frac{M}{T_{R,1}}, \quad (8)$$

the following expression is obtained:

$$F_{eff} = A_b(p_{b1} - 0.25p_{r1} - 0.5p_{b2} + 0.125p_{r2}). \quad (9)$$

For each of the 16 valve configurations shown in Figure 2, each of the pressure terms in Equation (9) is either equal to the source pressure or the tank pressure. Thus, each valve configuration corresponds to a unique effective force. Figure 6 shows uniform stepping of the effective force, which is a direct result of the conditions imposed on the transmission ratios (Equation (3)) and the area ratios (Equation (6)).

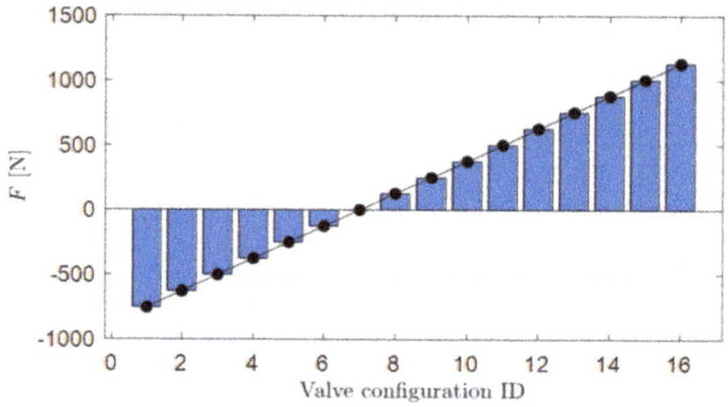

Figure 6. Effective force delivered by the hydraulic drive for different valve configurations.

3. Design Optimization Study

A major requirement of the exoskeleton device is that it should be compact and lightweight while providing its necessary function. To determine such a compact and lightweight design of the knee exoskeleton proposed by the authors, a design optimization study is conducted.

3.1. Objective Functions

For the user's comfort, it is important that the lateral width of the exoskeleton device (indicated as w in Figure 7) is as small as possible. In the exoskeleton design, this dimension is decided by the maximum distance between the guide curves of the two mechanisms. Thus, the first optimization objective is:

$$\text{minimize } w; w = \max|y_{B1}| + \max|y_{B2}|. \quad (10)$$

where B_1 and B_2 represent the guide curves of the two mechanisms.

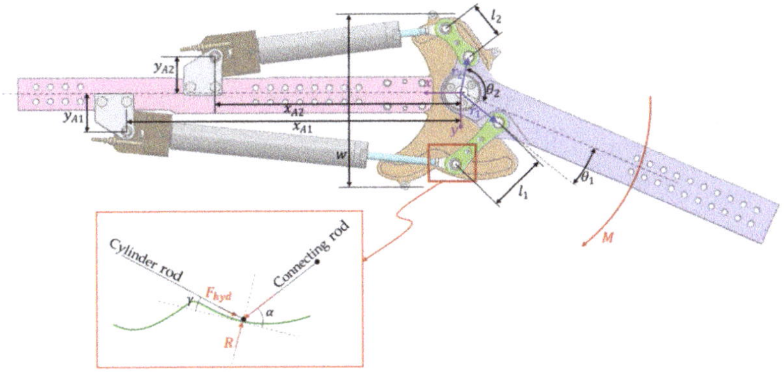

Figure 7. Illustration of critical geometrical dimensions used in the optimization problem. Inset shows the angles formed by the cylinder rod and connecting rod with the tangent of the guide curve at the guide pin.

The second optimization objective is related to loading at the guide curves. As shown in the inset of Figure 7, the guide curve needs to support the loads from the hydraulic cylinder and the connecting rod. Supporting higher loads will require thicker plates made of stronger material, which will lead to an increase in the weight of these plates. Thus, minimization of the load on the guide plate forms the second objective.

To determine the load on the guide plates and the contact force R, the gait cycle is divided into discrete time steps. For each time step in a gait cycle, from the knowledge of the knee torque requirement and the mechanism transmission ratio, the effective force requirement is determined (Equation (8)). The nearest discrete force level deliverable from the drive (shown in Figure 6) is chosen and the corresponding force from each cylinder (F_{hyd}) is determined. Figure 8 shows the forces for each cylinder, where the four levels of forces observed are the result of four pressurization cases: (a) $p_b = 0$, $p_r = p_S$; (b) $p_b = p_r = 0$; (c) $p_b = p_r = p_S$; (d) $p_b = p_S$, $p_r = 0$.

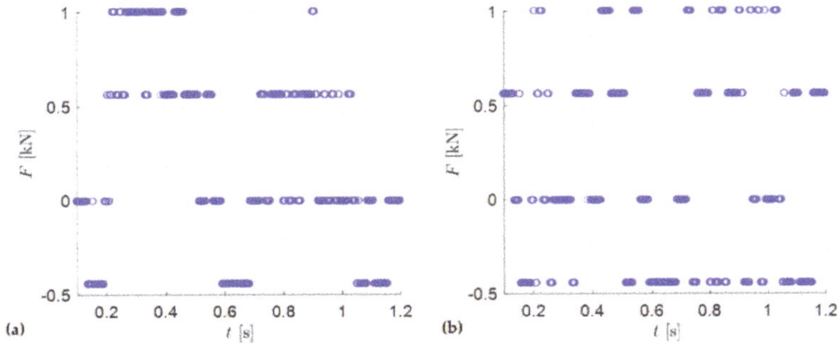

Figure 8. Force delivered by the hydraulic cylinders to support the required knee torque at different knee angles: (**a**) cylinder for mechanism 1; (**b**) cylinder for mechanism 2.

Next, the contact force at the guide plate is evaluated as:

$$R = F_{hyd}(\sin\gamma + \cos\gamma\tan\alpha), \tag{11}$$

where γ and α are the angles that the cylinder and connecting rod form with the tangent to the guide curve (as illustrated in the inset of Figure 7).

Thus, the second objective function is:

$$\text{minimize } R_{max}; R_{max} = \max(R_1, R_2). \tag{12}$$

3.2. Optimization Variables

The design variables for the optimization study are the dimensions r_1, r_2, l_1, l_2, x_{A1}, y_{A1}, x_{A2}, y_{A2}, θ_1, and θ_2, as indicated in Figure 7. Table 1 shows the bounds imposed on these variables in the optimization process. The lower bounds of the crank lengths (r_1 and r_2) are set to 20 mm to allow enough space at the knee joint for installation of an angle sensor (shown in Section 4.4). The lower bounds of the connecting rods (l_1 and l_2) are set to 20 mm to allow enough space for realization of the revolute joints at their ends. The upper bounds of these linkages are set to 60 mm since linkages longer than this will lead to designs with lateral width (w) much higher than the typical width of the knee.

Table 1. Lower (lb) and upper (ub) bounds of the design variables in the optimization problem.

	$r_1(mm)$	$r_2(mm)$	$l_1(mm)$	$l_2(mm)$	$x_{A1}(mm)$	$y_{A1}(mm)$	$x_{A2}(mm)$	$y_{A2}(mm)$	$\theta_1(°)$	$\theta_2(°)$
lb	20	20	20	20	180	30	130	30	0	30
ub	60	60	60	60	350	40	300	40	80	190

The distance of each cylinder base mounting from the knee joint (x_{A1} and x_{A2}) is influenced by the stroke of that cylinder. The stroke of each cylinder is determined from the transmission ratio of the respective mechanism and the angular span of the knee joint. For the knee angle range $\psi \in [75°, 180°]$, the stroke values of the two cylinders are determined to be 90 mm and 45 mm. The higher values in the bounds of x_{A1} and x_{A2} account for the extra distance due to the lengths of the cylinder base and cylinder rods in extension. The y-coordinates of the cylinder base mounting (y_{A1} and y_{A2}) are limited to [30, 40] mm range. The lower bound ensures that a realistic mounting of the cylinders on the shank part is achievable, whereas the higher bound ensures the compactness of the device.

3.3. Optimization Constraints

The following constraints are imposed on the optimization problem:

- The design must satisfy the transmission ratio requirements described in Section 2.2.
- The design must be kinematically feasible for the full range of knee motion.
- During the full range of knee motion, interference between different parts of the exoskeleton device must be avoided.
- It can be observed that the groove for mechanism 1 has a relatively sharp peak near the middle. If the curvature of the groove at this peak becomes too high, the knee motion may experience jerky behavior when the guide pin passes through this region. To avoid this, the curvature of the groove must be smaller than a certain value (170 m^{-1} is considered as an appropriate limit).

3.4. Optimization Procedure and Results

The multi-objective optimization problem described in the preceding subsections is solved using the NSGA II algorithm [37] in MATLAB environment. The initial design space is populated with 1000 designs and the optimization algorithm is executed for 100 generations.

Figure 9 shows the results obtained from the optimization study in terms of the approximate pareto-front between the two objectives (minimization of the device width and minimization of the contact force at the guide plates). The existence of the pareto-front indicates that there is a clear trade-off between these objectives and no single design performs the best with respect to both the objectives.

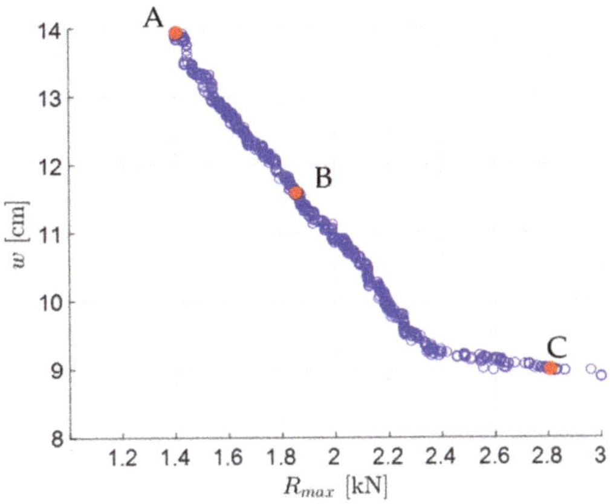

Figure 9. Approximate pareto-front obtained from the optimization study.

Three designs chosen from the pareto-front are shown in Figure 10. Design A has a smaller contact force (1.4 kN) but is much larger in size (14 cm width). On the other hand, design C has a smaller size (9 cm width) but exhibits high contact force (2.8 kN). Design B is a good compromise between the two extreme designs ($R_{max} = 1.9$ kN , $w = 11.6$ cm) and, hence, is chosen in this work. The illustrations of the exoskeleton device in Figures 1, 3, and Figure 7 show this chosen design. The values of the design variables for this design are reported in Table 2.

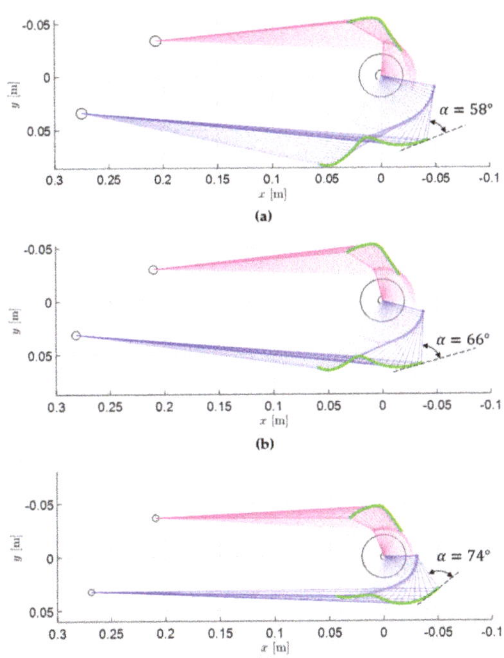

Figure 10. Three designs from the pareto-front: (**a**) design A; (**b**) design B; (**c**) design C. Mechanisms 1 and 2 are shown in blue and magenta, respectively. The guide groove profiles are shown in green.

Table 2. Values of the design variables for the chosen design (design B).

$r_1(mm)$	$r_2(mm)$	$l_1(mm)$	$l_2(mm)$	$x_{A1}(mm)$	$y_{A1}(mm)$	$x_{A2}(mm)$	$y_{A2}(mm)$	$\theta_1(°)$	$\theta_2(°)$
38.9	29.6	48.7	30.2	281.7	31.3	210.0	30.0	55.4	174.2

It is notable that the inverse relationship between the device size and contact force in Figure 9 is the result of the fact that, as the size of the device is reduced, the angle between the connecting rod and guide groove (α in Figure 7) needs to increase, resulting in the increase in the contact force at the guide groove. This observation is illustrated for the three designs in Figure 10, where the value of α for the highest knee angle configuration is reported.

4. Prototype Development

The exoskeleton design chosen from the optimization study in the previous section is realized in the form of a prototype. In the following subsections, key details related to the development/procurement of each component of the prototype and its integration in the prototype are presented.

4.1. Hydraulic Cylinders

The miniature hydraulic cylinders are obtained from HAWE Micro Fluid GmbH [38]. As shown in Figure 11, the cylinder design is based on the twin-tube principle, which allows location of both bore and rod side chamber ports at the cylinder base. The annular region between the outer and inner tubes allows fluid flow between the rod side chamber and the cylinder base.

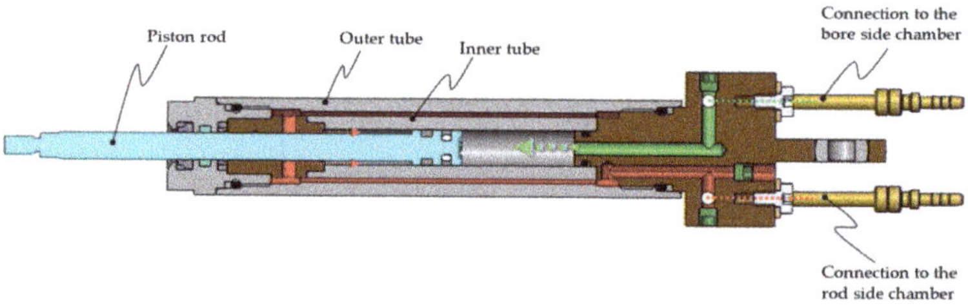

Figure 11. Sectioned view of the hydraulic cylinder from HAWE, illustrating the twin-tube design and the internal flow channels.

The strokes of the cylinders are 90 mm and 45 mm, which fulfils the requirements of the exoskeleton design. The bore and rod diameter of the cylinders are 8 mm and 6 mm, respectively, resulting in the area ratio of 2.29:1. As this cylinder is an off-the-shelf component from HAWE, the desired area ratio of 4:1 could not be fulfilled. This results in non-uniformity of the force steps deliverable by the hydraulic drive. This non-uniformity is visible in Figure 12, which shows the forces for different valve configurations. The interpretation of each valve configuration in terms of the state of pressurization/depressurization of each of the cylinder chambers is illustrated in Table 3. From Figure 12, the force step non-uniformity is moderate and, thus, is tolerable for the sake of the first prototype.

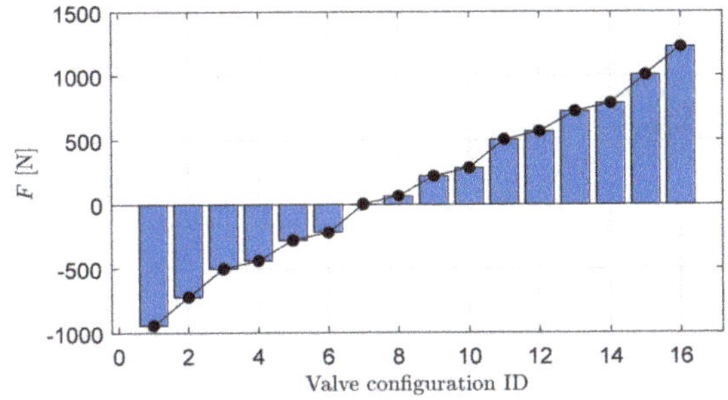

Figure 12. Effective force for different valve configurations using HAWE cylinders.

Table 3. Cylinder chamber state for different valve configurations (u). $V_{b1/2}$ and $V_{r1/2}$ are the bore and rod side chambers of cylinder 1/2. Chamber state of 1 and 0 indicate pressurized and depressurized, respectively.

u	**1**	**2**	**3**	**4**	**5**	**6**	**7**	**8**	**9**	**10**	**11**	**12**	**13**	**14**	**15**	**16**
V_{b1}	0	0	0	0	0	0	0	1	0	1	1	1	1	1	1	1
V_{r1}	1	1	0	1	0	1	0	1	0	1	0	1	0	1	0	0
V_{b2}	1	1	1	0	1	0	0	1	0	1	1	0	1	0	0	0
V_{r2}	0	1	0	0	1	1	0	0	1	1	0	0	1	1	0	1

4.2. Linkage Parts and Guide Plates

In the mechanical design of the exoskeleton device, the thigh part, the shank part, and the hydraulic cylinders are aligned on the same plane (referred to as the central plane in Figure 13). This prevents the rise of any lateral moment during exoskeleton operation.

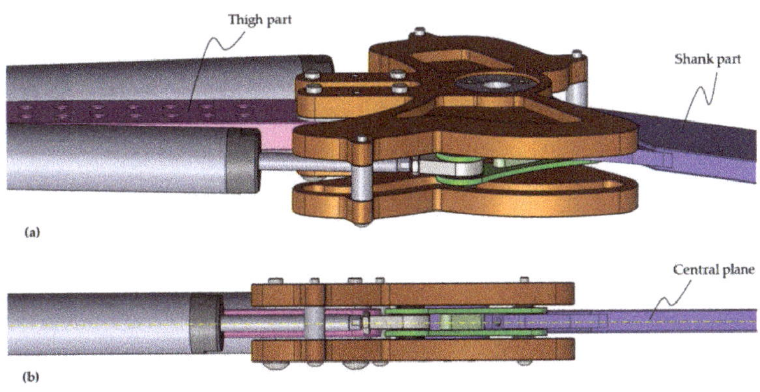

Figure 13. Illustration of the central plane and coplanarity of different linkages: (**a**) tilted view of the design; (**b**) bottom view of the design indicating the coplanarity of the thigh (magenta), cylinder (grey), and shank (blue) parts.

To link these coplanar cylinder rod and shank parts, the connecting rod part is realized via two individual parts that are interconnected via spacer elements (Figure 14). Similarly, there are two guide plates that are interconnected via spacer elements (Figure 13). On the upper side (mechanism 2 side) of the device, there are two such connections. In contrast, the lower side (mechanism 1 side) has only one interconnection due to the space constraint arising from the presence of the shank part on this side.

Figure 14. Realization of the connecting rod linkage in the prototype.

To determine the appropriate dimensions of the key loaded parts (i.e., connecting rod elements, guide plates, and shank) and their material type, stress analyses are conducted using Finite Element Analysis (FEA) tool in Solidworks Simulation environment [39]. Brief details of this study are presented in the following subsections.

4.2.1. Connecting Rod Linkage

From the knowledge of the force at the hydraulic cylinder (Figure 8) and the orientation of the cylinder and the connecting rod (inset of Figure 7), the force experienced by the connecting rod linkage is determined as:

$$F_{cr} = F_{hyd}\frac{\cos\gamma}{\cos\alpha} \tag{13}$$

Figure 15 shows the forces acting on the connecting rod linkages. Higher force level at higher knee angles is a direct consequence of transmission ratio amplification as well as the high knee torque demand at these knee angles.

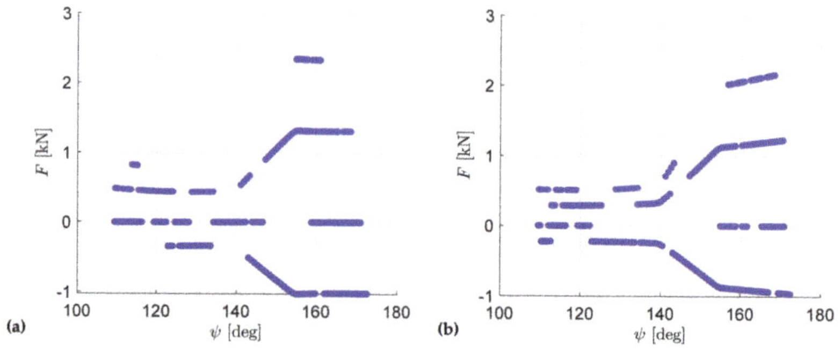

Figure 15. Forces (positive being compressive, negative being tensile) acting on the connecting rod linkages: (**a**) mechanism 1; (**b**) mechanism 2.

From Figure 15 and the fact that the connecting rod linkage is realized via two parts, the maximum load to be supported by each part is 1.2 kN force in compression and 0.5 kN force in tension. Figure 16 shows the stress experienced by these parts with 2 mm thickness. As the stress remains below 200 MPa, using steel E360 material (σ_{YS} = 360 MPa) provides a decent factor of safety (1.8) for prototype purposes.

4.2.2. Guide Plates

The guide plates need to support the load exerted by the guide pins of the two mechanisms. This contact force evaluation was explained in Section 3.1 (Equation (11)). Figure 17 shows the variation in these forces with the knee angle. The angle of 161° is the most severe position, where each of the two plates must support 1 kN load at the groove for mechanism 1 and 0.9 kN load at the groove for mechanism 2. These forces are acting outwards on the grooves from the knee joint perspective. The inwards forces are low (0.5 kN on mechanism 1 and 0.4 kN on mechanism 2) and compressive, and, thus, the material between the grooves and the knee joint can easily support them. Hence, the stress analysis is focused mainly on the outward loads.

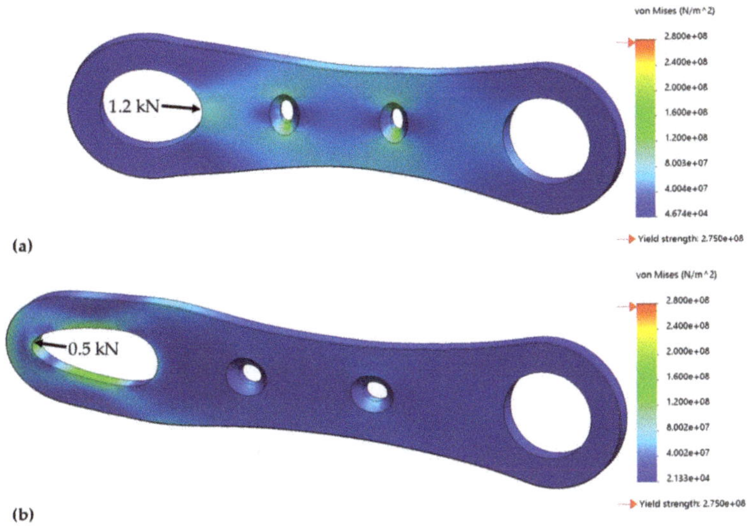

Figure 16. Stress analysis of the connecting rod elements in (**a**) compression and (**b**) tension.

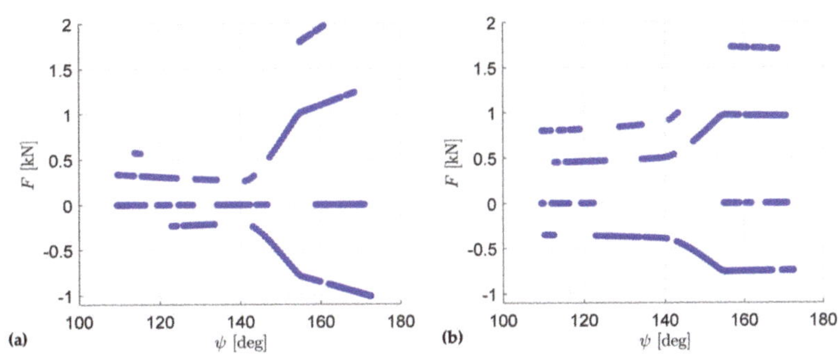

Figure 17. Forces (positive being outwards, negative being inwards) acting on the grooves of the guide plates: (**a**) grooves for mechanism 1; (**b**) grooves for mechanism 2.

Figure 18 shows that the maximum stress experienced by the guide plate is ~150 MPa. Thus, using steel E360 material ($\sigma_{YS} = 360$ MPa) provides a decent factor of safety (2.4). Furthermore, the maximum deformation of the guide groove remains below 40 µm (Figure 19), which is tolerable by the bearing at the guide pin.

4.2.3. Shank Part

The portion of the shank part near the knee joint is critical from the strength perspective since there are three linkage connections (connections with the thigh part and two connecting rods) next to each other. Via an iterative procedure (involving design modification and stress analysis), the design of the shank part is developed such that it can handle the stresses involved.

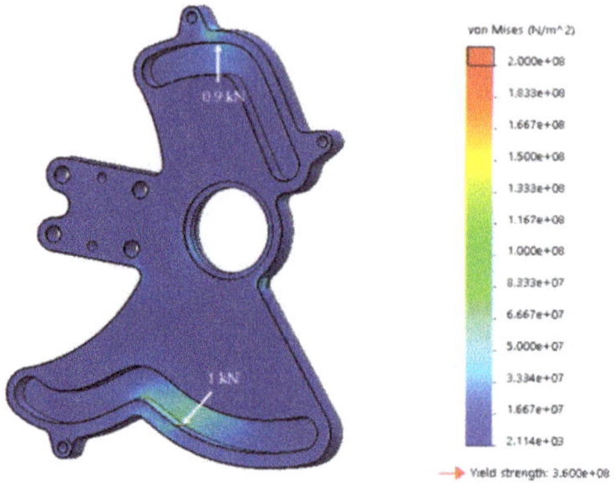

Figure 18. Stress analysis of the guide plate with the contact force exerted at the location of the guide pins for the knee angle of 161°.

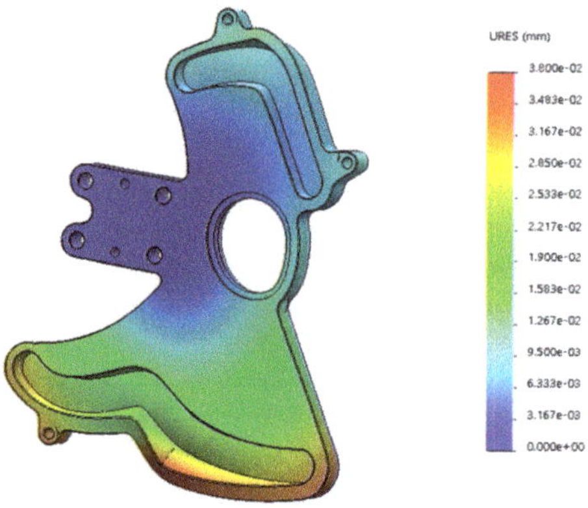

Figure 19. Deformation of the guide plate under the forces shown in Figure 18.

The shank part is most stressed at the highest knee torque demand scenario when connecting rod 1 pushes the shank with a force of 2.3 kN and connecting rod 2 pulls with the force of 0.9 kN (as per Figure 15). Another scenario where the shank is significantly stressed is when the torque demand is moderate, but, to match the demand, both cylinders apply a pushing force. From Equation (4), it can be understood how the difference in the two high forces scaled by the transmission ratios can match a desired low torque.

Figure 20 shows the chosen design of the shank where the stresses remain below 75 MPa for both extreme scenarios, and, thus, using aluminum AW 5083 material (σ_{YS} =193 MPa) allows a good factor of safety (2.6).

4.2.4. Bearings

At all the joints, plain bearings from Igus [40] are used. These bearings are made of special material Iglidur and are known for being extremely wear-resistant, robust, and self-lubricating [41]. With the compressive strength of 100 MPa, the bearings of appropriate dimensions manufactured from this material are able to sustain the loads experienced in the normal operation of the device.

Finally, at the guide pin, to allow the pin to move in the guide curve with little friction, needle bearings are used. Figure 21 shows the sectioned view of the design to reveal the bearings used.

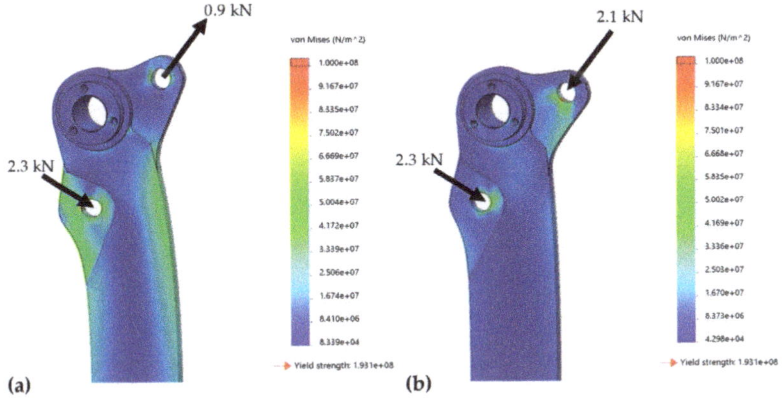

Figure 20. Stress analysis of the shank part: (**a**) high torque transfer case; (**b**) high internal force case.

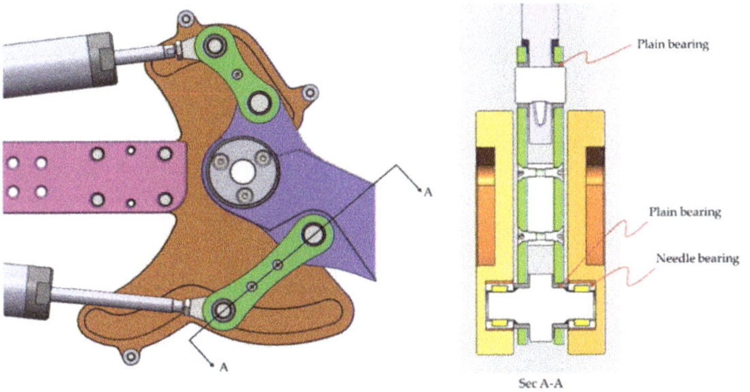

Figure 21. Illustration of the bearings (plain bearings from Igus and needle bearings) used in the design.

4.3. Valves

The miniature 2/2-way valves were obtained from researchers at Tampere University, Finland [42]. The eight valves are assembled in a compact fashion in a uniquely designed valve manifold, as shown in Figure 22. The flow capacity of each valve is 1.2 L/min at the pressure drop of 35 bar. The flows in the hydraulic drive during actuation of the gait cycle remain below 0.5 L/min except for one hydraulic line (connecting to the bore chamber of cylinder 1) in the swing phase, as shown in Figure 23, which goes up to 1.75 L/min. Thus, it is arguable that the pressure drop in this valve is too high for efficient operation and a larger valve would be preferable. However, this valve offers a key advantage of compactness, and such compact valves cannot be found elsewhere in the market.

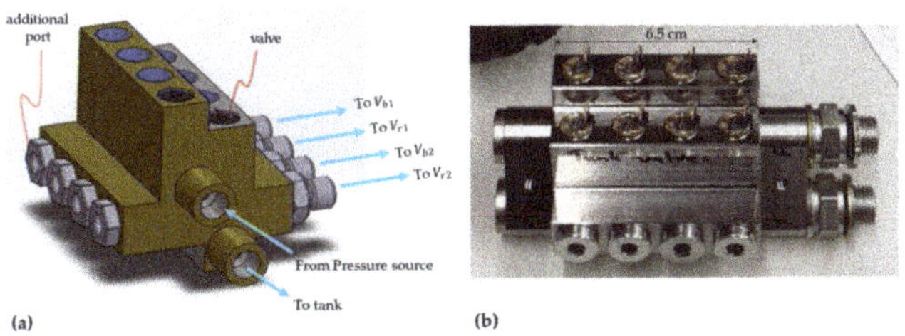

Figure 22. Valve manifold housing eight 2/2-way valves: (**a**) CAD; (**b**) procured valve block.

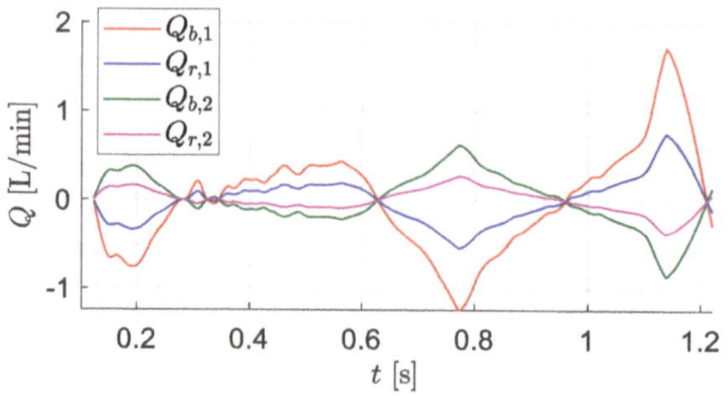

Figure 23. Flow in and out of the cylinder chambers over a gait cycle.

4.4. Sensors

For the control strategy described later in Section 5, the only information needed from the exoskeleton device is the knee angle. Thus, an angle sensor is integrated into the exoskeleton device. The hall effect sensor (model ETx25K) from Megatron is chosen due to its compact size [43]. The sensor's accuracy of $\pm 1^\circ$ is adequate for the controller's operation.

Figure 24 shows the integration of this sensor in the device. A magnet holder is fabricated out of a plastic material that has a circular groove for the sensor magnet. The magnet is inserted into this groove and then the holder with the magnet is press-fitted into the joint cover (which rotates with the shank). Next, a sensor holder is fabricated and mounted to the guide plate. Finally, the sensor is mounted on the sensor holder. This arrangement ensures a precise distance between the magnet and the sensor, which is important for sensor's proper functionality.

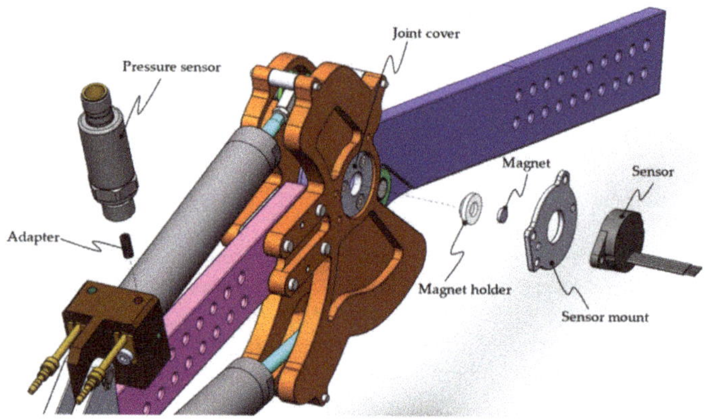

Figure 24. Sensor integration into the exoskeleton design.

During the testing phase, the pressure information from the cylinder chambers can be important. Thus, pressure sensors (AP019 from autosen [44]) are mounted on the cylinder base, as shown in Figure 24.

5. Exoskeleton Control Strategy

Control of an exoskeleton device depends on the functional requirements of the device. In this work, the authors consider the use case of patients suffering from paraplegia where they are unable to control their leg. Thus, the exoskeleton controller makes all the decisions in a gait cycle without any input from the user. The controller is tasked with following the characteristics of a typical gait cycle (i.e., track the knee angle observed in a typical gait cycle) and providing the necessary torque at the knee.

Two different strategies have been proposed by the authors for different phases in the gait cycle, a brief description of which is presented in the following subsections (a detailed description can be found in [34]).

5.1. Passive Control in the Stance Phase

From the typical knee motion characteristics in a gait cycle (Figure 4), it can be observed that the stance phase involves an inverse linear relationship between knee angle and knee torque. That is, as the knee angle decreases, the knee torque increases and vice versa. Thus, the knee motion in the stance phase can be realized via an elastic element in the knee joint. For the hydraulically driven knee exoskeleton proposed by the authors, the compressibility of the hydraulic oil can be exploited to achieve this elastic behavior.

As shown in Figure 25, during knee flexion, cylinder chambers V_{b1} and V_{r2} reduce in size. Thus, in the stance phase, both valves connecting to each of these chambers are turned off. As a result, when the knee flexes at the beginning of the stance phase, the hydraulic oil in these chambers becomes compressed. Consequently, the pressure in the chambers rises, resulting in an increase in the torque delivered by the drive. The opposite occurs in the knee extension at the latter part of the stance phase.

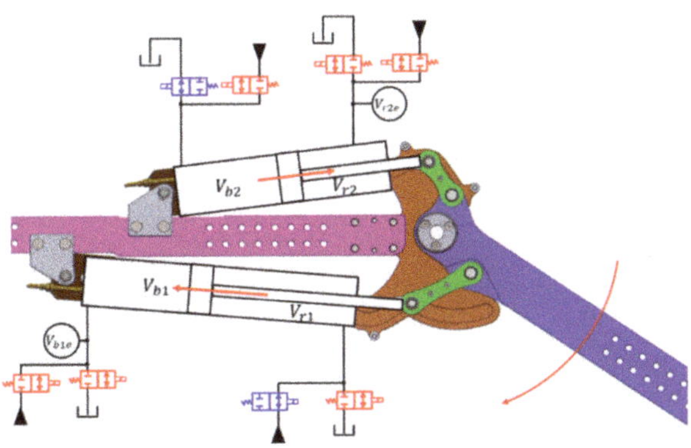

Figure 25. The valve configurations to realize the elastic behavior in the stance phase.

To ensure that the pressure in the cylinder chambers remains below their rated pressure (200 bar) in this elastic operation, extra volumes (V_{b1e} and V_{r2e}) are added to the isolated cylinder chambers.

The size of these volumes can be determined via simple analysis, starting from the pressure build up equation for a closed volume (consisting of the cylinder chamber of area A and length z, and the extra volume V_e):

$$\frac{dp}{dt} = -\frac{K}{V_e + Az} A \frac{dz}{dt} . \tag{14}$$

Here, K is the bulk modulus of the fluid. Integrating the equation from the beginning of the elastic phase (instant a) to the middle of the elastic phase (instant b) and simplifying, the following expression for V_e is obtained:

$$V_e = \frac{A\left(z_a e^{-\frac{p_b - p_a}{K}} - z_b\right)}{1 - e^{-\frac{p_b - p_a}{K}}} . \tag{15}$$

From the design of the mechanism, the piston positions (z_a, z_b) in the cylinder at different knee angles are known. Further, the mechanism is designed so that the cylinder pressure of 200 bar can supply the necessary peak torque in the stance phase. Thus, from instant a to b, the pressure should rise from 0 bar to 200 bar, i.e., in Equation (15),

$p_b - p_a = 200$ bar. For a hydraulic fluid with bulk modulus of 14,000 bar, the sizes of these volumes are determined to be V_{b1e} =85 mL and $V_{r2e} = 19$ mL. However, the presence of air in the fluid may lower the size requirements of these volumes. In the prototype, the valve block has additional ports (Figure 22), where these volumes of variable sizes (realized via appropriate fittings) can be mounted.

5.2. Model Predictive Control in the Swing Phase

In the swing phase, the exoskeleton device needs to track a desired knee motion typical of a gait cycle. To achieve this, a simplified form of model predictive control is employed.

The control strategy is presented concisely in Figure 26. At time step t_n, the controller uses the information of the current actual knee angle ($\psi_{a,n}$) and the desired knee angle at the next time step ($\psi_{d,n+1}$) to come up with a path (described by a 3rd order polynomial curve) to connect the current trajectory and the desired trajectory. This path is shown in blue in the figure. For this path, the knee angular position, velocity, and acceleration at the mid time ($t_m = 0.5(t_n + t_{n+1})$) are determined. This information is used in the mechanical model of limb motion (Appendix A) to determine the required knee torque. Next, from the transmission ratio information, the effective force requirement from the hydraulic drive is determined. Finally, a valve configuration (out of 16) is chosen that corresponds to the nearest digital force deliverable from the drive. This valve configuration is commanded by the controller at the current time step.

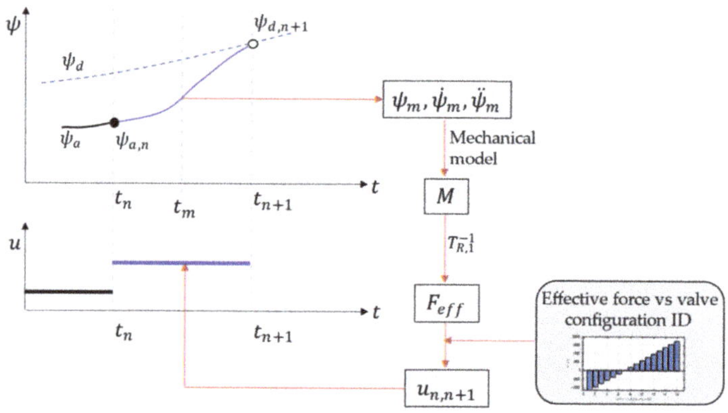

Figure 26. Illustration of the control strategy used in the swing phase of the gait cycle.

6. Experimental Tests

Testing of novel exoskeleton devices is typically carried out in two phases. The first phase involves testing the operation of the exoskeleton without human in the loop. This phase is important for safety purposes. Once the motion of the exoskeleton device is found to be satisfactory and safe, then the second phase of testing is conducted with human wearing the exoskeleton device.

In this work, first-phase tests are conducted. The details of the experimental setup and test results are presented in the following subsections.

6.1. Experimental Setup

The exoskeleton prototype is mounted on a test rig, as shown in Figure 27. Two weights (1 kg each) are added on each side of the shank part to partially simulate the weight of the shank and foot of the wearer. The angle sensor attached to the knee joint measures the knee angle and transmits it to the data acquisition and control system. As described in Section 5.2, the measured knee angle is used in the motion control strategy. The pressure sensors are also present in the test setup for monitoring purposes. The details of the sensors used in the test setup are reported in Table 4.

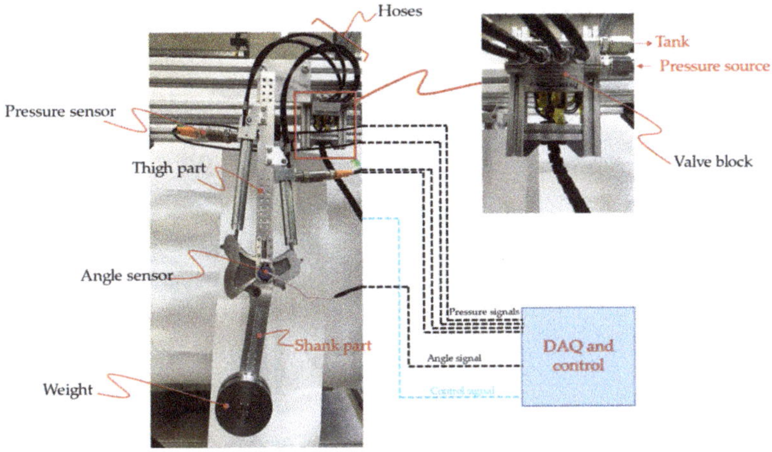

Figure 27. Experimental setup showing the exoskeleton device mounted on a test rig and connections to the data acquisition (DAQ) and control system. The inset indicates the valve block and connections to the pressure source and tank.

Table 4. Specification of the sensors used in the tests.

Sensor	Model	Accuracy	Measuring Range	Resolution
Angle sensor	Megatron ETx25K	±1%	0°–360°	0.07°
Pressure sensor	Autosen AP019	±0.2%	0–250 bar	-

The hydraulic power supply used in this work is a traditional supply system available to the authors that is capable of supplying flow at a preset pressure level. For an exoskeleton device that would eventually be worn by a user, the hydraulic power supply will comprise a miniature pressure-controlled electrohydraulic pump that could fit in the backpack of the user. Focus on this aspect of hydraulic power supply will be a task for the future.

Since the weights attached to the shank part are 2/5th of the typical weight of the shank and foot [35], the supply pressure is also set to 2/5th of the design pressure, i.e., 80 bar.

For data acquisition and control, a B&R X20 PLC system [45] is chosen. The system is flexible and can easily be adapted for any given I/O configuration, e.g., analog signals from the pressure sensors, bus connections, and driving the digital valves via "boost and hold" command [42].

The controller model described in Section 5.2 is implemented in MATLAB/Simulink environment. The Simulink model is then imported into the programming tool "Automation Studio" from B&R using "Automation Studio Target for Simulink" software (which allows automatic code generation from the Simulink model). Finally, "Automation Studio" establishes a TCP/IP connection with the B&R target system.

During tests, the measurements are recorded with a sampling time of 0.5 milliseconds, and postprocessing is completed in the MATLAB environment.

6.2. Test Trajectory

The goal of the tests in this work is to validate the motion control strategy proposed by the authors for the swing phase (Section 5.2) and the capability of the exoskeleton device to follow the desired motion. As shown in Figure 28a, the knee angle trajectory in the swing phase can be approximated via a cosine curve. The equation describing this curve is:

$$\psi[^\circ] = 141^\circ + 31^\circ \cos(11(t - 0.1)). \tag{16}$$

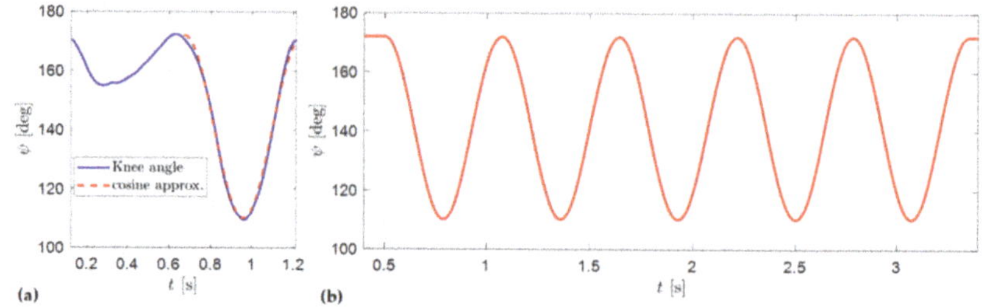

Figure 28. (**a**) Knee angle and a cosine curve approximation in the swing phase; (**b**) the repeating cosine curve used as the test trajectory.

This cosine curve is repeated to generate a test trajectory (Figure 28b), which is then used in the experiments.

6.3. Test Results

The experiments are conducted on the test setup using the control strategy described in Section 5.2. The only difference is the mechanical model, which, for the case of the test setup, is:

$$J\ddot{\psi} = M + mgl\sin\psi\,, \tag{17}$$

where J is the moment of inertia of the weights around the knee joint, M is the torque delivered by the hydraulic drive, m is the mass of the weights, g is the acceleration due to gravity, and l is the distance of the center of mass of the weights from the knee joint. Here, the inertia of the shank part is neglected since it is much lighter (~8 times) compared to the weights.

The exoskeleton device is asked to follow the test trajectory detailed in Section 6.2. The controller time step is set to 50 milliseconds.

Figure 29 shows the results from four sets of experiments. The conditions for the experiments are identical except for the initial knee angle and the pressure level of the source, which varies slightly between the experiments. The results indicate that the exoskeleton device is able to track the desired trajectory reasonably well. However, at some instances, it fails to reach the full knee extension as desired.

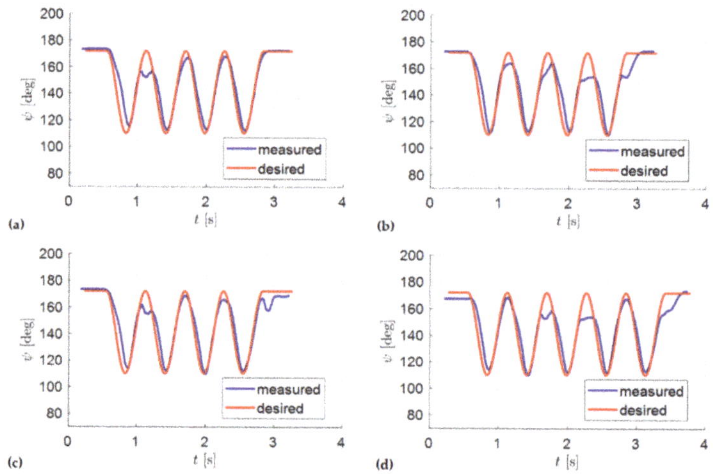

Figure 29. Results in the form of knee angle measured and desired (from the planned trajectory) for four sets of experiments: (**a**) Experiment A; (**b**) Experiment B; (**c**) Experiment C; (**d**) Experiment D.

For a quantitative analysis of the results, the error between the measured trajectory and desired trajectory is defined as:

$$e = |\psi_{meas} - \psi_d| \,. \tag{18}$$

The mean error (e_{mean}), its standard deviation (e_{std}), and maximum error (e_{max}) for the four sets of measurements are reported in Table 5. The error statistics are very similar across the experiments. Mean error of 5–6° is deemed to be tolerable by the user. However, the maximum error of 20–25° is of concern, which is caused by the inability of the device to reach full extension for some instances.

Table 5. Statistics for the error between the measured and desired trajectories.

Error Statistics	(a)	(b)	(c)	(d)
e_{mean}	4.6°	5.9°	5.1°	6.1°
e_{std}	5.4°	5.8°	4.9°	5.0°
e_{max}	24.5°	19.3°	21.7°	19.2°

A closer investigation into this discrepancy is illustrated via Figure 30. At $t = 1.596$ s, the controller commands the valve configuration of $u = 8$. As per Table 3, the state of the cylinder chambers should be: $\{V_{b1}, V_{r1}, V_{b2}, V_{r2}\} = \{1, 1, 1, 0\}$. However, as per Figure 30c, V_{b1} chamber starts to pressurize only at $t = 1.634$ s. Thus, for 38 ms, the state of the chambers is $\{0, 1, 1, 0\}$, which corresponds to $u = 1$. Consequently, as per Figure 12, instead of negligible force ($u = 8$), the drive delivers very high negative force, resulting in a sharp deceleration of the knee angle and a deviation from the desired trajectory.

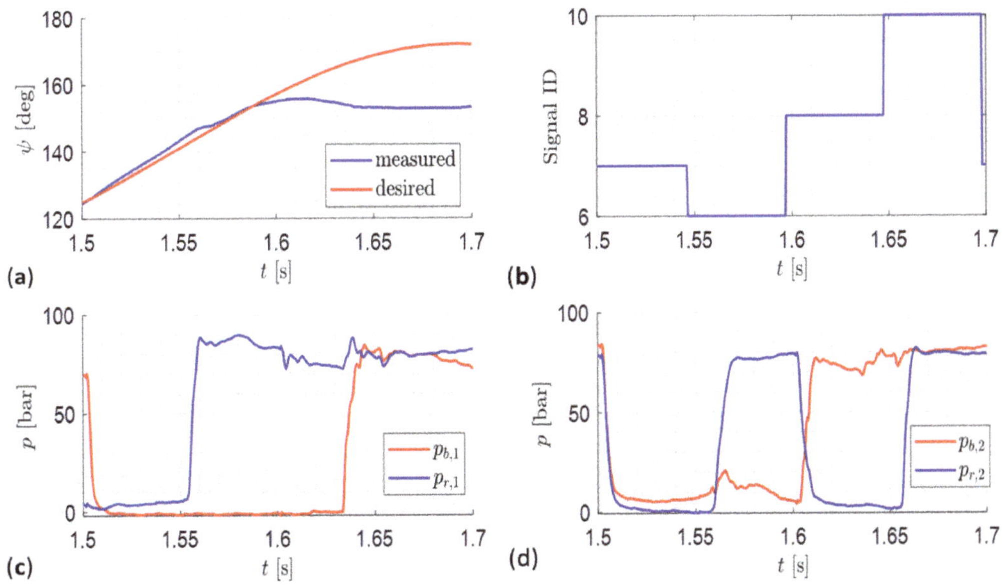

Figure 30. (**a**) Zoomed view of plot from Figure 29d; (**b**) valve configurations commanded by the controller; (**c**) pressure in cylinder 1 chambers; (**d**) pressure in cylinder 2 chambers.

The delayed pressurization of V_{b1} chamber is explained via the following arguments. During knee extension, V_{b1} chamber expands quickly. As gravity is helping the extension, the force needed from the drive is negative. From Figure 12, the valve configuration will remain below 7, i.e., V_{b1} chamber remains connected to the tank. Thus, V_{b1} is a fast-expanding chamber connected to tank, which needs high flow from the tank so that the chamber pressure can remain close to the tank pressure. However, the smallness of the valve (described in Section 4.3) restricts the amount of flow from the tank to the chamber. Consequently, the chamber pressure goes sub-atmospheric. Due the typical presence of entrained air in the fluid, its bulk modulus falls severely at such low pressures. Thus, when the chamber is connected to the pressure source, it takes more time to pressurize.

It is notable that this discrepancy in trajectory tracking only occurs in a specific situation near the end of knee extension when the controller commands a valve configuration that requires pressurized V_{b1}. At different instances, when the controller commands other valve configurations that do not require pressurization of this chamber, the discrepancy does not appear.

This issue can be addressed by using a larger valve that allows high enough flow to ensure that the pressure in V_{b1} chamber does not fall below the atmospheric pressure. However, the availability of larger valves in a compact arrangement remains a challenge, and, thus, other alternatives are needed to be investigated in the future.

7. Discussion

The digital hydraulically driven knee exoskeleton developed and tested in this work presents a significant advancement in terms of using hydraulics technology for actuation of exoskeleton devices. Compared to the traditional hydraulics technology used by past exoskeletons [8,12,14,15,18], the digital hydraulically driven exoskeleton device promises higher compactness, efficiency, and robustness. The compactness is illustrated in this work via the size of the valve block containing eight 2/2-way valves. Since the operations performed by the device in this work and the past devices in the literature are different, a direct efficiency comparison is not possible at this stage. However, the next phase of full gait cycle testing with user-in-loop will allow this direct efficiency comparison.

Another key advantage of the exoskeleton device presented in this work is that, since it is designed for paraplegic patients, it is capable of supplying 100% of the torque requirement at the knee joint. In contrast, most of the past works on exoskeleton development deal with partial disability or power augmentation. Thus, the devices developed are capable of supplying only up to 30% to 50% of the knee torque [15,23].

Since paraplegic patients are unable to control their leg, the controller of the exoskeleton device developed in this work makes all the decisions in the gait cycle. A simplified model predictive controller is proven (in this work and authors' past works [34]) to be sufficient for this task. In contrast, the devices developed for patients with partial disability and for power augmentation should account for the interaction between the force supplied by the human and the device. Thus, researchers working on such devices have used a variety of complex control strategies, ranging from cascade force control [14] to joint torque control [15] and repetitive learning control [16].

In the future, the authors would also like to extend the use case of their exoskeleton device to patients who can partially control their knee. A key challenge in this regard is identification of the motion intent of the wearer. Several approaches for motion intent identification have been proposed by other researchers [46]; however, one compatible with the drive system used in this work needs to be investigated.

The final discussion point is that of the prototype weight. The first-generation prototype developed in this work has been observed to be too heavy for users' comfort (2.8 kg). In contrast, Sun et al.'s LEHA weighs 2.5 kg (including the power supply) [29]. The reason behind this high weight (despite the optimized design) primarily lies in the twin-tube cylinder design. Such a design unnecessarily increases the amount of material used, thus increasing the weight (the cylinders weigh 0.62 kg and 0.54 kg, respectively). However, this is an obvious limitation of using off-the-shelf components. Sun et al. were able to achieve a low weight of their device by custom-designing the components. In the future, the prospects of inhouse manufacturing of cylinders will also be explored by the authors, which could address the weight issue and also enable an area ratio of 4:1, which is ideal for uniform force stepping of the digital drive. Furthermore, the simplicity of the guide plate design results in its high weight (0.4 kg each). In the next generation of the prototype, the guide plate design could be further optimized by developing a truss-frame-type structure, which will significantly lower its weight.

8. Conclusions

This article presents development and first-phase testing of a novel digital hydraulically driven knee exoskeleton prototype, the idea of which was conceived by the authors in their previous works. Via a multi-objective optimization study, a design is developed that is optimal with respect to its size and weight. The designs of the linkage parts are further refined while ensuring that each part can safely support the load exerted on it over a typical operational cycle. Subsequently, the prototype is manufactured and assembled with procured hydraulic cylinders, valves, and sensors. Finally, first-phase testing (without human) is conducted, where the prototype is mounted on a test stand and the controller is tasked with following a given trajectory that resembles the swing motion in a typical gait cycle. The results indicate that the prototype is able to track the desired motion, except with occasional discrepancies at full knee extension.

In the future, investigations to address the aforementioned discrepancy need to be conducted. These should be followed by testing the full gait cycle operation with a human in the loop.

Overall, the novel digital hydraulically driven knee exoskeleton device prototyped and tested in this work is a significant step towards bringing the advantages of digital hydraulics technology to the field of hydraulically driven exoskeleton devices. This can pave the way for widespread adoption of power-dense hydraulics technology in actuation of exoskeletons. The resulting compact lightweight exoskeleton devices will ultimately improve the comfort level of the wearers.

Furthermore, this work also showcases the gaps in the current hydraulics components' availability (e.g., miniature yet efficient valves and lightweight cylinders), which has historically hindered usage of hydraulics technology in exoskeleton actuation. This can be interpreted as an exciting opportunity for the industry and the research community for further research and development in this regard.

Author Contributions: Conceptualization, R.R., R.S. and P.L.; methodology, R.R., M.L. and A.P.; software, R.R. and A.P.; validation, R.R.; formal analysis, R.R., R.S. and M.L.; investigation, R.R.; resources, R.S. and P.L.; data curation, R.R.; writing—original draft preparation, R.R.; writing—review and editing, R.R. and R.S.; visualization, R.R. and M.L.; supervision, R.S. and P.L.; project administration, R.S. and P.L.; funding acquisition, R.S. All authors have read and agreed to the published version of the manuscript.

Funding: This work was completed in the framework of the COMET K2 Center on Symbiotic Mechatronics, which is funded by the Austrian Federal Government, the State Upper Austria, and by its Scientific and Industrial Partners.

Data Availability Statement: Not applicable.

Acknowledgments: Open Access Funding by the University of Linz. The authors thank Miika Paloniitty and Matti Linjama from Tampere University for providing the miniature hydraulic valves for this research work.

Conflicts of Interest: The authors declare no conflict of interest. The funders had no role in the design of the study; in the collection, analyses, or interpretation of data; in the writing of the manuscript; or in the decision to publish the results.

Nomenclature

A	Area
B	Guide curve
F	Force
g	Acceleration due to gravity
J	Moment of inertia
K	Bulk modulus of the fluid
K_E	Kinetic energy of the system
l	Length of the connecting rod
M	Knee torque
m	Mass
P	Pressure source
P_E	Potential energy
Q	Flow rate
R	Contact force at the guide curve
r	Crank length
T	Tank
t	Time
T_R	Transmission ratio
u	Valve configuration
V	Volume
v	Valve
W	Work
w	Lateral width of the exoskeleton device
x, y	Coordinate positions
z	Piston position in the cylinder
Greek letters	
α	Angle between the connecting rod and the tangent to the guide curve
γ	Angle between the hydraulic cylinder rod and the tangent to the guide curve
θ	Angle between the crank and the shank
σ_{YS}	Yield strength
ϕ	Angular position of limb part
ψ	Knee angle
Subscripts	
1,2	Mechanism identifier
a	Actual
b	Bore side of hydraulic cylinder
cr	Connecting rod
d	Desired
f	Friction
m	Middle
$meas$	Measured
n	Time step
r	Rod side of hydraulic cylinder
S	Source
Abbreviations	
eff	Effective
hyd	Hydraulic

Appendix A. Mechanical Model of Lower Limb Motion

The mechanical model of limb motion has been presented by the authors in previous works [33,34]. However, for the sake of completeness, the authors recapitulate the model in this appendix.

The lower limb motion is modelled via a planar system comprising foot, shank, and thigh. The motion of the upper body is until the thigh is considered to be known from the HuMoD database. Thus, the knee and ankle joint motions are the DOFs (degrees of freedom) considered in the model.

The kinetic and potential energy of the system is expressible as:

$$K_E = \frac{1}{2}\sum_{i=1}^{2}\left(m_i\dot{x}_{S_i}\cdot\dot{x}_{S_i} + J_i\dot{\phi}_i^2\right), \tag{A1}$$

$$P_E = g\sum_{i=1}^{2} m_i x_{S_i}\cdot e_y\,. \tag{A2}$$

where m_i is the mass, x_{S_i} is the position of the center of mass, J_i is the moment of inertia, ϕ_i is the angular position of the ith limb part, and e_y is the unit vector in vertical direction.

Next, the virtual work due to torques at the joints (M_{Gi}) and the ground reaction force (F_G) is:

$$\delta W = \sum_{i=2}^{3} M_{Gi}(\delta\phi_i - \delta\phi_{i-1}) + F_G^T\delta x. \tag{A3}$$

Substituting these expressions into the Lagrange equation of motion and simplifying, the following expression for the knee torque is obtained:

$$M = m_{eff,1}\ddot{\phi}_1 + m_{eff,2}\ddot{\phi}_2 - kB_F F_G - kV_{f0}. \tag{A4}$$

where $m_{eff,1}$ and $m_{eff,2}$ are the effective masses of foot and shank, $kB_F F_G$ is the ground reaction term, and kV_{f0} contains the terms related to the centripetal force, gravity, and the force from thigh. The complete expressions are lengthy, and, hence, are not shown here.

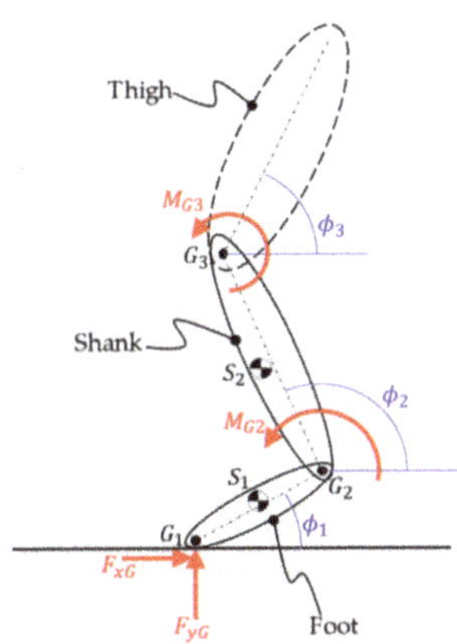

Figure A1. Planar view of limb with variables present in the mechanical model.

References

1. Li, S.; Francisco, G.E.; Zhou, P. Post-stroke hemiplegic gait: New perspective and insights. *Front. Physiol.* **2018**, *9*, 1021. [CrossRef] [PubMed]
2. Nilsson, A.; Vreede, K.S.; Häglund, V.; Kawamoto, H.; Sankai, Y.; Borg, J. Gait training early after stroke with a new exoskeleton—The hybrid assistive limb: A study of safety and feasibility. *J. NeuroEng. Rehabil.* **2014**, *11*, 92. [CrossRef] [PubMed]
3. Voilqué, A.; Masood, J.; Fauroux, J.C.; Sabourin, L.; Guezet, O. Industrial Exoskeleton Technology: Classification, Structural Analysis, and Structural Complexity Indicator. In Proceedings of the 2019 Wearable Robotics Association Conference (WearRAcon), Scottsdale, AZ, USA, 25–27 March 2019; pp. 13–20.
4. Proud, J.K.; Lai, D.T.H.; Mudie, K.L.; Carstairs, G.L.; Billing, D.C.; Garofolini, A.; Begg, R.K. Exoskeleton application to military manual handling tasks. *Hum. Factors* **2022**, *64*, 527–554. [CrossRef] [PubMed]
5. Osipov, A. Fire exoskeleton to facilitate the work of the fireman. *E3S Web. Conf.* **2019**, *126*, 00015. [CrossRef]
6. De la Tejera, J.A.; Bustamante-Bello, R.; Ramirez-Mendoza, R.A.; Izquierdo-Reyes, J. Systematic review of exoskeletons towards a general categorization model proposal. *Appl. Sci.* **2021**, *11*, 76. [CrossRef]
7. Scheidl, R. Digital Fluid Power for Exoskeleton Actuation—Guidelines, Opportunities, Challenges. In Proceedings of the Ninth Workshop on Digital Fluid Power, Aalborg, Denmark, 7–8 September 2017.

8. Zoss, A.; Kazerooni, H.; Chu, A. On the Mechanical Design of the Berkeley Lower Extremity Exoskeleton (BLEEX). In Proceedings of the 2005 IEEE/RSJ International Conference on Intelligent Robots and Systems, Edmonton, AB, Canada, 2–6 August 2005; pp. 3465–3472.
9. Zoss, A.B.; Kazerooni, H.; Chu, A. Biomechanical design of the Berkeley Lower Extremity Exoskeleton (BLEEX). *IEEE/ASME Trans. Mechatron.* **2006**, *11*, 128–138. [CrossRef]
10. Kazerooni, H.; Racine, J.-L.; Huang, L.; Steger, R. On the Control of the Berkeley Lower Extremity Exoskeleton (BLEEX). In Proceedings of the 2005 IEEE International Conference on Robotics and Automation, Barcelona, Spain, 18–22 April 2005; pp. 4353–4360.
11. Xie, H.; Li, X.; Li, W.; Li, X. The Proceeding of the Research on Human Exoskeleton. In *Advances in Intelligent Systems Research*; Atlantis Press: Amsterdam, The Netherlands, 2014; pp. 754–758.
12. Karlin, S. Raiding Iron Man's Closet [Geek Life]. *IEEE Spectr.* **2011**, *48*, 25. [CrossRef]
13. Huo, W.; Mohammed, S.; Moreno, J.C.; Amirat, Y. Lower limb wearable robots for assistance and rehabilitation: A state of the art. *IEEE Syst. J.* **2016**, *10*, 1068–1081. [CrossRef]
14. Chen, S.; Chen, Z.; Yao, B.; Zhu, X.; Zhu, S.; Wang, Q.; Song, Y. Adaptive robust cascade force control of 1-DOF hydraulic exoskeleton for human performance augmentation. *IEEE/ASME Trans. Mechatron.* **2017**, *22*, 589–600. [CrossRef]
15. Kim, H.; Shin, Y.J.; Kim, J. Design and locomotion control of a hydraulic lower extremity exoskeleton for mobility augmentation. *Mechatronics* **2017**, *46*, 32–45. [CrossRef]
16. Yang, Y.; Dong, X.; Liu, X.; Huang, D. Robust repetitive learning-based trajectory tracking control for a leg exoskeleton driven by hybrid hydraulic system. *IEEE Access* **2020**, *8*, 27705–27714. [CrossRef]
17. Chen, S.; Han, T.; Dong, F.; Lu, L.; Liu, H.; Tian, X.; Han, J. Precision interaction force control of an underactuated hydraulic stance leg exoskeleton considering the constraint from the wearer. *Machines* **2021**, *9*, 96. [CrossRef]
18. Ouyang, X.; Ding, S.; Fan, B.; Li, P.Y.; Yang, H. Development of a novel compact hydraulic power unit for the exoskeleton robot. *Mechatronics* **2016**, *38*, 68–75. [CrossRef]
19. Kosaki, T.; Li, S. A water-hydraulic upper-limb assistive exoskeleton system with displacement estimation. *J. Robot. Mechatron.* **2020**, *30*, 149–156. [CrossRef]
20. Kaminaga, H.; Ono, J.; Nakashima, Y.; Nakamura, Y. Development of Backdrivable Hydraulic Joint Mechanism for Knee Joint of Humanoid Robots. In Proceedings of the 2009 IEEE International Conference on Robotics and Automation, Kobe, Japan, 12–17 May 2009; pp. 1577–1582.
21. Kaminaga, H.; Amari, T.; Niwa, Y.; Nakamura, Y. Electro-Hydrostatic Actuators with Series Dissipative Property and Their Application to Power Assist Devices. In Proceedings of the 2010 3rd IEEE RAS & EMBS International Conference on Biomedical Robotics and Biomechatronics, Tokyo, Japan, 26–29 September 2010; pp. 76–81.
22. Kaminaga, H.; Tanaka, H.; Nakamura, Y. Mechanism and Control of Knee Power Augmenting Device with Backdrivable Electro-Hydrostatic Actuator. In Proceedings of the 13th World Congress in Mechanism and Machine Science, Guanajuato, Mexico, 19–23 June 2011.
23. Kaminaga, H.; Amari, T.; Niwa, Y.; Nakamura, Y. Development of Knee Power Assist Using Backdrivable Electro-Hydrostatic Actuator. In Proceedings of the 2010 IEEE/RSJ International Conference on Intelligent Robots and Systems, Taipei, Taiwan, 18–22 October 2010; pp. 5517–5524.
24. Tanaka, H.; Kaminaga, H.; Nakamura, Y. Pressure feedback control based on singular perturbation method of an electro-hydrostatic actuator for an exoskeletal power-assist system. *J. Robot. Mechatron.* **2012**, *24*, 354–362. [CrossRef]
25. Kurosawa, H.; Walker, P.S.; Abe, S.; Garg, A.; Hunter, T. Geometry and motion of the knee for implant and orthotic design. *J. Biomech.* **1985**, *18*, 487–499. [CrossRef]
26. Lee, D.; Song, B.; Park, S.Y.; Baek, Y.S. Development and control of an electro-hydraulic actuator system for an exoskeleton robot. *Appl. Sci.* **2019**, *9*, 4295. [CrossRef]
27. Jiang, J.; Wang, Y.; Cao, H.; Zhu, J.; Zhang, X. A novel pump-valve coordinated controlled hydraulic system for the lower extremity exoskeleton. *Trans. Inst. Meas. Control* **2020**, *42*, 2872–2884. [CrossRef]
28. Lee, T.; Lee, D.; Song, B.; Baek, Y.S. Design and control of a polycentric knee exoskeleton using an electro-hydraulic actuator. *Sensors* **2020**, *20*, 211. [CrossRef]
29. Sun, M.; Ouyang, X.; Mattila, J.; Chen, Z.; Yang, H.; Liu, H. Lightweight electrohydrostatic actuator drive solution for exoskeleton robots. *IEEE/ASME Trans. Mechatron.* **2022**, 1–12. [CrossRef]
30. Scheidl, R.; Linjama, M.; Schmidt, S. Is the future of fluid power digital? *Proc. Inst. Mech. Eng. Part I J. Syst. Control. Eng.* **2012**, *226*, 721–723. [CrossRef]
31. Cao, H.; Ling, Z.; Zhu, J.; Wang, Y.; Wang, W. Design Frame of a Leg Exoskeleton for Load-Carrying Augmentation. In Proceedings of the 2009 IEEE International Conference on Robotics and Biomimetics (ROBIO), Guangxi, China, 13–19 December 2009; pp. 426–431.
32. Holl, E.; Scheidl, R.; Eshkabilov, S. Simulation Study of a Digital Hydraulic Drive for a Knee Joint Exoskeleton. In Proceedings of the ASME/BATH 2017 Symposium on Fluid Power and Motion Control, Sarasota, FL, USA, 4 December 2017.
33. Rituraj, R.; Scheidl, R.; Ladner, P.; Lauber, M. A Novel Design Concept of Digital Hydraulic Drive for Knee Exoskeleton. In Proceedings of the ASME/BATH 2021 Symposium on Fluid Power and Motion Control, Virtual, 13 December 2021.

34. Rituraj, R.; Scheidl, R. Advancements in the Control Strategy for Digital Hydraulically Driven Knee Exoskeleton. In Proceedings of the ASME/BATH 2022 Symposium on Fluid Power and Motion Control, Bath, UK, 19 September 2022.
35. HuMoD. Available online: https://www.sim.informatik.tu-darmstadt.de/res/ds/humod/ (accessed on 30 March 2022).
36. Wojtusch, J.; Von Stryk, O. HuMoD—A Versatile and Open Database for the Investigation, Modeling and Simulation of Human Motion Dynamics on Actuation Level. In Proceedings of the 2015 IEEE-RAS 15th International Conference on Humanoid Robots (Humanoids), Seoul, Republic of Korea, 3–5 November 2015; pp. 74–79.
37. Deb, K.; Pratap, A.; Agarwal, S.; Meyarivan, T. A fast and elitist multiobjective genetic algorithm: NSGA-II. *IEEE Trans. Evol. Comput.* **2002**, *6*, 182–197. [CrossRef]
38. HAWE Micro Fluid GmbH. Available online: www.hawe.com (accessed on 22 February 2022).
39. SOLIDWORKS Simulation. Available online: https://www.solidworks.com/domain/simulation (accessed on 1 June 2022).
40. Igus. Available online: https://www.igus.co.uk/ (accessed on 3 July 2022).
41. Iglidur. Available online: https://www.igus.co.uk/info/plain-bearings-x-material-data (accessed on 3 July 2022).
42. Linjama, M.; Paloniitty, M.; Tiainen, L.; Huhtala, K. Mechatronic design of digital hydraulic micro valve package. *Procedia Eng.* **2015**, *106*, 97–107. [CrossRef]
43. Megatron Kit Encoder Etx25K. Available online: https://www.megatron.de/en/products/angle-sensors/encoder-etx25k-with-3d-hall-as-kit-versions.html (accessed on 30 July 2022).
44. Autosen AP019 Pressure Sensor. Available online: https://autosen.com/en/Process-Sensors/Pressure-sensors/Electronic-pressure-sensor-G1-4M-AP019 (accessed on 7 August 2022).
45. B&R X20 System. Available online: https://www.br-automation.com/en/products/plc-systems/x20-system/ (accessed on 7 August 2022).
46. Yang, J.; He, Y.; Shi, P.; Yu, H. A review on human intent understanding and compliance control strategies for lower limb exoskeletons. *Proc. Inst. Mech. Eng. Part I J. Syst. Control. Eng.* **2022**, *236*, 1067–1086. [CrossRef]

energies

MDPI

Article

Energy Savings in Hydraulic Hybrid Transmissions through Digital Hydraulics Technology

Israa Azzam, Keith Pate, Jose Garcia-Bravo and Farid Breidi *

School of Engineering Technology, Purdue University, West Lafayette, IN 47906, USA; iazzam@purdue.edu (I.A.); pate5@purdue.edu (K.P.); jmgarcia@purdue.edu (J.G.-B.)
* Correspondence: breidi@purdue.edu; Tel.: +1-(765)-496-5140

Abstract: Hydraulic hybrid drivetrains, which are fluid power technologies implemented in automobiles, present a popular alternative to conventional drivetrain architectures due to their high energy savings, flexibility in power transmission, and ease of operation. Hydraulic hybrid drivetrains offer multiple environmental benefits compared to other power transmission technologies. They provide heavy-duty vehicles, e.g., commercial transportation, construction equipment, wagon handling, drilling machines, and military trucks, with the potential to achieve better fuel economy and lower carbon emissions. Despite the preponderance of hydraulic hybrid transmissions, state-of-the-art hydraulic hybrid drivetrains have relatively low efficiencies, around 64% to 81%. This low efficiency is due to the utilization of conventional variable displacement pumps and motors that experience high power losses throughout the drive cycle and thus fail to maintain high operating efficiency at lower volumetric displacements. This work proposes and validates a new methodology to improve the overall efficiency of hydraulic hybrid drivetrains by replacing conventional pump/motor units with their digital counterparts. Compared to conventional pump/motors, the digital pump/motor can achieve higher overall efficiencies at a wide range of operating conditions. A proof-of-concept digital pump/motor prototype was built and tested. The experimental data were integrated into a multi-domain physics-based simulation model of a series hydraulic hybrid transmission. The proposed methodology permits enhancing the overall efficiency of a series hydraulic hybrid transmission and thus allows for energy savings. Simulating the system at moderate load-speed conditions allowed achieving a total efficiency of around 89%. Compared to the average efficiency of the series hydraulic hybrid drivetrains, our simulation results reveal that the utilization of the state-of-the-art digital pump enables improving the total efficiency of the series hydraulic hybrid drivetrain by up to 25%.

Keywords: fluid power; hydraulic hybrid drivetrains; digital hydraulics; energy savings; efficiency

Citation: Azzam, I.; Pate, K.; Garcia-Bravo, J.; Breidi, F. Energy Savings in Hydraulic Hybrid Transmissions through Digital Hydraulics Technology. *Energies* **2022**, *15*, 1348. https://doi.org/10.3390/en15041348

Academic Editors: Paolo Casoli, Massimo Rundo and Helena M. Ramos

Received: 6 January 2022
Accepted: 10 February 2022
Published: 13 February 2022

1. Introduction

Fluid power has a vast area of engineering technology, where it uses pressurized fluids to generate, monitor and control, and transmit power [1]. Hydraulic oil is typically used due to its low freezing point, high boiling point, relatively large bulk modulus, and self-lubrication characteristics [2]. Fluid power has a well-defined research area and scholarly activities, which serves widespread industries like agriculture, construction, transportation, aerospace, marine, manufacturing, and many other industries that require high power to weight ratios [3]. In addition to its capability of achieving high power densities, fluid power is characterized by its flexibility in transmission, ease of operation, relatively low cost, controllability, and manageability [4]. These advantages make fluid power technology convenient for performing crucial operations that require high power density and reliability, like heavy-duty applications, e.g., construction vehicles, material handling equipment, and military operations [5].

The fluid power sector is a Workhorse of the United States economy [6]. According to the National Fluid Power Association (NFPA), thousands of companies in the United

States utilize hydraulic power systems. Such companies are renowned as the top industrial U.S. firms, involving more than 845,000 employees for an annual payroll of more than $60 billion [7]. The fluid power sector has multiple application areas. Four main application areas comprise mobile hydraulics, industrial hydraulics, pneumatics, and aerospace [8].

Hydraulic hybrid technology, a subset of mobile hydraulics, is considered one of the pioneer frameworks in automobiles. It is a desirable research topic widely examined by scholars to improve the overall efficiency of power transmission between a prime mover and the wheels. This technology allows for high fuel efficiency and offers energy savings and multiple environmental benefits, becoming a popular alternative to conventional drivetrain configurations [9]. Hydraulic hybrid drivetrains offer multiple benefits over the conventional manual/automatic transmissions and the continuously variable transmissions (CVTs). Unlike conventional drivetrains and CVTs, hydraulic hybrid drivetrains comprise two power sources: the prime mover (an internal combustion engine or electric motor) and a positive variable displacement pump/motor installed with a high-pressure storage device (accumulator) [10]. The variable displacement pump allows achieving variable speeds by regulating the flow rate, which prevents the gearshift discontinuities in conventional drivetrains (manual/automatic) during speed regulation. CVTs do not experience excessive gearshift discontinuities during speed regulation. However, CVTs have limited belts, also known as chain strengths, which cause torsional moments, generating excessive frictional forces and torsional moments [11]. Besides this advantage, utilizing a hydraulic accumulator in hydraulic hybrid drivetrains allows for better power management [12]. The accumulator enables storing and discharging energy and thus achieving higher overall efficiencies and energy savings while attaining lower carbon emissions and better fuel economy [13].

In addition to its preponderance over manual/automatic drivetrains and CVTs, the hydraulic hybrid drivetrains have many advantages over electric hybrid transmission [14,15]. Although electric hybrid systems also utilize a transmission source and an energy storage source (batteries), the batteries in electric systems can only charge and discharge around 20% of energy available during braking [16]. However, the accumulators employed in the hydraulic hybrid transmissions can store and deliver up to 80% of the braking energy [17], depending on the duty cycle. This operation is desirable during vehicle acceleration/deceleration to improve fuel economy and reduce emissions [18,19]. It permits transmitting a large amount of energy rapidly, thus attaining higher power densities and power to weight ratios [20].

These advantages make the hydraulic hybrid technology more convenient for heavy-duty operations [21], like heavy equipment manufacturing, road appliances, heavy load material handling equipment, and many other types of equipment that require high power densities and energy capture. For instance, commercial vehicles, such as heavy load vehicles undergoing high dynamic-drive-cycle operations, have started to rely on this transmission technology. They carry bulky loads while experiencing continual stops, requiring more braking and high energy capture [15]. Besides heavy-duty vehicles, some utility vehicles designed to carry light weights also rely on hydraulic hybrid technology. For example, a sport utility vehicle was developed in a research laboratory to demonstrate the concept of a hydraulic hybrid transmission in a passenger vehicle. The vehicle was designed to operate the system in four different modes: hydrostatic driving mode, hybrid driving mode, a combination of hydrostatic and hybrid driving modes, and braking mode [22].

This paper is organized as follows: Section 2 exhibits a review of the currently existing hydraulic hybrid transmissions. It introduces the different available hydraulic hybrid drivetrain configurations: series, parallel, and series-parallel, and it discusses the overall operation of each. Section 3 presents a series of notable applications on hydraulic hybrid drivetrains. Section 4 formulates the problem statement of this work, elucidating the inadequacies of the currently existing hydraulic hybrid drivetrains. Moreover, it highlights and introduces the adopted methodology to overcome such inadequacies. Section 5 gives an overview of digital hydraulics technology in fluid power and presents the design of the state-of-the-art digital pump, its basic operation, digital motoring/pumping data, and the capability to achieve high efficiencies at different pump displacements. Section 6 demonstrates the hydraulic hybrid transmission efficiency improvement after integrating digital hydraulics by simulating a physical series hydraulic hybrid architecture using the intended digital motoring/pumping data. It presents the adopted circuit schematic, the comprehensive simulation model, and the resulting outcomes. Section 7 concludes with a brief discussion of the main ideas learned from the work.

2. Existing Hydraulic Hybrid Transmission Configurations

Three main architectures of hydraulic hybrid transmission drivetrains are currently available: the parallel configuration, series configuration, and the series-parallel configuration. These hydraulic hybrid transmission configurations employ a variable displacement pump/motor, a high-pressure reservoir for energy storage (typically a nitrogen accumulator), and a low-pressure reservoir (fluid tank) [23]. The variable displacement pump/motor acts as a motor when converting the hydraulic energy into mechanical energy to drive the vehicle. It acts as a pump when converting the mechanical energy into hydraulic energy stored in the accumulator to be delivered when needed.

The following sections exhibit a brief review of existing relevant hydraulic hybrid drivetrains (parallel, series, and series-parallel) and discuss their overall operation.

2.1. Parallel Hydraulic Hybrid Configuration

A parallel hydraulic hybrid has a simple architecture, design, and implementation compared to other configurations [24]. It involves two sub-systems: a conventional mechanical transmission directly connected to the output shaft via a differential gearbox and a hydraulic transmission [25]. The sub-systems are coupled in parallel using a mechanical power transmitter like a gearbox, a chain, or a belt. Figure 1 shows a parallel hydraulic hybrid configuration utilizing a belt mechanical power transmitter. As shown in Figure 1, the drivetrain consists of mechanical transmission (7), a clutch (2), and the prime mover (1), typically an internal combustion engine. The mechanical transmission (7) is directly connected to the engine (1) via the clutch component (2) and to the differential gearbox (8) on the drive shaft (9) to provide the needed power to each wheel. The hydraulic drivetrain comprises a variable displacement pump/motor (4), an accumulator (5), a fluid tank (6), and other hydraulic components, e.g., valves and fittings. The hydraulic pump/motor (4) is connected in parallel to the mechanical transmission (7) via the belt linkage (3). The clutch (2) allows engaging and disengaging the engine shaft from the rest of the system based on the accumulator's stored energy, thus controlling the engine on/off signal [26]. For instance, decoupling the engine from the system occurs when enough energy is in the accumulator to boost the wheels during acceleration. However, engaging the engine shaft to the transmission shaft is required to drive the wheels at low accumulator storage levels.

After introducing the basic concept beyond the parallel hybrid configuration, now, we briefly discuss its overall operation. The transmission duty cycle constitutes three main stages: acceleration, cruising, and braking. During acceleration, the accumulator acts as a hydraulic launch assist. It discharges its pressurized fluid to drive the hydraulic pump/motor (4) (operating as an energy source) and thus transforms the high-pressure energy into mechanical energy at the motor's shaft. The mechanical belt (3) transmits the generated mechanical power to the drive shaft (9), thus boosting the motion at the wheels. At the start of the cruising phase, the accumulator's pressure decreases below its pre-charged pressure, forcing the clutch to engage the engine shaft with the rest of the system to power the drive shaft (9) and drive the wheels. During the last stage, i.e., braking, the clutch (2) disengages the engine shaft from the drive shaft (9), the vehicle slows down. The rotary motion at the wheels is transmitted via the belt (3) to drive the hydraulic pump/motor (4) that will start acting as a pump to charge the accumulator. This cycle repeats every braking operation.

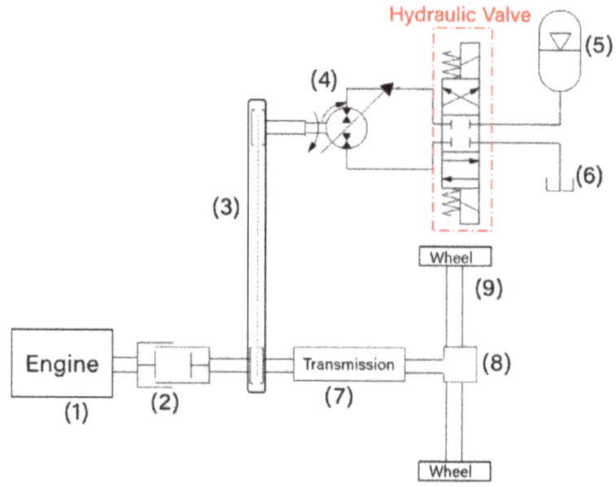

Item	Description
1	Engine
2	Mechanical Clutch
3	Mechanical Transmission Belt
4	Variable Displacement Motor/Pump
5	Gas Accumulator
6	Reservoir
7	Mechanical Transmission
8	Differential Gear Box
9	Output Shaft

Figure 1. Schematic for the parallel hydraulic hybrid configuration.

2.2. Series Hydraulic Hybrid Configuration

Unlike the parallel hydraulic hybrid configuration, the series hydraulic hybrid configuration does not utilize a mechanical transmission. Instead, it employs a variable displacement pump directly connected to the prime mover via a clutch [27]. As shown in Figure 2, in the series configuration, the mechanical transmission is replaced with a hydraulic variable displacement pump (3) connected in series to the hydraulic drivetrain consisting of a variable displacement pump/motor (6), accumulator (5), fluid tank (4), and other hydraulic components.

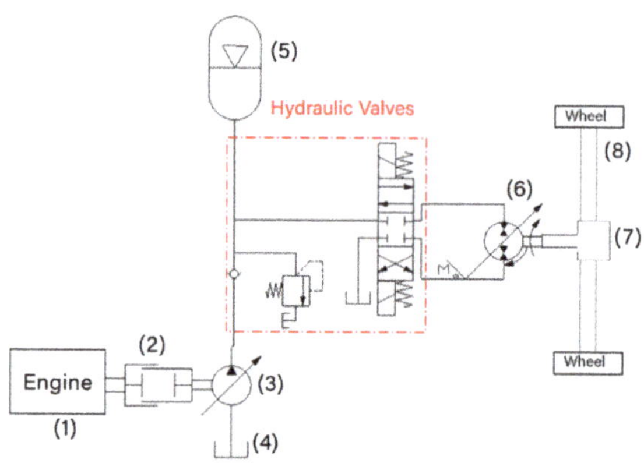

Item	Description
1	Engine
2	Mechanical Clutch
3	Variable Displacement Pump
4	Reservoir
5	Gas Accumulator
6	Variable Displacement Motor/Pump
7	Differential Gear Box
8	Output Shaft

Figure 2. Schematic for the series hydraulic hybrid configuration.

Similar to the parallel architecture, the operation comprises three main phases: acceleration, cruising, and braking. During the acceleration phase, the clutch (2) disengages the engine shaft from the pump shaft, and the pressurized hydraulic fluid stored in the accumulator (5) is delivered into the hydraulic pump/motor (6), allowing it to act as a motor. This operation converts the hydraulic energy in the accumulator into mechanical energy at the output shaft (8). After acceleration, cruising starts, and the fluid inside the accumulator drops to its minimum level. Its pressure decreases below the nitrogen's pre-charge pressure, forcing the engine (1) to operate and the clutch (2) to engage. Accordingly, the engine's energy is transmitted to the hydraulic pump (3), allowing the pump to deliver the fluid from the low-pressure reservoir (4) to the hydraulic pump/motor (6) to drive the wheels. The excess high-pressure fluid produced by the hydraulic pump is stored in the accumulator to be delivered when needed. During braking, the engine shuts down, the clutch disengages, and the hydraulic pump/motor (6) starts acting as a pump to push the fluid from the low-pressure reservoir into the accumulator. Thus, the kinetic energy from the wheels is converted into hydraulic energy stored in the accumulator. This stored energy will be used to accelerate the vehicle when needed.

2.3. Series-Parallel Hydraulic Hybrid Configuration

The series-parallel hydraulic hybrid system, also known as power-split, comprises the series and the parallel hydraulic hybrid layouts [28]. It integrates the two hydraulic hybrid configurations in one design [29]. Like the parallel configuration, the power-split architecture, presented in Figure 3, utilizes two sub-systems coupled via a mechanical transmitter, for example, a belt (3): a conventional mechanical drivetrain and a hydraulic drivetrain. The mechanical drivetrain employs a mechanical transmission (8) directly connected to the output shaft (10) via a differential gearbox (9) and the engine (1) via a mechanical clutch (2). The hydraulic drivetrain comprises two variable displacement pump/motors (4 & 5), accumulator (6), low-pressure reservoir (7), and other hydraulic components. Similar to the series architecture, the hydraulic transmission is directly connected to the output shaft (10).

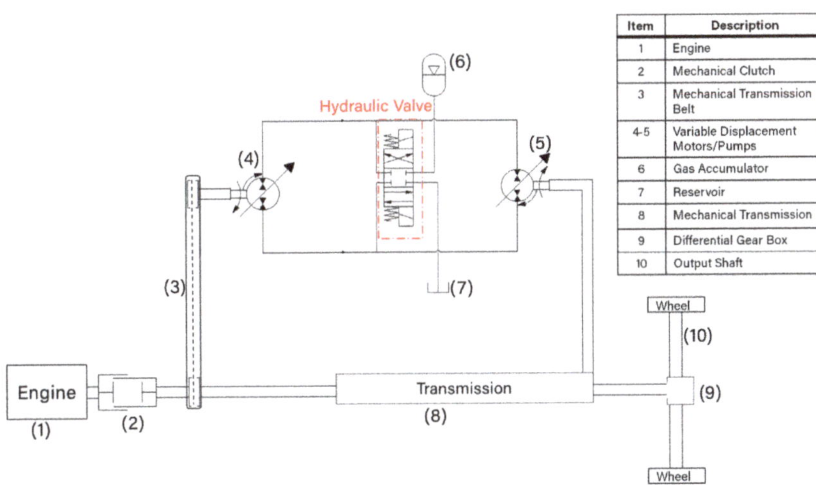

Figure 3. Schematic for the series-parallel hydraulic hybrid configuration.

This configuration enables operating the system in two different modes: pure hydraulic mode and hybrid mode. The pure hydraulic mode offers a simultaneous operation of the variable displacement pump/motors (4 & 5). However, the hybrid mode allows a seamless operation of the mechanical transmission (8) and the hydraulic drivetrain [30]. The seamless integration between the mechanical drivetrain and the hydraulic system allows transmitting the power from the engine to the wheels while maintaining optimal engine operation. This mechanism permits achieving high mechanical transmission performance in addition to the high-efficiency energy storage and regeneration, and thus operating the system at higher overall efficiency and lower fuel consumption. However, the design comes at the expense of increased system complexity, weight, and cost.

3. Summary of Notable Mobile Applications Using Hydraulic Hybrid Drivetrains

This section briefly reviews notable applications and proposed models for hydraulic hybrid drivetrains.

A comprehensive work presented by Stecki et al. [31] presented a review of the most notable computer simulation models of hydraulic hybrids available in the literature before 2005 and summarized the resulting simulation outcomes. The comprehensive report addresses the design, implementation, and simulation of more than eight prototype vehicles using CVT transmissions combined with hydrostatic regeneration. Most of the vehicles studied were on-road vehicles. The authors mentioned ADVISOR (National Renewable Energy Laboratory, Golden, CO, USA), the Advanced Vehicle Simulator developed by the National Renewable Energy Laboratory based in the US Department of Energy's (DOE's), as a simulation tool for developing hybrid drivetrain models. One of the vehicles discussed focused on a tactical vehicle, developed in MATLAB/Simulink which is created by MathWorks (an American privately held corporation based in California) and implemented in ADVISOR [32]. In this work, a Permo-Drive regenerative energy management system for an army tactical vehicle was modeled and simulated on MATLAB/Simulink. The model was then incorporated into the vehicle simulator (ADVISOR), a simulation tool for developing hybrid drivetrain models. The authors concluded their review by stating that extra fuel trails are required for examining the accuracy of the adopted fuel map and thus providing further validation of the model. Thy concluded that the accumulators might present a safety issue and recommended that their implementation require further accumulator technologies.

Besides the work of Stecki et al., the work of Stelson et al. [10] presents a simulation study of various hydraulic hybrid drivetrains comparing the three hydraulic hybrid configurations (series, parallel, and power-split). The prime mover is emulated as a combustion engine throughout their simulation studies. A dynamic programming approach is adopted, in which the fuel consumption was used as the optimization variable. The developed models reveal that the three configurations of the presented on-road vehicles (series, parallel, and power-split) improve fuel consumption for the two cycles: urban drive cycle and highway drive cycle. For the urban drive cycle, the three hydraulic hybrid configurations (series, parallel, and power-split) show an improvement of 14.7%, 53.2%, and 10.0%, respectively.

Regarding the highway drive cycle, the series, parallel, and power-split architectures were enhanced by 8.9%, 11.9%, and 8.5%, respectively. Furthermore, throughout the simulation, the employed pump/motors were oversized. This forced the units to operate at partial displacements and thus led to a low overall hydraulic efficiency.

In 2004, Wu et al. [33] utilized the dynamic programming optimization methodology to optimize the accumulator's charging/discharging process for a Class VI parallel hydraulic hybrid delivery truck. The researchers used a 99.9 L accumulator operating at a maximum pressure of 36 MPa throughout this work. The proposed control algorithm involves two optimization parameters: the fuel consumption variable as the primary control parameter and the state of charge of the accumulator (SOC) as the secondary optimization parameter. The researchers estimated the fuel economy between 28% and 47%, depending on the efficiency of the hydraulic components. In 2009, Van Batavia [34] proposed a new control approach to monitor and regulate the hydraulic launch assist operation. The adopted strategy relies on operating the engine at its peak Torque-RPM performance. An approximate range of operation comprised between 1000 RPM to 5500 RPM and 125 N.m to 200 N.m for a Diesel engine rated for 110 kW capacity. The work claims that the proposed control strategy improved the fuel economy by 50% and reduced emissions by 40%. The proposed configuration used a variable displacement bent axis pump and a separate variable displacement bent axis motor. The bent axis units are known to have high efficiencies at 100% displacements of up to 98% [35], yet this efficiency is only achievable at high displacements and is substantially lower at low displacements.

Buchwald et al. [36] introduced a Hydraulic—Internal combustion hybrid concept and four topologies for a passenger bus in Denmark. Besides the latter work, in 2008, Eaton Hydraulics [37] presented its hydraulic launch assist system for parallel drivetrain configurations. The objective was to improve power savings by supplementing engine power while launching the vehicle from a stopped position. The system was conceived for vehicles operating with heavy stop-and-go cycles, like refuse trucks, delivery vehicles, and buses. This system, developed in collaboration with the U.S. Environmental Protection

Agency, depended on using a state-of-the-art pump/motor unit specifically designed for a parallel drivetrain configuration using an axial piston design.

Tollefson et al. [38] examined a new configuration of internal combustion engine/hydropneumatic accumulator energy storage for automobiles. The proposed architecture requires one single hydrostatic pump-motor (P/M) unit. Computer-based simulations were conducted over a wide range of parameter variations using a suitable control strategy. The simulation results revealed that the novel concept allowed significant fuel savings. The proposed hydraulic accumulator passenger car could achieve up to 65 mpg during city driving conditions, unlike the present automobiles that can attain only 25–30 mpg under similar conditions.

Similarly, Kepner [39] presents a Hydraulic Power Assist (HPA) demonstration on a full-size sport utility vehicle. It exhibits a description of the vehicle, the resulting fuel economy data, and performance and noise outcomes. The vehicle is constructed as a collaboration between Ford Motor Company, United States Environmental Protection Agency (EPA), and Advanced Technology Division, in the context of a Cooperative Research and Development Agreement (CRAD). It utilizes a valve block provided by Ifield Technologies, a pump/motor connected in parallel to the driveshaft with the conventional powertrain, and carbon fiber accumulators acting as HPA. Results show that the utilization of the HPA improved the Environmental Protection Agency (EPA) and the Heavier Acceleration city cycles by 23.6% and 35.5%, respectively. Bozic [40] presented a hybrid configuration comprising a combustion engine as the prime mover, hydrostatic transmission, and a coexisting vehicle electric generation. According to the literature consulted, it is evident that the implementation of hybrid powertrains significantly contributes to energy savings, in some cases, as high as 50% [33,34]. Yet, the system's efficiency may be compromised by the efficiency of the heart of the hybrid transmission; that is, the pump, the motor, or the pump/motor units. These units can operate at an efficiency as high as 98%, provided the units are operating at 100% displacement, but this may be the case of only a fixed segment of a duty cycle and may not be true at all operating conditions.

4. Inadequacies of the Existing Hydraulic Hybrid Drivetrains

Despite the prominence of fluid power transmission, the existing state-of-the-art fluid power drivetrains are ineffective [3]. A study published by the United States Department of Energy shows that, among all industries, fluid power drives have low efficiencies between 9% and 60%, depending on the application, with an average efficiency not exceeding 22% [8]. The insufficiency in the fluid power transmissions' efficiencies is notably associated with the performance of the employed hydraulic units. The variable displacement pump/motor is an essential hydraulic unit employed in all hydraulic applications; it is the heart of fluid power systems [41]. Its efficiency significantly affects the performance and overall efficiency of fluid power transmission [42,43].

The conventional variable displacement pump/motors, i.e., existing state-of-the-art hydraulic pumps, are efficient only when operating at full displacements, whereas at partial displacements, their performance decreases until operating at low efficiencies down to 30% [42,44,45]. The insufficiency of the conventional pumps at partial or low displacements affects the system's overall efficiency due to the high pressure that may remain in the displacement chamber, in addition to leakage and other mechanical losses [46]. Given the limited efficiency of the conventional variable displacement pump/motors, the overall efficiency of the hydraulic hybrid drivetrains is relatively low, around 64–81% [47]. This critical problem has motivated many researchers in multiple fluid power applications to propose, design, and implement new techniques and methodologies to improve the performance of hydraulic hybrid transmissions by enhancing the maximum efficiency of variable displacement vane and piston pumps [48–55]. For instance, the work in [49] addresses the pre-defined problem by proposing a dynamic seven-stage model that predicts the thickness of the fluid film between the valve plate and cylinder block in axial piston pumps to get accurate information on lubrication. In [51], relying on a coating methodology, the authors improved the overall efficiency map of axial piston pumps by minimizing the friction losses by applying nanocoatings to the pump's slippers. Besides these works, the work in [54] provides research results for a previously executed research study. The resulting outcomes can be used to enhance the efficiency of many technological systems in power engineering and transport involving hydraulic pumps and motors. Similarly, the work in [55] focuses on improving the numerical algorithm used for solving the optimal control problem for the three-dimensional heat equation, which can improve many technological systems like hydraulic units.

All the proposed techniques in the pre-listed works still result in low overall efficiencies when operating at partial volumetric displacements. There is no noticeable improvement in the pumps' lowest efficiency at low volumetric displacements [56]. Accordingly, the work presented in [42] proposed integrating digital technologies into fluid power systems through designing digital variable displacement pump/motors that allow maintaining high efficiencies over a wide range of displacements.

Building on the results from [41,42,57,58], this work aims at enhancing the performance and efficiency of a hydraulic hybrid transmission by simulating the implementation of an experimentally-validated digital pump/motor into a hydraulic hybrid transmission. As will be proved through a physical-based simulation, unlike the ordinary variable displacement pump/motors, the adopted digital pump/motors pursue to overcome the efficiency limitation in hydraulic hybrid drivetrains by increasing the maximum overall efficiency of the transmission.

5. Digital Hydraulics

Digital Hydraulics is a pioneering sub-field of fluid power, which relies on advanced on/off control technologies to enhance hydraulic systems' performance, reinforce advancements in fluid power applications, and minimize energy consumption [59]. This technology can replace conventional fluid power components, such as directional control valves, with multiple switching on/off valves in a parallel configuration to improve efficiency [41]. Such valves' configurations promote a superior design in this application because they use a poppet-style actuation that does not allow leakage across the valve, commonly seen in spool valves. In turn, this configuration reduces the system's losses, enables more control, increases the performance and efficiency, and reduces the cost of maintaining components [60–62]. This technology has also been utilized in various other configurations to improve the efficiency of conventional systems, including replacing traditional check valves on fixed displacement pumps. It also enables the opportunity to limit and divert flow in conventional fixed displacement pumps allowing variable displacements [44].

Some conventional piston pump/motors, e.g., reciprocating piston pumps, utilize mechanical check valves on the inlet and outlet ports of the pump to separate the system's high- and low-pressure sides. During pumping, this configuration allows fluid to enter through the low-pressure check valve when the pump is on its intake stroke. When the pump exhausts its cylinder, pressure closes the intake check valve and opens the exhaust check valve, displacing fluid to the high-pressure side of the system. This configuration limits the pump to fixed displacements since the pressure differential in the chamber is the only method of controlling the actuation of the intake and exhaust valves. Thus, the work in [42] proposes an electrically actuated digital inline 3-piston pump that replaces the conventional valves with digital switching on/off valves. The proposed digital pump/motor is a 28cc/rev unit comprising three displacement chambers, each with two high-speed switching on/off valves. Figure 4 shows the pump/motor's configuration for a single displacement chamber. The figure shows that Valve 1 controls the flow at the low-pressure side, and Valve 2 controls the flow at the high-pressure side. The proposed configuration allows opening or closing the valves at any time throughout the pumping/motoring cycle. This configuration adds a degree of controllability and flexibility to the system, enabling various digital pumping/motoring techniques, referred to as flow diverting and limiting operating strategies [44,63].

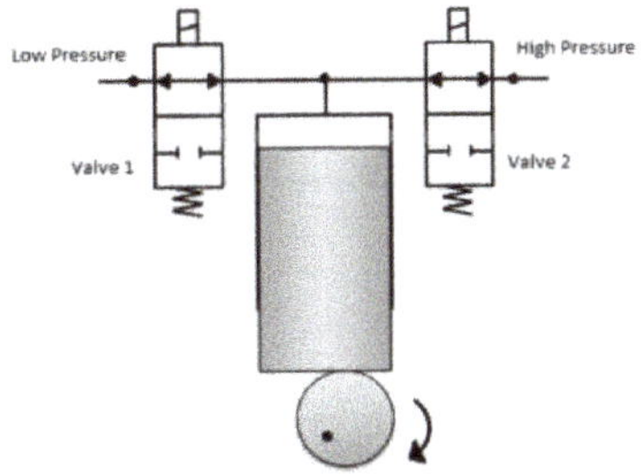

Figure 4. Single-piston digital pump/motor configuration [64].

Besides the electrically actuated digital pump, our team designed and imple-mented a mechanically actuated digital pump, which can also achieve high efficiencies over a wide range of displacements. The following sub-sections exhibit an overview of the internal mechanical structure of the mechanically actuated digital pump, intro-duce the electrically actuated pump's most efficient operating strategy, and present the experimental testing and validation on the electrically actuated pump.

5.1. Internal Mechanical Structure of the Mechanically Actuated Digital Pump

This subsection gives a brief introduction to the mechanical structure of the mechanically actuated digital pump, which can achieve efficiencies higher than the ones reported in this paper. The internal mechanical structure of the intended digital pump, shown in Figure 5, comprises three main sub-systems: (1) gear-cam unit, (2) cam-follower unit, and the (3) valve manifold. The gear-cam unit utilizes three main half-masking cams, where each cam consists of two sub-cams that are phased relative to one another to allow achieving different cam profiles. This unit is responsible for the phasing mechanism of the camshafts, where it allows varying the cams' orientation relative to the shaft during rotation. The cam-follower unit consists of the three followers, their mount bracket, and the three valves to be actuated via the cam-follower mechanism. This unit allows converting the rotational motion at the cam to a linear motion at the follower and thus controlling the valves' opening and closing. Therefore, varying the cam profile using the phasing mechanism permits varying the opening and closing of the valves and thus enables achieving variable pump displacement. A detailed description of the mechanical structure and the pumping mechanisms is presented in [65].

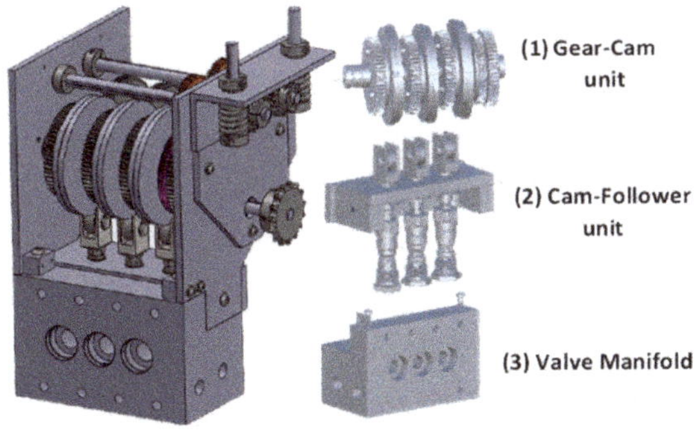

Figure 5. CAD drawing showing the basic units of the mechanically actuated digital pump.

5.2. Operating Strategies of the State-of-the-Art Electrically Actuated Digital Pump

The electrically actuated digital variable displacement pump/motor [42] relies on the control flow entering and leaving the displacement chambers. The pump/motor displacement variation can be achieved by controlling the actuation of two on/off valves on the intake and exhaust ports of the pump. The control and timing of the valves' actuation enable various pumping/motoring operating strategies. Four different operating strategies have been explored and tested [42] utilizing the state-of-the-art digital pump. The four different operating modes are Partial Flow-Diverting (PFD), Partial Flow-Limiting, Sequential Flow-Diverting (SFD), and Sequential Flow-Limiting (SFL). The most efficient operating strategy discovered is the SFL [63]. SFL is a technique that permits the states of each cylinder to be enabled or disabled, sending either 100% or 0% displacement to the output valve, respectively. This method can be used for both pumping and motoring cycles.

An example of a 100% displacement pumping cycle is shown in Figure 6. This cycle achieves similar functionality as the mechanical check valve. During pumping, the intake valve is opened for the entire stroke (A) and closed when the exhaust stroke begins (B). The exhaust valve is then opened and held until the exhaust stroke is completed (C). This transfers 100% of the fluid entering the piston chamber on the intake stroke, disregarding the dead volume in the cylinder.

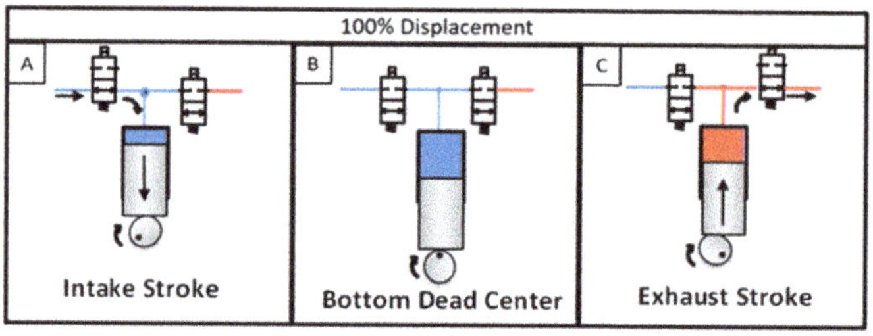

Figure 6. Enabled chamber operating at 100% displacement during SFL mode.

An example of 0% displacement for pumping is shown in Figure 7. During this cycle, all intake and exhaust valves remain closed for the entire process, allowing the fluid to be compressed and decompressed. As the cylinder begins its intake stroke, the piston starts to decompress the fluid in the dead volume of the cylinder (A) until the cylinder reaches the Bottom Dead Center (B), which allows depressurizing the fluid and creating a partial vacuum in the chamber. Once the piston begins the exhaust stroke (C), the fluid is compressed back to its original state. Since the intake and exhaust valves remained closed during this cycle, 0% displacement is delivered by the cylinder. Thus, this cycle allows the chamber's dead volume fluid to be depressurized and pressurized, and no fluid is delivered to the system.

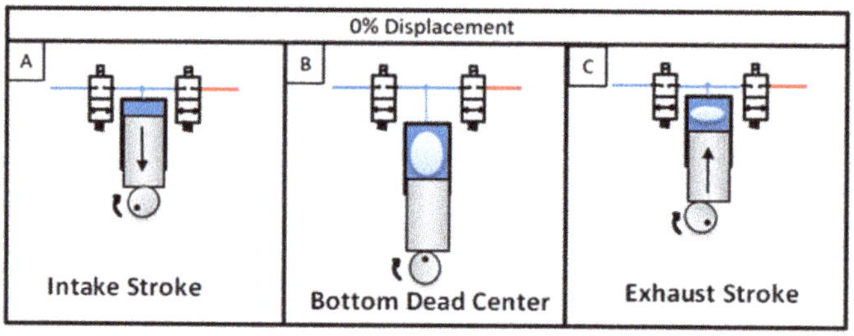

Figure 7. Disabled chamber operating at 0% displacement during SFL mode.

The partial vacuum created during the intake stroke is called chamber voiding. The chamber voiding phenomenon was experimentally inspected by [66]. It was experimentally proved that the gas created during the chamber voiding phase, i.e., (B), would dissolve entirely as the fluid is re-compressed.

The central concept of this strategy relies on varying the state of the chambers sequentially to change the 'ump's displacement. To illustrate, a three-piston pump configuration is shown below in Figure 8. To achieve 66% pumping displacement, two chambers (1 & 2) would operate at 100% displacement, while the third chamber (3) would run at 0% displacement.

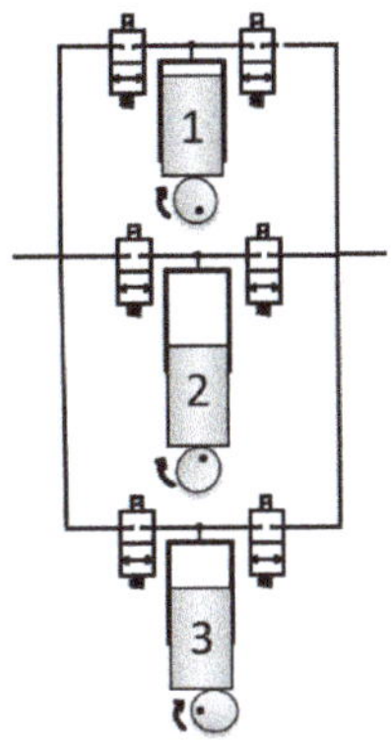

Figure 8. Three-piston digital inline piston pump configuration.

Sequencing is averaged over various pump revolutions to achieve a wide range of displacements between 0, 33%, 66%, and 100% [42]. As a case in point, to achieve 11.11% displacement, the first cycle of the pump would turn on one cylinder (1) and turn off two cylinders (2 & 3). The following two pump revolutions would sequence all cylinders (1, 2 & 3) off. This operation achieves 1/9th, or 11%, displacement over the three cycles. A ~1% displacement resolution can be achieved by averaging the sequences over 32 shaft revolutions.

5.3. Experimental Validation of the Electrically Actuated Digital Pump/Motor

The experimental test stand used for testing the SFL operating strategy is shown in Figure 9. The test stand involves the intended 28 cc/rev digital pump, two accumulators (one at the suction side and one at the discharge side), manifold housing with six on/off valves (two for each piston chamber), and other hardware used to complete the hydraulic circuit, data acquisition, and control. The two accumulators were added to the system to reduce pressure and flow ripples and avoid cavitation problems during the suction phase. For instance, if the hydraulic system experienced sudden stops during suction, rapid changes in the system pressure will generate pressure ripples causing shock waves. In this case, the system's pressure will go above the accumulator's pre-charged pressure. The accumulator at the discharge port will act as a shock cushioning. It will allow the excessive flow to enter, compressing the gas cushion (nitrogen bladder) and thus cushion and minimize the shock. The second accumulator (one located at the suction port) is required because the pump is bi-directional, i.e., the pump can discharge fluid from either port. A detailed description of the hardware used is presented in [66].

The experimental testing of the SFL pumping/motoring strategy was conducted under steady-state conditions at a temperature T = 50 °C and different pump displacements, 25%, 50%, 75%, and 100% displacement. Moreover, the SFL pumping and motoring cycles were examined at different system pressures ranging from 4 MPa to 18 MPa and shaft speeds up to 700 RPM. Operating the digital pump at shaft speeds beyond 700 RPM requires high-speed switching valves, affecting the valves' variability and causing non-linearity and errors in the valves' timing (opening and closing time). A timing valve correction algorithm was developed in [67] and utilized throughout the experimental testing to overcome this problem and account for the valve timing error. The developed algorithm obtains the delay in the valve timing using the pressure readings at the low- and high-pressure ports, thus regulating the valves' opening and closing time. Although the adopted algorithm enabled high-speed switching, the valves' opening and closing speeds are not high enough. Thus, the maximum operating shaft speed was limited to 700 RPM throughout the research study.

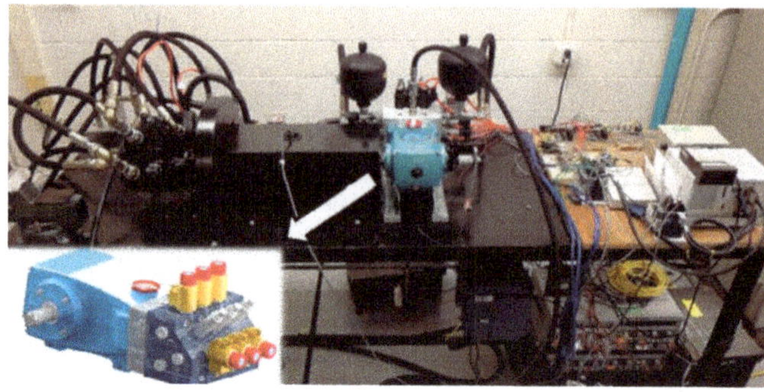

Figure 9. The experimental test stand for testing different operating strategies of the electrically actuated digital 3-piston pump [42].

After executing the experimental testing, the required data (torque, pressure, flow, etc.) were acquired at various pump displacements (25%, 50%, 75%, and 100% displacement), system pressures (4 MPa, 10 MPa, and 18 MPa), and shaft speeds (300 RPM, 500 RPM, and 700 RPM). Then, the efficiencies were calculated according to the obtained data.

The experimental outcomes resulting from operating the SFL strategy at 100% displacement during the pumping and motoring cycles are shown in Figures 10 and 11. Figure 10 illustrates the calculated efficiency curves across various system pressures and shaft speeds at 100% displacement. As shown, at high shaft speeds and low pressures, the pump can achieve efficiencies of at least 78%, while at low shaft speeds and high pressures above 16 MPa, the pump can achieve high efficiencies up to 92%. Furthermore, the system pressure significantly impacts the overall pump's efficiency, where the pump's efficiency significantly increases as pressure increases.

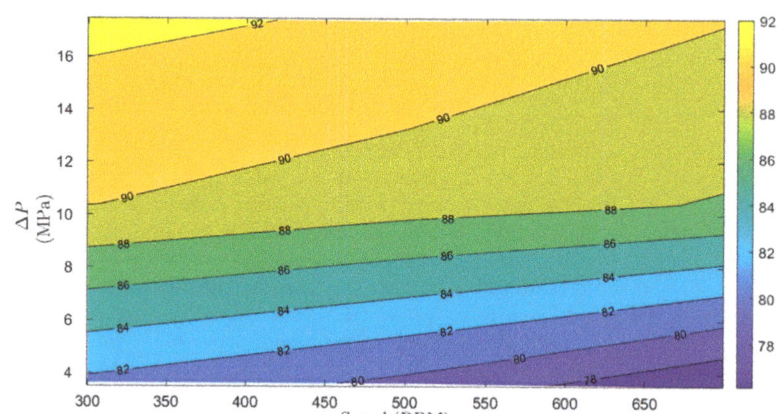

Figure 10. Efficiency map of the electrically actuated digital pump/motor operated in the SFL pumping strategy at 100% displacement.

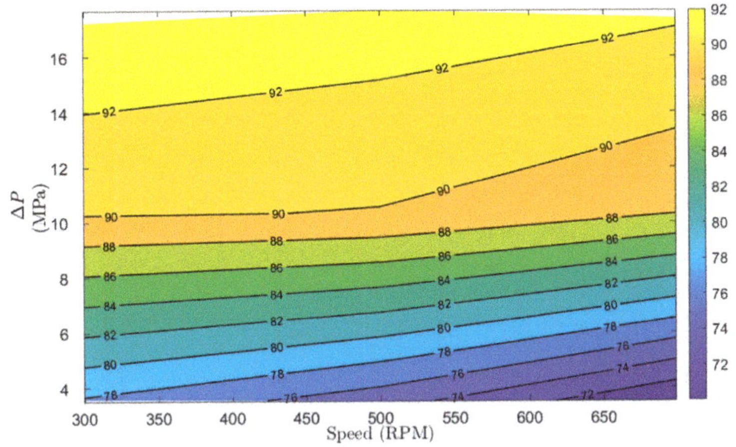

Figure 11. Efficiency map of the electrically actuated digital pump/motor operated in the SFL motoring strategy at 100% displacement.

Similarly, the efficiency map corresponding to operating the pump with the SFL strategy at 100% displacement during the motoring mode is shown in Figure 11. As shown in the figure, the motoring mode enables efficiencies of at least 72% and up to 92%. Unlike the results of the pumping mode, in motoring mode, the efficiency curves are almost parallel to the x-axis, which means that the shaft's speed has a negligible impact on the pump's efficiency. For instance, we can still achieve efficiencies from 72% to 92% at any RPM, depending on the system's pressure.

6. Simulation Model: Hydraulic Hybrid Drivetrains Utilizing Digital Pump/Motors

This section presents a physics-based series hydraulic hybrid drivetrain simulation conducted on MATLAB Simscape, a graphical programming software created by MathWorks [68]. The simulated MATLAB Simscape model mimics regular terrain hydraulic vehicles' series hydraulic hybrid transmission. The goal of the model is to investigate the improvement of the hydraulic hybrid transmission performance after replacing the conventional pump/motors with the digital pump/motor units. The intended state-of-the-art digital units are emulated by applying the digital motoring/pumping data acquired at the tested displacements (25%, 50%, 75%, and 100% displacement), pressures (4 Mpa, 10 MPa, and 18 MPa), and shaft speeds (300 RPM, 500 RPM, and 700 RPM) to the physical simulation components (pump and motor). The digital data are applied to the physical components of the simulation model by converting it into lookup tables imported to the physical pump and motor, employed throughout the simulation.

The following sub-sections present the designed circuit schematic, display the simulation model, design, and functionality during the three transmission stages (acceleration, cruising, and braking), and discuss the resulting outcomes.

6.1. Circuit Schematics

The adopted series hydraulic transmission, illustrated in Figure 12, comprises a variable displacement pressure compensated pump (5) driven by an electric motor (2), a mechanical clutch (3) to engage and disengage the motor's shaft from the rest of the system, a pressurized gas accumulator (1) for energy storage-boost purposes, a low-pressure reservoir (8), and a variable displacement pump/motor (11) for power transmission. The pump/motor (11) operates as a motor during acceleration and cruising, i.e., motoring mode, and works as a pump during braking, i.e., pumping mode. A check valve (6) is added to block the backflow (flow from and into the accumulator) from entering the pump's discharge port. Two 4/3 way (9 & 10) and one 3/2 way (7) directional flow control valves allow a seamless operation during the motoring and pumping modes. The 4/3 way valves (9 & 10) are closed-center valves with spring return having two directly actuated solenoid coils, while the 3/2 way valve (7) is a spring return valve with one actuated solenoid coil. As demonstrated through the simulation model, valve (9) is employed during the motoring mode, i.e., acceleration and cruising, where the user controls the valve based on the direction of motion, i.e., forward or backward drive. However, valve (10) is used during the pumping mode, i.e., regenerative braking, and it is controlled based on the direction of rotation received by the unit's shaft (11). Valve (7) allows switching between valves (9) and (10) based on the system's operation mode. During the motoring mode, i.e., acceleration and cruising, valve (7) is set at its un-actuated position, allowing pumping the fluid into the motor (11) through valve (9). However, during the pumping mode, i.e., at braking, the mechanical braking signal will be converted into an electrical signal at valve (7), forcing it to switch to the other position and thus allowing the motor (11) to pump the flow from the tank into the accumulator through the valve (10).

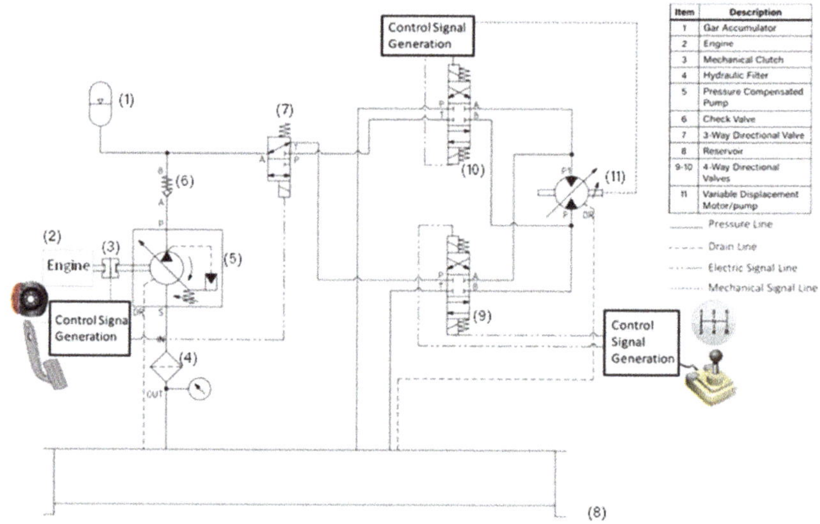

Figure 12. A circuit illustrating the simulated series hydraulic hybrid transmission.

6.2. Simulation Model, Design, and Operation

Figure 13 exhibits the physical model for the series hydraulic hybrid transmission illustrated in Figure 12. The adopted system is modeled and simulated using Simscape Fluids and Drivetrains libraries on MATLAB-Simulink. The simulation model is constructed by integrating the required numerical block diagrams with the physical paradigms. All the hydraulic physical components, i.e., pump, motor, valves, and other hydraulic units, are compiled using interfaceable physical connections. The figure shows that the model involves a pressure compensated pump and a variable displacement motor that operates as a motor when pumping high-pressure fluid into one of its ports and as a pump when receiving a mechanical signal at its output shaft. Furthermore, the model involves physical hydraulic sensors to sense the flow and the pressure at specific points in the system. For instance, pressure-flow sensors are installed at the pump's discharge port, accumulator's port, and motor's inlet and outlet ports. The accumulator's flow sensor is installed into the accumulator. Then, a positive flow rate at the accumulator's port designates flow into the accumulator; during charging. However, a negative flow denotes flow out of the accumulator; during discharging. Besides the pressure-flow sensors, mechanical sensors, e.g., rotational motion and torque sensors, are added at the pump's and motor's shaft to measure the resulting shaft's speed and torque. Acquiring this data (torque and speed) allowed us to compute the input and output mechanical powers. The power data was then exported into the MATLAB workspace for calculating the average input and output powers. The average input and output powers were individually calculated over the entire cycle using the mean function in MATLAB.

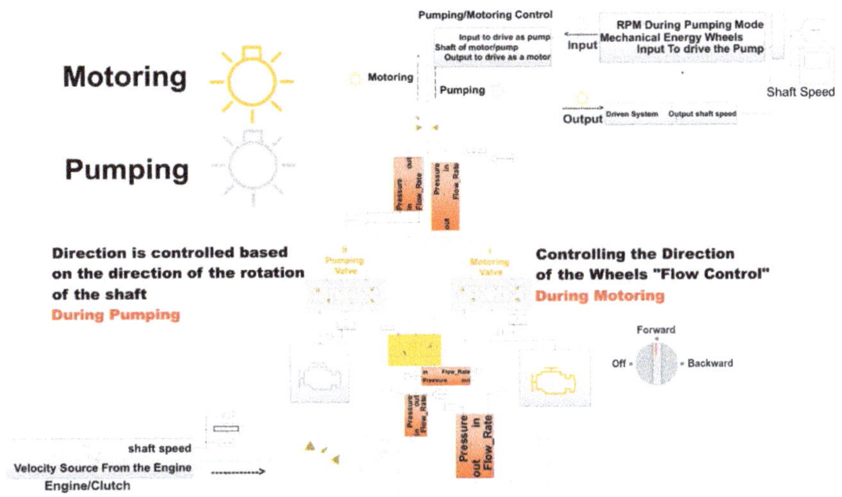

Figure 13. A MATLAB Simscape simulation model for the series hydraulic hybrid transmission.

The digital pumping/motoring data presented in Section 5.2 are converted into lookup tables and then imported to the physical pumping/motoring units (the utilized pressure-compensated pump and the variable displacement motor) to incorporate the pre-introduced digital pump/motor. The pumping data is imported into the physical motor (11) during braking as a lookup table. Also, it is imported into the physical pump (5) during cruising, knowing that the pump will be off during acceleration. Similarly, the motoring data is applied to the physical motor as a lookup table during acceleration and cruising. Thus, the physical pump and motor (pressure-compensated pump and the variable displacement motor) are sized based on the imported lookup tables, which allows setting the maximum and minimum pressures in the system in addition to the required flow rate. Then, the accumulator and other hydraulic components, e.g., valves, fittings, hoses, etc., are sized and selected based on the acquired system's pressure and flow rate. Based on the digital pumping/motoring data, the hydraulic units (5) and (11) will operate between 300 RPM and 700 RPM with a maximum displacement of 28 cc/rev. Furthermore, the hydraulic units will work under a volumetric efficiency up to 94% and overall efficiency of at least 78% and up to 90% during pumping. However, unit (11) will operate during motoring under volumetric and total efficiencies up to 99% and 92%, respectively. Furthermore, the simulation model accounts for the system's internal losses, e.g., losses in hydraulic pipelines, fittings, and valves. These internal losses are simulated by adding hydraulic resistance, capacitance, and inductance to the distributed parameter model using the simscape block models for hydraulic pipelines, fittings, and hoses.

During the series hydraulic hybrid transmission, the gas accumulator serves as a hydraulic lunch assist device to provide auxiliary flow during high demand requirements (within the acceleration stage), forcing the gas inside the accumulator to expand and compress under adiabatic conditions. Based on the intended application, the accumulator is sized using the following equations [69]:

$$P_1 V_1^k = P_2 V_2^k \tag{1}$$

$$V_1 = \frac{V_x (P_3/P_1)^{1/k}}{1 - (P_3/P_2)^{1/k}} \quad (2)$$

where P_1, P_2, and P_3 in Pa denote the accumulator's pre-charged pressure, system maximum operating pressure, and system minimum operating pressure, respectively, set based on the experimentally validated data of the digital pump. V_1 and V_2 in L designate the nitrogen's volume at P_1 and P_2, respectively, V_x in L denotes the required fluid volume, and $k = 1.4$ is the nitrogen constant during adiabatic operation. To prevent the accumulator's poppet (poppet that opens and closes, allowing fluid to enter and leave) from fully closing during the expansion of the nitrogen bladder, the accumulator's pre-charged pressure P_1 has to be set between 60% to 80% of the minimum system pressure [70,71]. Thus, for P_2 = 1.9e + 7 Pa, P_3 = 5.3e + 6 Pa and $P_1 = 0.75P_3$, we get V_1 = 36.3 L and V_x = 18 L.

The adopted physical model comprises two emulation options, options one and two shown in Figure 14, for simulating the intended system. Option one allows for a user interface simulation. The user can interact with the transmission model by switching between the motoring and the pumping modes, thus mimicking the driving and braking situations. Given this option, the user generates the duty cycle by choosing when to drive and brake. On the other hand, option two enables a complete auto-simulation over the three transmission stages (acceleration, cruising, and braking). The simulation will be done automatically based on the provided duty cycle for this option.

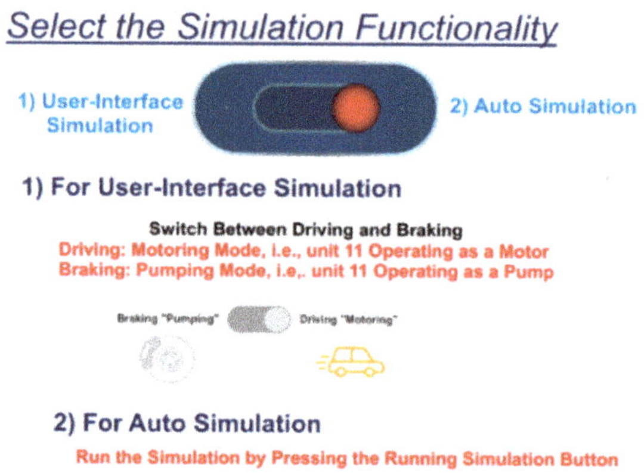

Figure 14. User interface illustrating the functionality of the simulation model.

Relying on option two, several drive cycles were adopted to test the simulation model's capability over the three transmission stages. The total time frame of the tested drive cycles ranged from 200 s/cycle to 300 s/cycle, where the acceleration and braking phases required long times up to 60 s and 40 s, respectively. The acceleration and braking timing is due to the accumulator's charging and discharging time, which was selected based on the experimentally acquired pump data that was limited with a maximum and minimum pressure around 19.1 MPa and 5.3 MPa, respectively.

One of the tested duty cycles is shown in Figure 15. The figure shows that the physical system is simulated for 400 s for two consecutive cycles. Each cycle comprises the three stages of the series hydraulic hybrid transmission (acceleration, cruising, and braking). During the first cycle, the acceleration phase occurs between 0 and 60 s, the cruising starts at time = 60 s and continues until time = 140 s, and the re-regenerative braking takes place within 140 and 180 s. The second cycle starts 20 s after the first cycle, which accounts for the complete stopping of the vehicle. Similarly, the three transmission stages (acceleration, cruising, and braking) occur for 60, 80, and 40 s, respectively, during the second cycle.

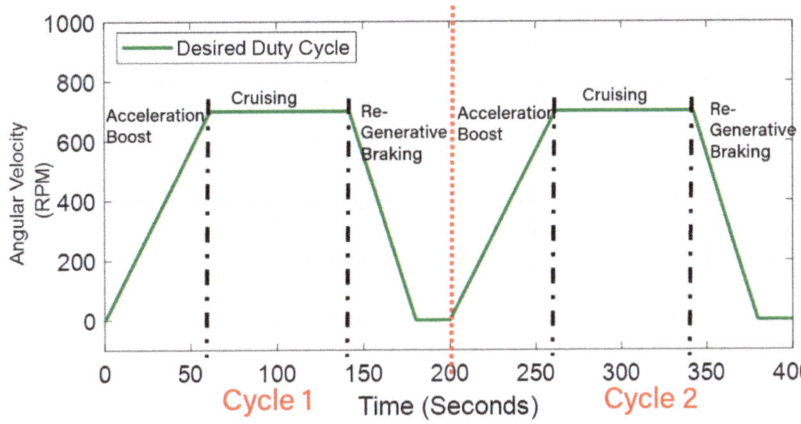

Figure 15. The desired duty cycle.

To fully simulate the hydraulic hybrid series transmission system for achieving the desired duty cycle in Figure 15, it is required to apply the desired operating angular velocity at the system's input port and specify its load. The system's operating RPM is simulated by applying an input signal builder using a physical velocity source. The transmission's total mass was included within the physical paradigms of the utilized hydraulic units and all components in the Simscape environment. The transmission's load is simulated using four Simscape Tire blocks, which simulate the tires by their radius and ideal inertia. In [72], Valeski approximates the average tire's weight and diameter to be around 10 kg and 40 cm, respectively. Thus, the calculated moment of inertia on each tire was around 0.37 kg·m^2. Given the calculated moment of inertia on each tire, a moderate physical torque source with an ideal moment of inertia of 1.5 kg·m^2 was applied at the motor's shaft. Now, we explain the operation of the simulation model during each stage.

6.2.1. Acceleration

During acceleration, the accumulator is fully-charged with pressurized fluid, and its pre-charged pressure is greater than the system's operating pressure. At this phase, the pump is off, and thus the accumulator acts as an auxiliary power supply. The accumulator discharges its stored energy into the motor (11), operating in its motoring mode, to accelerate the vehicle. The valves' opening/closing operation is controlled based on the accumulator's pre-charged pressure. For instance, during this phase, because the pre-charged pressure of the nitrogen inside the accumulator is greater than the system's pressure, no signal is generated at the valve (7). Then, valve (7) is kept at its un-actuated position to pass the flow through the valve (9) and block it at valve (10).

6.2.2. Cruising

At the end of the acceleration phase, the fluid inside the accumulator starts decreasing, and the pre-charged pressure becomes below the system's pressure. Namely, the accumulator discharged almost all its stored energy, and thus the system requires an additional hydraulic power supply to drive the wheels. At this stage (cruising), the drop in the pre-charged pressure triggers an electric signal at the engine and forces the clutch to engage the engine to the rest of the system. The pump starts pumping the fluid from the low-pressure reservoir (hydraulic tank) into the hydraulic motor (11) that is still operating as a motor to execute the motion at the wheels. Like the acceleration phase, the fluid is delivered to the motor through the valve (9), where valve (10) is still closed.

6.2.3. Braking

During braking, the motor (11) switches to the pumping mode, and the clutch decouples the engine from the rest of the system, forcing the pump (5) to shut down. The pre-charged pressure of the nitrogen gas inside the accumulator drops below the system's pressure. At this stage, an electric signal is triggered at valve (7), forcing it to switch to the second position and thus pumping the fluid through valve 10 and blocking it at valve 9. The motor utilizes the braking kinetic energy at its output shaft to pump fluid from the tank into the accumulator converting the mechanical energy into hydraulic energy stored in the accumulator to be delivered during acceleration.

7. Resulting Outcomes

After introducing the simulation model and discussing its fundamental operation, throughout the following sections: Sections 7.1 and 7.2, we present the simulation outcomes resulting from applying the digital motoring/pumping data into units (5) and (11) at moderate load-speed conditions (1.5 kg·m^2 and 700 RPM).

7.1. Pressure-Flow Outcomes

Figures 16 and 17 illustrate the variations in the system's pressure and flow throughout the proposed duty cycle. They show the pressure-flow outcomes at different ports in the system during the three transmission stages (acceleration, cruising, and braking). As shown in the figures, during the acceleration phase (0 to 60 s in cycle one and 200 to 260 s in cycle two), the pump's shaft angular velocity is kept at zero, then triggered after 40 s. The pump started operating after 40 s because the accumulator had enough stored energy at the start of the acceleration phase (from 0 to 40 s). Thus, the clutch disengages the engine's shaft from the pump's shaft, causing a zero RPM at the pump's input shaft. The accumulator acts as a hydraulic launch assist and discharges its stored energy into the motor. Thus, the flow rate at the accumulator's port flows in the negative direction, i.e., from the accumulator into the system, as shown in Figure 16. A pressure drop of about 10 MPa is created between the motor's inlet and outlet ports, as shown in Figure 17, allowing it to convert the hydraulic energy into mechanical energy enabling the wheels to accelerate until the output shaft speed reaches 600 RPM. As the accumulator delivers the auxiliary power, the fluid's pressure at the accumulator's port, shown in Figure 16, decreases because the accumulator's poppet is still open to deliver fluid to the system. After the accumulator discharges about 80% to 85% of its stored energy, i.e., at time = 40 s, the pump starts operating via the clutch-engine mechanism maintaining the system's output angular velocity at about 600 RPM, where its discharge pressure, captured in Figure 16, increases up to 6–7.5 MPa. After 20 s, at time = 60 s, the vehicle undergoes the cruising stage. The pump operates at its maximum speed (700 RPM) with an operating pressure of around 7.5 to 8 MPa. The motor's output shaft operates at almost moderate speed (500 RPM). During this stage, the pressure at the accumulator's port, shown in Figure 16, is the same as the system's pressure because the accumulator's poppet is closed. Thus, the accumulator is neither charging nor discharging. After 80 s, the vehicle undergoes the braking phase, in which the clutch starts disengaging the engine shaft from the pump shaft resulting in a decrease in its RPM. The motor (11) switches to its pumping mode during this phase, pumping the fluid from the tank into the accumulator. The pressure at the motor's outlet port, shown in Figure 17, increases up to 17.5 MPa causing a high pressure at the accumulator's port and the pump's discharge port, shown in Figure 16. The accumulator starts charging, where its flow rate, shown in Figure 16, flows in the positive direction, i.e., from the system into the accumulator.

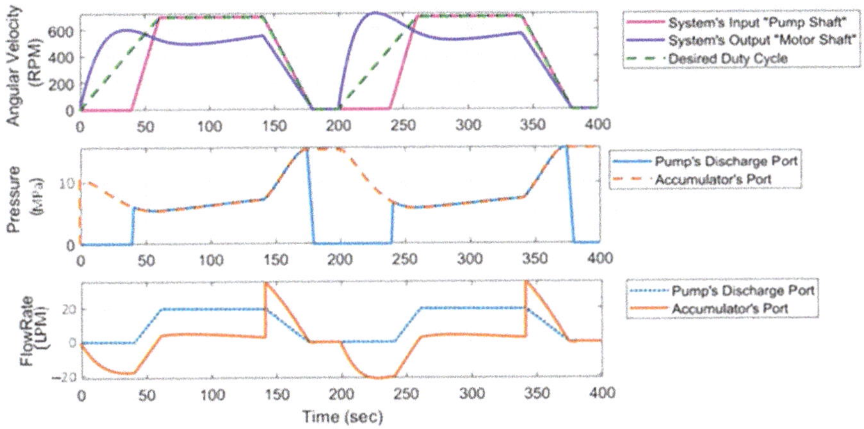

Figure 16. The resulting pressure and flow simulation outcomes at the pump's discharge and accumulator's ports throughout the desired duty cycle.

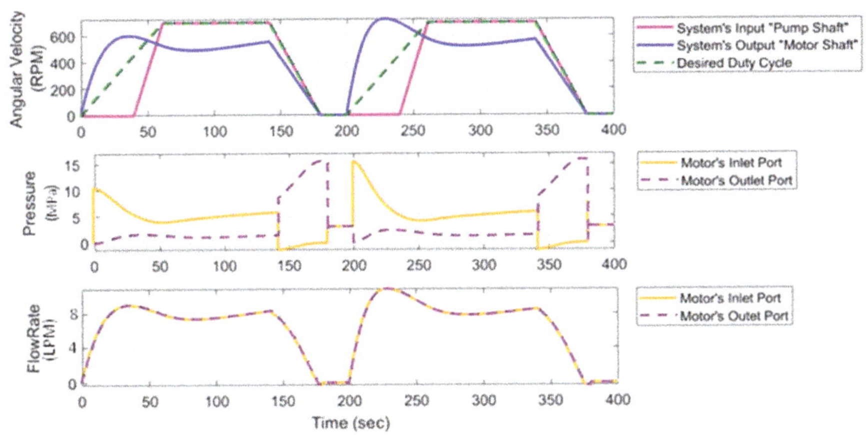

Figure 17. The resulting pressure and flow simulation outcomes at the Motor's inlet and outlet ports throughout the desired duty cycle.

Figures 18 and 19 show the variations in the accumulator's fluid volume and the motor's pressure drop, respectively, throughout the desired duty cycle for the three transmission stages (acceleration, cruising, and braking). As discussed previously, during the acceleration phase, the accumulator discharges its stored energy into the motor's inlet port to accelerate the motion at the wheels. Thus, the motor operates in its motoring mode. The accumulator's discharging process causes a decrease in fluid volume, shown in Figure 18, and a positive pressure drop of about 10 MPa at the motor, shown in Figure 19. As shown in Figure 19, the 10 MPa pressure drop across the motor slightly decreases until 5–6.5 MPa. Then, because the accumulator is neither charging nor discharging during the cruising phase, its fluid volume, shown in Figure 18, is kept almost constant. During this phase, the pressure drop across the motor, shown in Figure 19, is maintained at around 7.5 MPa. During braking, the accumulator's fluid volume, shown in Figure 18, increases because the accumulator is charging. A negative pressure drop is created across the motor, as shown in Figure 19, where the motor is operating in its pumping mode.

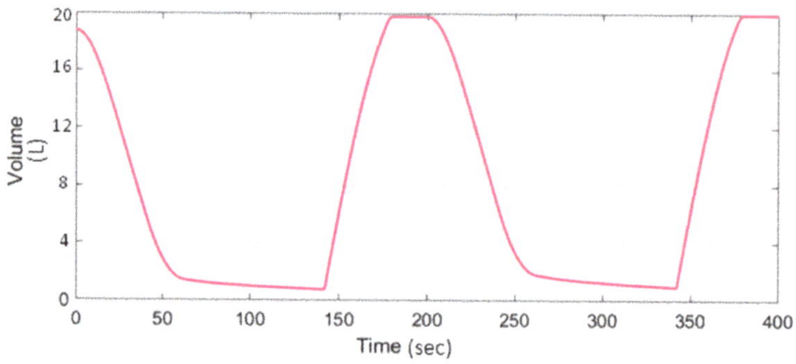

Figure 18. Simulation outcomes for the fluid volume inside the accumulator.

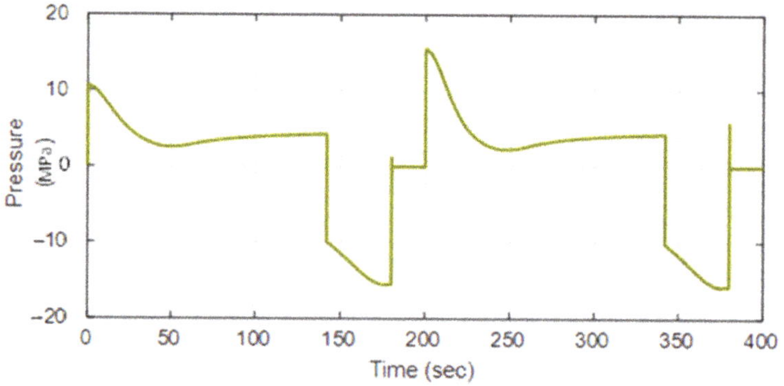

Figure 19. Simulation results for the pressure drop across the motor, i.e., the pressure difference between the inlet and outlet motor's pressure.

7.2. Power-Efficiency Outcomes

Figure 20 shows the load-speed outcomes and the resulting mechanical power at the pump and motor shafts, i.e., the system's input and output shafts, respectively, throughout the three transmission stages (acceleration, cruising, and braking). During acceleration, the pump is not operating, and thus the accumulator is the only power supply in the system. The accumulator discharges its stored energy into the motor. The motor converts the hydraulic energy into mechanical energy resulting in an output mechanical power of around 1300 W in cycle one and about 2000 W in cycle two. The difference in the mechanical powers in cycles one and two is due to the accumulator's initial stored energy. During cruising, the pump starts operating at around 700 RPM, producing an input mechanical power of about 2100 W. This results in an output mechanical power of around 1000 W at the motor's shaft due to the 1.5 kg·m^2 inertia added at the wheels. During braking, the motor switches to its pumping mode. It utilizes the braking energy and converts it into hydraulic energy stored in the accumulator.

Simulating the system to achieve the desired duty cycle at moderate load-speed conditions (1.5 kg·m^2 and 700 RPM) results in average input power of 876 W at the system's input shaft. This results in an average output power of 780 W at the system's output shaft, which outputs an overall efficiency of around 89%. This result reveals a significant improvement in the efficiency of the series hydraulic hybrid drivetrains. Due to the rated efficiency of the conventional hydraulic pump/motors, the total average efficiency of the series hydraulic hybrid transmission is around 64% to 81%, which is relatively low [47]. Therefore, the utilization of the state-of-the-art multi-piston digital pump reinforced the system's performance by increasing the overall efficiency at least 8% and up to 25%.

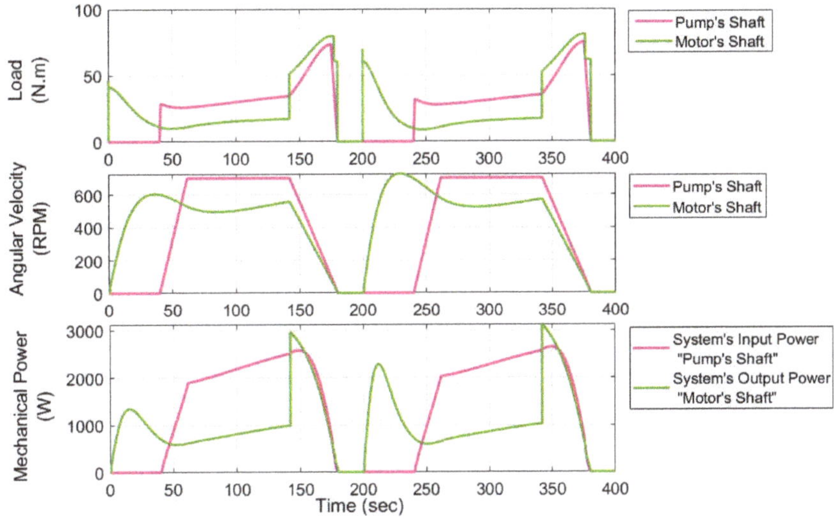

Figure 20. Load-Speed and mechanical power simulation results.

8. Conclusions

The overall average efficiency of series hydraulic hybrid drivetrains is comparatively low, between 64% and 81%, due to the utilization of conventional variable displacement pumps that become inefficient when operating at partial displacements. This work estimates and addresses the efficiency improvement of a series hydraulic hybrid transmission by integrating hydraulic hybrid technology with digital hydraulics. It proposes replacing the conventional piston pump/motor employed in series hydraulic hybrid drivetrains with the multi-piston digital pump designed and implemented in preceding works. The proposed methodology is tested by conducting a physical-based simulation of a series hydraulic hybrid drivetrain on MATLAB Simscape. The state-of-the-art multi-piston digital pump is simulated by applying its digital motoring/pumping data into the power supply units, i.e., pump and motor utilized throughout the simulation model. The adopted simulation model permitted evaluating the performance of the series hydraulic hybrid drivetrain, and it allowed examining the improvement of the system's overall efficiency. The final simulation outcomes resulted in 89% overall efficiency and thus exceeded the conventional series hydraulic hybrid drivetrain (64% to 81%). This resulting efficiency reveals that the utilization of the state-of-the-art digital pump enables enhancing the total efficiency of the hydraulic hybrid drivetrains by at least 8% and up to 25%. The attained 89% efficiency can also be better enhanced using a more refined digital unit. The digital pump utilized throughout the simulation model was a proof of concept prototype limited to 700 RPM shaft speed. Thus, a refined one would run at a broader range of inputs and achieve even better results. As future work, we aim to transform the proof-of-concept digital pump into a higher-speed refined unit to test the proposed transmission experimentally.

Author Contributions: Conceptualization, F.B. and I.A.; Investigation, J.G.-B. and F.B.; Methodology, I.A. and F.B.; Supervision, J.G.-B. and F.B.; Validation, J.G.-B.; Visualization, I.A. and F.B.; Writing—original draft, I.A., K.P. and J.G.-B.; Writing—review & editing, I.A. and F.B. All authors have read and agreed to the published version of the manuscript.

Funding: This research received no external funding.

Data Availability Statement: The data that support the findings of this study are available on request from the corresponding author, F.B.

Conflicts of Interest: The authors declare no conflict of interest.

References

1. McCloy, D.; Martin, H.R. *Control of Fluid Power: Analysis and Design*, 2nd ed.; Ehlh: Chichester, UK, 1980. Available online: https://ui.adsabs.harvard.edu/abs/1980ehlh.book.....M/abstract (accessed on 10 October 2021).
2. Durfee, W.; Sun, Z. *Fluid Power System*; A National Science Foundation Engineering Research Center: Minneapolis, MN, USA, 2009; pp. 1–48.
3. Vacca, A. Energy Efficiency and Controllability of Fluid Power Systems. *Energies* **2018**, *11*, 1169. [CrossRef]
4. Konami, S.; Nishiumi, T. *Hydraulic Control Systems*; John Wiley & Sons: Singapore, 2016; pp. 187–299. [CrossRef]
5. Fabis-Domagala, J.; Domagala, M.; Momeni, H. A Concept of Risk Prioritization in FMEA Analysis for Fluid Power Systems. *Energies* **2021**, *14*, 6482. [CrossRef]
6. What is Fluid Power? National Fluid Power Association. 2021. Available online: https://www.nfpa.com/home/AboutNFPA/What-is-Fluid-Power.htm (accessed on 1 December 2021).
7. U.S. Fluid Power Industry Brief. 2021. Available online: https://www.nfpa.com/home/industry-stats/Industry-Brief.htm (accessed on 1 December 2021).
8. Love, L.; Lanke, E.; Alles, P. Estimating the Impact (Energy, Emissions and Economics) of the US Fluid Power Industry. In Oak Ridge National Laboratory (Issue December). 2012. Available online: http://www.osti.gov/servlets/purl/1061537/ (accessed on 1 December 2021).
9. Kwon, H.; Sprengel, M.; Ivantysynova, M. Thermal modeling of a hydraulic hybrid vehicle transmission based on thermodynamic analysis. *Energy* **2016**, *116*, 650–660. [CrossRef]
10. Stelson, K.A.; Meyer, J.J.; Alleyne, A.G.; Hencey, B. Optimization of a passenger hydraulic hybrid vehicle to improve fuel economy. *Proc. JFPS Int. Symp. Fluid Power* **2008**, *2008*, 143–148. [CrossRef]
11. Chen, J.-S. Energy Efficiency Comparison between Hydraulic Hybrid and Hybrid Electric Vehicles. *Energies* **2015**, *8*, 4697–4723. [CrossRef]
12. Gary, J. Hydrid Transmission. *Acad. Manag. Rev.* **2006**, *31*, 386–408.
13. Zhang, Z.; Chen, J.; Wu, B. The control strategy of optimal brake energy recovery for a parallel hydraulic hybrid vehicle. *Proc. Inst. Mech. Eng. Part D J. Automob. Eng.* **2012**, *226*, 1445–1453. [CrossRef]
14. Rydberg, K.-E. Energy Efficient Hydraulic Hybrid Drives. In Proceedings of the 11th Scandinavian International Conference on Fluid Power, SICFP'09, Linköping, Sweden, 2–4 June 2009.
15. Tvrdić, V.; Podrug, S.; Šuljić, I.; Matić, B. Hydraulic hybrid vehicle configurations and comparison with hybrid electric vehicle. In Proceedings of the Comtemporary Issues in Economy & Technology—CIET 2018, Split, Croatia, 1–2 June 2018; pp. 548–556.
16. Florida, N.; López, C.; Pocomucha, V. CORE View metadata, citation and similar papers at core. ac. uk. *J. Dedik. Pendidik.* **2012**, *2*, 35–43.
17. Shan, M. Modeling and Control Strategy for Series Hydraulic Hybrid Vehicles. Ph.D. Thesis, The University of Toledo, Toledo, OH, USA, December 2009.
18. Tavares, F.; Johri, R.; Salvi, A.; Baseley, S.; Filipi, Z. *Hydraulic Hybrid Powertrain-In-the-Loop Integration for Analyzing Real-World Fuel Economy and Emissions Improvements*. SAE Technical Paper. 2011. Available online: https://doi.org/10.4271/2011-01-2275 (accessed on 20 October 2021).
19. Midgley, W.; Cebon, D. Comparison of regenerative braking technologies for heavy goods vehicles in urban environments. *Proc. Inst. Mech. Eng. Part D J. Automob. Eng.* **2012**, *226*, 957–970. [CrossRef]
20. Li, C.-T.; Peng, H. Optimal configuration design for hydraulic split hybrid vehicles. In Proceedings of the 2010 American Control Conference, Baltimore, MD, USA, 30 June–2 July 2010; pp. 5812–5817. [CrossRef]
21. Kim, N.; Rousseau, A. A Comparative Study of Hydraulic Hybrid Systems for Class 6 Trucks. *SAE Tech. Pap.* **2013**, *2*. [CrossRef]
22. Sprengel, M.; Bleazard, T.; Haria, H.; Ivantysynova, M. Implementation of a Novel Hydraulic Hybrid Powertrain in a Sports Utility Vehicle. *IFAC-PapersOnLine* **2015**, *48*, 187–194. [CrossRef]
23. Bottiglione, F.; Mantriota, G.; Valle, M. Power-Split Hydrostatic Transmissions for Wind Energy Systems. *Energies* **2018**, *11*, 3369. [CrossRef]
24. Jae, Y.; Filipi, Z.; Kim, Y.J.; Filipi, Z. Simulation Study of a Series Hydraulic Hybrid Prop System for a Light Truck. *SAE Trans.* **2021**, *116*, 147–161.
25. Hui, S.; Junqing, J. Research on the system configuration and energy control strategy for parallel hydraulic hybrid loader. *Autom. Constr.* **2010**, *19*, 213–220. [CrossRef]
26. Zhou, S.; Walker, P.; Tian, Y.; Zhang, N. Mode switching analysis and control for a parallel hydraulic hybrid vehicle. *Veh. Syst. Dyn.* **2020**, *59*, 928–948. [CrossRef]
27. Sim, T.P.; Li, P.Y. Analysis and Control Design of a Hydro-Mechanical Hydraulic Hybrid Passenger Vehicle. In Proceedings of the Dynamic Systems and Control Conference, Hollywood, CA, USA, 12–14 October 2009; pp. 667–674. [CrossRef]
28. Cheong, K.L. Design and Analysis of Hydraulic Hybrid Passenger Vehicles. Ph.D. Thesis, University of Minnesota, Minneapolis, MN, USA, 2015.
29. Ramdan, M.I.; A Stelson, K. Optimal design of a power-split hybrid hydraulic bus. *Proc. Inst. Mech. Eng. Part D J. Automob. Eng.* **2016**, *230*, 1699–1718. [CrossRef]
30. Sprengel, M.; Ivantysynova, M. Recent Developments in a Novel Blended Hydraulic Hybrid Transmission. *SAE Tech. Pap.* **2014**, *2014*. [CrossRef]

31. Stecki, J.; Matheson, P. Advances In Automotive Hydraulic Hybrid Drives. *Proc. JFPS Int. Symp. Fluid Power* **2005**, *2005*, 664–669. [CrossRef]
32. Matheson, P.; Stecki, J. Development and simulation of a hydraulic-hybrid powertrain for use in commercial heavy vehicles. *SAE Tech. Pap.* **2003**, *112*, 114–123. [CrossRef]
33. Wu, B.; Lin, C.-C.; Filipi, Z.; Peng, H.; Assanis, D. Optimal Power Management for a Hydraulic Hybrid Delivery Truck. *Veh. Syst. Dyn.* **2004**, *42*, 23–40. [CrossRef]
34. Van Batavia, B.L. *Hydraulic Hybrid Vehicle Energy Management System*; SAE Technical Paper; SAE International: Warrendale, PA, USA, 2009. [CrossRef]
35. Achten, P.A.J.; Schellekens, M.P.A.; Murrenhoff, H.; Deeken, M. Efficiency and Low Speed Behavior of the Floating Cup Pump. *J. Commer. Veh.* **2004**, *113*, 366–376. [CrossRef]
36. Buchwald, P.; Christensen, G.; Larsen, H.; Sunn Pedersen, P. Improvement of citybus fuel economy using a hydraulic hybrid propulsion system—A theoretical and experimental study. *SAE Tech. Pap.* **1979**, *88*, 1042–1056. [CrossRef]
37. Hydraulic Launch Assist. 2013. Available online: https://www.eaton.com/SEAsia/ProductsSolutions/Hydraulics/ProductsServices/HydraulicLaunchAssist/index.htm (accessed on 1 October 2021).
38. Tollefson, S.; Beachley, N.H.; Fronczak, F.J. Studies of an Accumulator Energy-Storage Automobile Design with a Single Pump/Motor Unit. *SAE Trans.* **1985**. [CrossRef]
39. Kepner, R.P. Hydraulic power assist—A demonstration of Hydraulic Hybrid Vehicle regenerative braking in a road vehicle application. *SAE Tech. Pap.* **2002**, *111*, 826–833. [CrossRef]
40. Bozic, A. Introducing Hydraulic-Electric Synergy into Hybrid Transmission Using the Free-piston Engine Technology. In Proceedings of the Commercial Vehicle Engineering Congress and Exhibition and Powertrain & Fluid Systems Conference, Rosemont, IL, USA, 29 October–1 November 2007. [CrossRef]
41. Breidi, F.; Garrity, J.; Lumkes, J. Design and Testing of Novel Hydraulic Pump/Motors to Improve the Efficiency of Agricultural Equipment. *Trans. ASABE* **2017**, *60*, 1809–1817. [CrossRef]
42. El-Breidi, F. Investigation of digital pump/motor control strategies. Ph.D. Thesis, Purdue University, West Lafayette, IN, USA, 2016.
43. Yang, H.-Y.; Pan, M. Engineering research in fluid power: A review. *J. Zhejiang Univ. A* **2015**, *16*, 427–442. [CrossRef]
44. Merrill, K.J.; Breidi, F.Y.; Lumkes, J. Simulation based design and optimization of digital pump/motors. In *Fluid Power Systems Technology*; American Society of Mechanical Engineers: New York, NY, USA, 2013.
45. Li, M.; Foss, R.; Stelson, K.A.; Van De Ven, J.D.; Barth, E.J. Design, Dynamic Modeling, and Experimental Validation of A Novel Alternating Flow Variable Displacement Hydraulic Pump. *IEEE/ASME Trans. Mechatron.* **2019**, *24*, 1294–1305. [CrossRef]
46. Wilhelm, S.R.; Van De Ven, J.D. Synthesis of a Variable Displacement Linkage for a Hydraulic Transformer. In Proceedings of the International Design Engineering Technical Conferences and Computers and Information in Engineering Conference, Washington, DC, USA, 28–31 August 2011. [CrossRef]
47. Chen, Y.L.; Liu, S.A.; Jiang, J.H.; Shang, T.; Zhang, Y.K.; Wei, W. Dynamic analysis of energy storage unit of the hydraulic hybrid vehicle. *Int. J. Automot. Technol.* **2013**, *14*, 101–112. [CrossRef]
48. Ivantysynova, M. Innovations in pump design-what are future directions? *Proc. JFPS Int. Symp. Fluid Power* **2008**, *2008*, 59–64. [CrossRef]
49. Zhang, C.; Huang, S.; Du, J.; Wang, X.; Wang, S.; Zhang, H. A new dynamic seven-stage model for thickness prediction of the film between valve plate and cylinder block in axial piston pumps. *Adv. Mech. Eng.* **2016**, *8*. [CrossRef]
50. Shang, L.; Ivantysynova, M. A temperature adaptive piston design for swash plate type axial piston machines. *Int. J. Fluid Power* **2016**, *18*, 38–48. [CrossRef]
51. Rizzo, G.; Massarotti, G.; Bonanno, A.; Paoluzzi, R.; Raimondo, M.; Blosi, M.; Veronesi, F.; Caldarelli, A.; Guarini, G. Axial piston pumps slippers with nanocoated surfaces to reduce friction. *Int. J. Fluid Power* **2015**, *16*, 1–10. [CrossRef]
52. Inaguma, Y.; Hibi, A. Reduction of friction torque in vane pump by smoothing cam ring surface. *Proc. Inst. Mech. Eng. Part C J. Mech. Eng. Sci.* **2007**, *221*, 527–534. [CrossRef]
53. Wieczorek, U.; Ivantysynova, M. Computer Aided Optimization of Bearing and Sealing Gaps in Hydrostatic Machines—The Simulation Tool Caspar. *Int. J. Fluid Power* **2002**, *3*, 7–20. [CrossRef]
54. Sazonov, Y.A.; Mokhov, M.A.; Gryaznova, I.V.; Voronova, V.V.; Tumanyan, K.A.; Frankov, M.A.; Balaka, N.N. Development and Prototyping of Jet Systems for Advanced Turbomachinery with Mesh Rotor. *Emerg. Sci. J.* **2021**, *5*, 775–801. [CrossRef]
55. Kostikov, Y.A.; Romanenkov, A.M. Approximation of the Multidimensional Optimal Control Problem for the Heat Equation (Applicable to Computational Fluid Dynamics (CFD)). *Civ. Eng. J.* **2020**, *6*, 743–768. [CrossRef]
56. Manring, N.D. Valve-Plate Design for an Axial Piston Pump Operating at Low Displacements. *J. Mech. Des.* **2003**, *125*, 200–205. [CrossRef]
57. Helmus, T.; Breidi, F.; Lumkes, J. Simulation of a variable displacement mechanically actuated digital pump unit. In Proceedings of the Eight Workshop on Digital Fluid Power, Tampere, Finland, 24–25 May 2016; pp. 95–106.
58. Chehade, A.; Breidi, F.; Pate, K.S.; Lumkes, J. Data-driven Adaptive Thresholding Model for Real-time Valve Delay Estimation in Digital Pump/Motors. *Int. J. Fluid Power* **2019**, *20*, 271–294. [CrossRef]

59. Winkler, B. Recent Advances in Digital Hydraulic Components and Applications. In Proceedings of the The Ninth Workshop on Digital Fluid Power, Aalborg, Denmark, 7–8 September 2017. Available online: https://www.et.aau.dk/digitalAssets/377/377525_id117-recent-advances-in-digital-hydraulic-components-and-applications.pdf (accessed on 1 December 2021).
60. Brandstetter, R.; Deubel, T.; Scheidl, R.; Winkler, B.; Zeman, K. Digital hydraulics and "Industrie 4.0". *Proc. Inst. Mech. Eng. Part I J. Syst. Control. Eng.* **2016**, *231*, 82–93. [CrossRef]
61. Peng, S.; Branson Iii, D.T.; Guglielmino, E.; Boaventura, T.; Caldwell, D.G. Performance Assessment of Digital Hydraulics in a Quadruped Robot Leg. 2012. Available online: http://asmedigitalcollection.asme.org/ESDA/proceedings-pdf/ESDA2012/44861/227/4451304/227_1.pdf?casa_token=fWco0NtKGfUAAAAA:9c3YrOxP0rRp-PmWLcv8DUFtas4Qe6HnyCq02vGig-lSARq94FAcQnGOpL9W-XpcM-sdwGk (accessed on 5 November 2021).
62. Pinto, L.P.G.; Belan, H.C.; Locateli, C.C.; Krus, P.; De Negri, V.J.; Lantto, B. New perspectives on digital hydraulics for aerospace applications. In Proceedings of the Aerospace Technology Congress, Solna, Stockholm, 11–12 October 2016.
63. Breidi, F.; Helmus, T.; Lumkes, J. The impact of peak-and-hold and reverse current solenoid driving strategies on the dynamic performance of commercial cartridge valves in a digital pump/motor. *Int. J. Fluid Power* **2015**, *17*, 37–47. [CrossRef]
64. Breidi, F.; Chehade, A.; Lumkes, J. Monitoring Digital Technologies in Hydraulic Systems Using CUSUM Control Charts. In *Fluid Power Systems Technology*; American Society of Mechanical Engineers: New York, NY, USA, 2019.
65. Helmus, T. Investigation and Implementation of Mechanically Actuated Valves for Digital Hydraulic Units. Ph.D. Thesis, Purdue University, West Lafayette, IN, USA, 2017.
66. Holland, M.A. Design of Digital Pump/Motors and Experimental Validation of Operating Strategies Doctor of Philosophy. 2012. Available online: http://www.purdue.edu/policies/pages/teach_res_outreach/c_22.html (accessed on 1 December 2021).
67. Breidi, F.; Garrity, J.; Lumkes, J., Jr. Investigation of a real-time pressure based valve timing correction algorithm. In *Fluid Power Systems Technology*; American Society of Mechanical Engineers: New York, NY, USA, 2017.
68. Simscape—MATLAB & Simulink. 2021. Available online: https://www.mathworks.com/products/simscape.html (accessed on 6 October 2021).
69. Sizing Tool—Reasontek Corp. 2021. Available online: https://www.reasontek.com/sizing-tool/ (accessed on 23 November 2021).
70. Cheema, J. Air and Oil Do Mix: Selecting the Right Accumulator Is Critical for Wind Farm Applications. Hydraulics and Pneumatic. 2013. Available online: http://ezproxy.derby.ac.uk/login?url=http://search.ebscohost.com/login.aspx?direct=true&db=edsbl&AN=RN332050940&site=eds-live (accessed on 28 November 2021).
71. Xu, Z.; Li, W.; Liu, X.; Chen, Z. Dynamic characteristics of coupling model of valve-controlled cylinder parallel accumulator. *Mech. Ind.* **2019**, *20*, 306. [CrossRef]
72. Valeski, B. Average Tire Weight (With 10 Examples)—Survival Tech Shop. 2021. Available online: https://www.survivaltechshop.com/tire-weight/ (accessed on 1 December 2021).

MDPI
St. Alban-Anlage 66
4052 Basel
Switzerland
Tel. +41 61 683 77 34
Fax +41 61 302 89 18
www.mdpi.com

Energies Editorial Office
E-mail: energies@mdpi.com
www.mdpi.com/journal/energies

www.ingramcontent.com/pod-product-compliance
Lightning Source LLC
Chambersburg PA
CBHW061822210926
51850CB00006B/776
* 9 7 8 3 0 3 6 5 7 8 6 6 8 *